THE BAKHSHĀLĪ MANUSCRIPT

THE BAKHSHĀLĪ MANUSCRIPT

A Study in Medieval Mathematics

Parts I, II & III (Bound in one)

Part I : Introduction
Part II : The Text (includes 42 plates)
Part III : The Text Re-Arranged

G.R. KAYE

ADITYA PRAKASHAN
NEW DELHI

First published: 1927
Reprint: 2004

ISBN 81-7742-058-5

₹2500.00

Published by P.K. Goel for Aditya Prakashan, F-14/65, Model Town II, Delhi – 110 009.
Printed at D.K. Fine Art Press (P) Ltd., Delhi – 110 052.

ARCHÆOLOGICAL SURVEY OF INDIA

New Imperial Series. Vol. XLIII

Parts I & II

THE BAKHSHĀLĪ MANUSCRIPT

A Study in Mediæval Mathematics

BY

G. R. KAYE

CALCUTTA: GOVERNMENT OF INDIA
CENTRAL PUBLICATION BRANCH
1927

CONTENTS.

PREFACE.

In order to correct an impression that certain passages in this volume might convey unless distinctly qualified, I must here refer to my indebtedness to the late Dr. Hoernle. Indeed, a considerable part of the analysis of the MS. is really his work,* and by his preliminary survey of the manuscript my task was considerably lightened. It was at Dr. Hoernle's special request that I undertook to carry on the work he had started, and he handed over to me most of the material he had himself prepared. Had he lived a little longer I should, no doubt, have had the benefit of further help from him, and this volume might have been issued as our joint work. Dr. Hoernle's lamented death prevented that plan being carried out; and unfortunately my views are so often opposed to those that where held by Dr. Hoernle that it would hardly be proper to make him a participator in them.

I am much indebted to Bodley's Librarian for special facilities that enabled me to examine the original manuscript under the most favourable conditions; to the Oxford University Press for their most excellent work in preparing the photo graphs of the manuscript and the collotype reproductions of the text; and to the Manager, Government of India Press, Calcutta, for the care and skill with which the transliteration has been printed.

G. R. KAYE.

BANHAM,
Attleborough,
Norfolk.

* Sections B, G, H, K and L are almost wholly the work of Dr. Hoernle, who also transiterated about half of the leaves of the MS. References to his published papers on the MS. are given on page 2.

ERRATA.

On the plate facing page 4 read '51 RECTO B' for '51 VERSO B'.

In table iv, 4 (Part ii) the fifteenth ligature is *sva*, not *sya*.

On p. 41 in the second example x_1, x_2 and x_3 have been wrongly interchanged in the solution and answer.

In the table on p. 14 'folio 55' should come before 'folio 49'. That is M 9 is folio 55 and M 10 is folio 49.

p. 20, § 52. Add "on folio 49 is a phrase *śūkhyair yajamti* &c."

p. 28, foot-note g : for $2\frac{17}{11}$ read $^{2}\frac{17}{11}$.

p. 29, penultimate line : for $\frac{36}{4}$ read $\frac{35}{4}$.

p. 62, § 109. Add 'the terms *prasṛiti* and *khārī* occur on folio 34, but with doubtful application.'

p. 63. Omit the last line of the capacity table.

p. 127. Fol. 29r(*a*). Second line. Read *guṇabhayāso*.

(*d*). Second line. Read *dya* for *sya*.

Fol. 29v.(*d*). Second line. Read *tṛitīya* for *tritīya*.

p. 130. Fol. 33v. Fourth line. Read *pādavyase* for *pāgadhyase*.

p. 131. Fol. 35r. (*b*). Second line. Read *yogosyā* for *yogaṁsvā*.

p. 136. Fol. 40v. (*d*). Second line. Read *prathamaṁ* for *prathamā*.

p. 137. Fol. 42v. Third line. Read *viṁśanām* for *viṁśanoṁ*.

p. 139. Fol. 44v. Second line. Read *bha 8* for *bhā 8*.

p. 139. Fol. 45v. First line. Read *akṛi* for *asya*.

p. 142. Fol. 49r. Second line. Read *danāstimūdṛi* for *pādanastimūdṛi*.

p. 145. Fol. 54r. (*b*). Read *sūtraṁ* (*c*) *jātaṁ*.

p. 147. Fol. 56r. Penultimate line. Read *yāramarjaye* for *yāvarjaye*.

p. 147. Fol. 57r. Last line. Read *ekorāśi* for *ekonarāśi*.

THE BAKHSHĀLĪ MANUSCRIPT.

PART I.—INTRODUCTION.

CHAPTER I.

1. In 1881 a mathematical work written on brich-bark was found at Bakhshālī near Mardān on the north-west frontier of India. This manuscript was supposed to be of great age and its discovery aroused considerable interest. Part of it was examined by Dr. Hoernle, who published a short account together with a translation of a few of the leaves in 1888. Dr. Hoernle had intended, in due course, to publish a complete edition of the text, but was unable to do so. The present volume gives the complete text.

2. Bakhshālī, or Bakhshalai, as it is written in the official maps, is a village of the Yusufzai sub-division of the Peshawar district of the North-West Frontier Province of India. It is situated on, or near, the river Mukhām, which eventually joins the Kābul river near Nowshera, some twenty miles further south. Six miles W.N.W. of Bakhshālī is Jamalgarhī, twelve miles to the west is Takht-i-Bhai, and twenty-five miles W.S.W. is Chārsada—famous for their Indo-Greek art treasures.

Bakhshālī is about 150 miles from Kābul, 160 from Srinagar, 50 from Peshawar, 350 from Balkh and 70 from Taxila. It is in the trans-Indus country and in ancient times was within the Persian boundaries—in the Arachosian satrapy of the Achaemenid kings. It is within that part of the country to which the name Gandhāra has been given, and was subject to those western influences which are so bountifully illustrated in the so-called Gandhāra art.

3. The only authentic record of the discovery of the manuscript appears to be contained in the following letter, dated the 5th of July 1881,[1] from the Assistant Commissioner at Mardān.

"In reply to your No. 1306, dated 20th ultimo, and its enclosures, I have the honour to inform you that the remains of the papyrus MS. referred to were brought to me by the Inspector of Police, Mian An-Wan-Udin. The finder, a tenant of the latter, said he had found the manuscript while digging in a ruined stone enclosure on one of the mounds near Bakhshāli.[2] These mounds lie on the west side of the Mardān and Bakhshālī roads and are evidently the remains of a former village. Close to the same spot the man found a triangular-shaped 'diwa,' a soap-stone pencil, and a large lota of baked clay with a perforated bottom. I had a further search made but nothing else was found.

"According to the finder's statement the greater part of the manuscript had been destroyed in taking it up from the place where it lay between stones. The remains when brought to me were like dry tinder, and there may be about fifty pages left some of which would be certainly legible to any one who knew the characters. The letters on some of the pages are very clear and look like some kind of *Prakrit*, but it is most difficult to separate the pages without injuring them. I had intended to forward the manuscript to the Lahore Museum in the hope that it might be sent on thence to some scholar, but I was unable to have a proper tin box made for it before I left Mardān. I will see to this on my return from leave. The papyrus will require very tender manipulation. The result will be interesting if it enables us to judge the age of the ruins where the manuscript was found."

---

[1] Apparently the manuscript was found in May 1881.

[2] General Cunningham in a private letter to Dr. Hoernle, dated Simla, 5th June 1882, says: "Bakhshālī is 4 miles north of Shāhbāzgarhī. It is a mound with the village on the top of it. The birch-bark manuscript was found in a field near a well without trace of any building near the spot, which is outside the mound village........"

This account is very unsatisfactory and there are indications that it is not altogether reliable. It was written, apparently from memory, some month or so after the discovery of the manuscript. The "ruins," it appears, were the creation of the writer's imagination, and the statement generally does not give the impression of exactitude.

4. In the meantime notices of the discovery had found their way into the Indian newspapers. Professor Bühler, who had read of the discovery in the "Bombay Gazette"[1], communicated the announcement to Professor Weber, who brought it to the notice of the fifth International Congress of Orientalists then assembled in Berlin.[2] In Bühler's letter to Weber it was stated that the manuscript had been found "carefully enclosed in a stone chamber," and it was thought that the newly discovered manuscript might prove to be "one of the Tripitakas which Kanishka ordered to be deposited in Stūpas."

There is nothing whatever in the record of the find to justify Bühler's statement, which seems to have originated in a rather strange interpretation of the words "while digging in a stone enclosure" that occur in the letter quoted above, and which are themselves of doubtful reliability. And Bühler's views would hardly have been worth recording here had it not been that their effect was altogether disproportionate to their value. Perhaps the exaggeration of the value of the find, however, served a useful purpose at the time; but now it has become embarrassing—for refuting it makes the present editor of the manuscript appear to be decrying his wares.

5. The manuscript was subsequently sent to the Lieutenant-Governor of the Punjab, who, on the advice of General Cunningham, directed it to be transmitted to Dr. Hoernle, then head of the Calcutta Madrasa, for examination and publication. In 1882 Dr. Hoernle gave a short description of the manuscript before the Asiatic Society of Bengal, and this description was published in the Indian Antiquary of 1883.[3] At the seventh oriental conference held at Vienna in 1886 he gave a fuller account which was published in the proceedings of the conference,[4] and also, with some additions, in the Indian Antiquary of 1888.[5] In 1902 Dr. Hoernle presented the manuscript to the Bodleian Library.

---

1 This account appears in the BOMBAY GAZETTE of Wednesday, August 13th, 1881, and is as follows :—

"The remains of a very ancient papyrus manuscript have been found near Baskhālī, in the Mardān tahsil, Peshawar District. On the west side of the Mardān and Baskhālī road are some mounds, believed to be the remains of a former village, though nothing is known with any certainty regarding them, and it was while digging in a ruined stone enclosure on one of these mounds the discovery was made. A triangular-shaped stone 'diwa', and a soap stone pencil, and a large lotah of baked clay, with a perforated bottom, were found at the same place. Much of the manuscript was destroyed by the ignorant finder in taking it up from the spot where it lay between the stones; and the remains are described as being like dry tinder, in some of the pages. However, the character, which somewhat resembles *Prakrit*, is clear, and it is hoped it may be deciphered when it reaches Lahore, whither we understand it is shortly to be sent."

2 The official record is given in the report of the Congress (Part I, p. 79) as follows :—

Der Präsident (Weber) verlas darauf aus einem Briefe von Prof. Bühler folgende hochwichtige Mittheilung :

"Ein Penjâbi Landmann soll beim Steinegraben einen alten Papyros gefunden haben, der sorfaltig in einen Steinkammer eingeschlossen war. Derselbe soll sehr umfangreich gewesen, doch Vieles vom Finder durch Unvorsichtigkeit zerstort sein. Die sehr bedeutenden Reste sind nach Lahore gebracht. Ganz Seiten sollen lesbar sein und die Schrift 'wie Prâkrit' aussehen. Es konnte dies wohl eines der Tripiṭaka's sein, die Kanishka in Stûpa's niederlegen liess. Ich habe gleich an Cunningham geschreiben und um ein Stuck wenigstens in Photographie gebeten."

3 Vol. xii, pp. 89-90.

4 Verhandlungen des vii Internationalen Orientalisten-Congresses, Arische Section, 127 *seq.*

5 Vol. xvii, pp. 33-48 and 275-279.

CHAPTER II.

6. The manuscript consists of some 70 leaves of birch-bark, but some of these are mere scraps. The largest leaf measures about 5·75 by 3·5 inches or 14·5 by 8·9 centimetres. The leaves, which are numbered according to the Bodleian Library arrangement from 1 to 70, may be classified according to their size and condition as follows :—

In fair condition but broken at the edges—size, not less than 5 by 3 inches (13 by 8 cm.)

1, 3, 4, 5, 6, 7, 8, 9, 10, 11, 12, 13, 14, 15, 16, 17, 18, 20, 22, 23, 24, 32, 33, 34, 36, 37, 43, 44, 47, 49, 59, 60, 61, 62, 63—total 35.

Rather more damaged but otherwise in fair condition—not less than $4\frac{3}{4}$ by 2 inches (12 by 5 cm.)

2, 25, 26, 42, 45, 46, 48, 50, 51, 52, 55, 56, 57, 58, 64, 69—total 16.

Much damaged.

21, 31, 35, 41, 53, 66, 68—total 7.

Scraps.

27, 28, 29, 30, 35, 38, 39, 40, 54, 68, 70—total 11.

One folio (19) is entirely blank.

Certain folios consist of two leaves stuck together, namely 7, 31, and 65, and possibly others. It would not be difficult to separate these double leaves without damaging the manuscript.

7. The eleven folios classified as " scraps " at first appeared completely refractory but the exercise of considerable patience has led the following to submit to some arrangement. Here the letters a, b, c, etc., refer to the scraps in the order in which they appear on the right sides of the leaves as illustrated and the revised arrangement is indicated below :—

Revised order.	References to the plates.
A1	$40^{a} + 39^{a} + 39^{b1} + 39^{c} + 38^{b}$.
A2	$39^{b} + 40^{c} + 38^{e} + 40^{b} + 39^{d}$.
A3	$40^{d} + 39^{e} + 38^{a}$.
A6	$51^{b} + 35^{a}$.
A9	$29^{d} + 29^{b} + 29^{c}$.
A10	$27 + 29^{a}$.

Of these rearrangements one side of each of A1, A6 and A9 are shown in the illustration facing p. 4. These rearrangements were made from the reproductions —a much more difficult task than working with the originals.

8. Dr. Hoernle's first estimate of the original size of the leaves was 7 by $8\frac{1}{4}$ inches, and this estimate was based upon " the well-known fact that the old birch-bark manuscripts were always written on leaves of a squarish size ", and upon the

obvious incompleteness of folio 17 (xliii),[1] but this estimate he himself corrected some twelve years later.[2]

Although no leaf of the manuscript is now complete there are 35 leaves in comparatively fair condition, and these vary in size[3] from about $5\frac{1}{4}$ by $3\frac{1}{4}$ inches to 6 by $3\frac{3}{4}$ inches. There are some lines of writing quite complete, though not many—and these complete lines measure from $5\frac{1}{2}$ to 6 inches in length. There is also evidence of a small margin on either side, and the actual space occupied is just over 6 inches or $15\frac{1}{2}$ centimetres. I estimate that the original length was from about $6\frac{1}{2}$ to 7 inches.

With regard to the depth I estimate from 4 to $4\frac{1}{2}$ inches. In a few cases we have the complete set of lines (generally from 10 to 11 to the page) and in some other cases the amount missing can be roughly estimated, while the actual maximum depth that occurs is $3\frac{3}{4}$ inches. My estimate of the original size of the leaves is therefore $6\frac{1}{2}$ to 7 inches long by 4 to $4\frac{1}{2}$ inches deep or 16·6 to 17·8 centimetres by 10·2 to 11·5 centimetres. Dr. Hoernle's final estimate was 7 by 4 inches.

8 (*a*). In ordinary Sanskrit manuscripts it is the custom to number the leaves (not the pages) in order, generally in the left hand margin of the reverse.[4] No single leaf of our manuscript has this part intact and there is no evidence of such numbering.

9. The leaves are now mounted between sheets of mica and placed within an album. The mica sheets are about 7·4 by 4·6 inches and are fixed together by strips of gummed paper at the edges leaving a clear area of $6\frac{1}{2}$ by $3\frac{3}{4}$ inches. The general arrangement is shown in Plate I (Part II). Some other method of mounting (*e.g.*, between glass plates) might be safer. It should be possible to separate the leaves now stuck together, and the possibility of thereby discovering new material would justify considerable trouble being taken in this matter.

10. Birch-bark is an outer bark of the Silver Birch (*Betula utilis, Betula bhojpatra,* or the *Bhurja tree,* as it is variously called) which flourishes in the Himalayas from Kashmir to Sikkim. It grows on all the higher ranges of the Kashmir hills from a height of about 6,000 feet to 12,000 feet. The forests in the Gurais district supply most of the *bhojpatra* that is sold in Srinagar. The bark is used chiefly for the roofing of houses, for wrapping up things, for lining baskets, etc., and the villagers still use it as a writing material.

To obtain the bark from the tree a deep cut is made vertically down a clean piece of bole, and the bark is then peeled off by the hand. The operation is very much the same as that employed in "girdling" pine trees except that the upper and lower cuts are not made. The paper bark appears to be thrown off by an under red bark and apparently one layer is produced each year. If all the laminæ are stripped off from a tree, it either dies, or, if it survives, it does not give good bark a second time. The most suitable size of tree is from 2 feet to $4\frac{1}{2}$ feet in girth. In

---

[1] *Indian Antiquary,* xvii, 1888, p. 33. The 'well-known fact' is not a fact at all.

[2] JASB, LIX, 1900, p. 126.

[3] The reproductions are of the same size as actual leaves.

[4] This is the general practice in Northern India. In the South the number is usually given on the obverse.

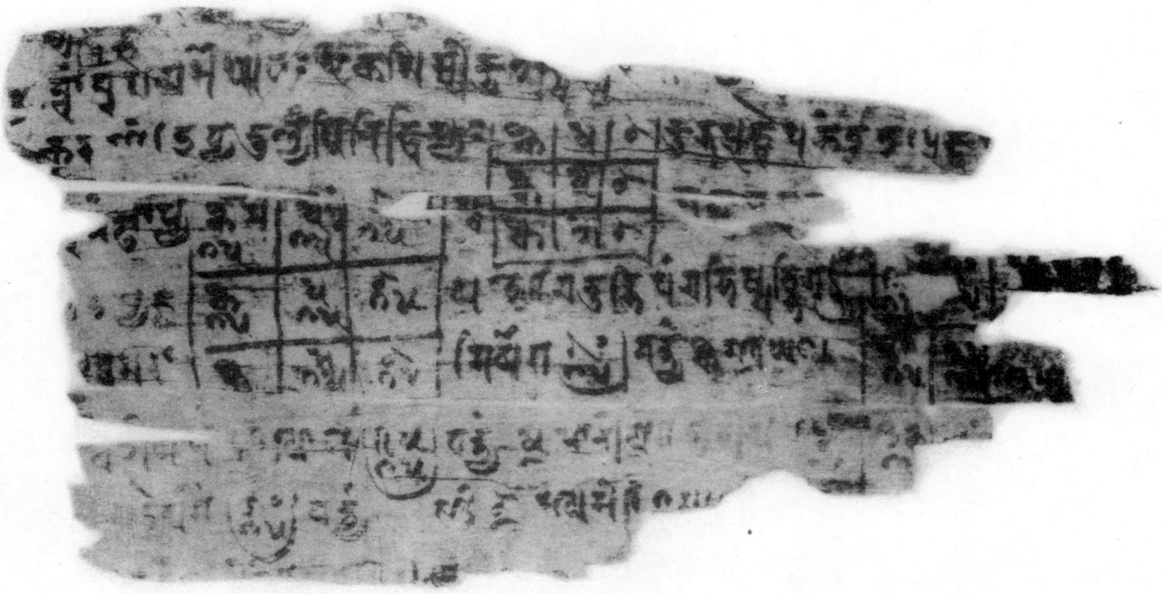

40 RECTO, E

40 RECTO, D

39 RECTO, E

38 RECTO, A

29 RECTO, D

29 RECTO, B

29 RECTO, C

51 VERSO, B

35 RECTO, A

RE-ARRANGED FRAGMENTS

larger trees the paper bark of the bole is rough and lignified and is of no use as a writing material.[1]

11. Each layer of bark is white or pinky-white on the outer side, but is a reddish or yellowish buff on the inner side. The number of layers varies and I have counted 47 in a strip taken from an old tree. A marked feature of the bark is the existence of numerous lenticels (glands) from 1 to about 6 cm. in length and from about 1 to 3 mm. in depth. These lenticels are reddish-brown in colour[2] and of darker shade than the natural reverse of the lamina, and each of them is continued throughout the several laminæ. On the natural obverse they appear much more distinct by contrast with the lighter back-ground, and in the reproductions of manuscripts this contrast sometimes appears to be emphasised. On the bole of the tree the lenticels are *horizontal* (*i.e.*, they are always at right angles to the axis of the bole or branch). On the older specimens they are slightly convex on the obverse and concave on the reverse. Traces of these lenticels can be seen in almost all the reproductions of our manuscript (see Fig. 1 and Plate XLI ; but they show much clearer in some of the Bower manuscript plates). The lenticels are of importance from the point of view of the scribe because they are of different structure from the rest of the bark, and they sometimes break away (see folio 16). There is a sort of grain running parallel with the lenticels and the bark tears easily in that direction.

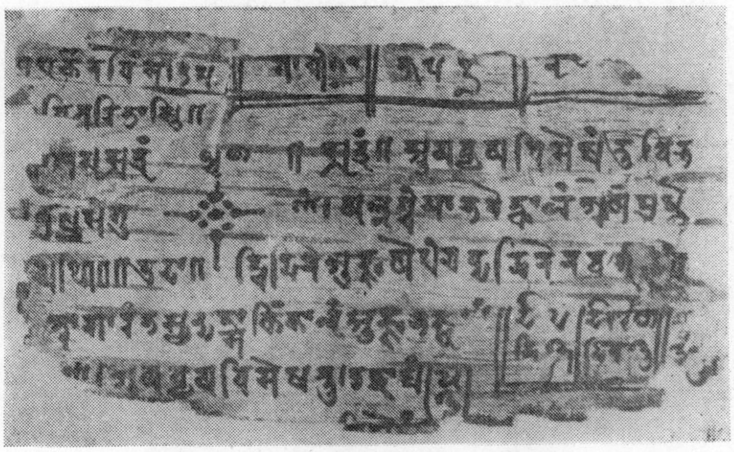

Fig. 1.

[1] *See* notes by Mr. E. Radcliffe in the *Indian Forester* (xxviii, 1902, pp. 25-27). I am much indebted to Lt.-Col. A. Gage, Director of the Botanical Survey of India, Mr. P. H. Clutterbuck, Inspector-General of Forests, and Mr. B. Coventry, Conservator of Forests, Kashmir, for information very kindly supplied by them.

The following particulars of specimen strips of birch-bark taken from trees of various ages, and kindly sent to me by Mr. B. Coventry, are perhaps worthy of record.

Girth of tree.		Size of Strip.				Number of laminæ.
		Length.*		Breadth.*		
Ft.	Inches.	Ft.	Inches.	Ft.	Inches.	
..	6	2	...	...	6	20
1	...	2	10	...	10	19
2	...	3	4	1	6	29
3	...	4	7	2	...	Shows previous stripping.
3	...	3	7	2	2	19
4	...	2	10	2	6	27 Bottom of trunk woody.
5	...	3	6	4	...	33 ,, ,, ,,
6	...	3	7	1	10	47 Taken from an upper branch.

* Here length means vertical length measured at right angles to the lenticels and breadth means the measurement *parallel* to the lenticels.

[2] On the outside of the tree they are sometimes black, and on very old trees they form woody excrescences.

12. The art of preparing the bark for writing upon appears to be lost, but Albīrūnī tells us that the strips were rubbed with (?) oil and polished. The manuscripts preserve no evidence of either of these processes. All that they tell us is something of the process of sub-division, and arrangement. Each leaf of the Bakhshālī manuscript appears to consist of half of the original thickness of the strip, *i.e.*, the original strip was divided into layers each of which consisted of some six laminæ. In the Bower manuscript the number of laminæ to a leaf varies from two to six while one leaf consists of at least twelve liminæ. The Kashmirian Arthava Veda exhibits rather more elaboration : the process of sub-division is carried to the extreme limits and each leaf consists of two single laminæ pasted together.

13. Possibly the original strip of birch-bark from which the leaves of the Bakhshālī manuscript were taken was roughly of the shape of the annexed diagram and was cut up into the oblongs indicated. If A, B, C, etc., represent the upper layer, and A^1, B^1, C^1, etc., the lower layer, then, according to the evidence of the leaves themselves, they were arranged for purposes of writing upon in the order A, A^1 ; B, B^1 ; C, C^1 ; etc., or A, A^1 ; D, D^1 ; etc.

A	B	C
D	E	F
G	H	I
J	K	L
M	N	O
P	Q	R

Fig. 2.

14. Birch-bark, even after preparation for writing upon, retains the natural marks of the wood showing the lenticels and occasionally knots. Such knots are like thumb marks in their individuality, so that, if a knot recurs, it can generally be identified. Returning to our diagram—it is obvious that if a knot occurs on, say, D its doublet will occur on D^1. In the Bower manuscript three knots that occur on folio II, 7 recur on II, 8 ; the knot on III, 8 recurs on III, 9 ; several knots on II, 23 recur on II, 24 ; one on II, 31 recurs on II, 32 ; and so on ; and in no case does any particular knot occur on more than two leaves. In the Bakhshālī manuscript the order of the leaves was uncertain but it was noticed that folios 12 and 13 had a common knot, and so had folios 32 and 36, folios 44 and 49, folios 51 and 52, folios

53 and 66. If our scheme of the arrangement of leaves also holds good here, then each of these pairs of leaves should consist of consecutive leaves, and the final order, based on other considerations, places them thus—

Bodleian order.	Final order.
12	G3
13	G4
32	M3
36	M4
44	M10
49	M9
51	A8, 9
52	A10
53	E2
66	E1

15. It should be noted also that where the reproductions are of double leaves stuck together the two sides should have no common natural markings (see, for example, folio 65); and, conversely, when the two sides have no natural marks in common there is ground for suspicion (*e.g.*, folios 46 and 3). In this matter too much reliance should not be placed on the reproductions.

Format.

16. Of the very few early birch-bark manuscripts the following are perhaps the best known :—

(*a*) The Kharoshṭhī Dhammapada from Khotan. Size 8 by 36[1] inches, (Ratio 0·2)[1]. Period 2nd century A.D.

(*b*) The Bower manuscript from Khotgarh. Size 9 by 2 to 11½ by 2½ inches, (Ratio 4·5). Period 5th century A.D.

(*c*) The Bakhshālī manuscript (Oxford). Size 7 by 4 inches, (Ratio 1·7). Period ?

(*d*) The Kashmirian Arthava Veda. Size 9·8 by 7·8 inches, (Ratio 1·25). Period 15th century A.D.

(*e*) The Deccan College Rig Veda (Poona). Size 10¼ by 9½ inches, (Ratio 1·08). Period ?

---

[1] The sizes here given are, of course, only approximations to the average sizes. See (*a*) *Journal Asiatique*, 1898, pp. 192 ff. (*b*) *The Bower Manuscript*, by A. F. R. Hoernle. (*d*) *The Kashmirian Arthava Veda*, Bloomfield & Garbe. (*e*) *Cat. MSS. of the Deccan College, Poona*, Vol. I, p. 1.

The shapes and relative sizes of these manuscripts are represented in the accompanying diagram—

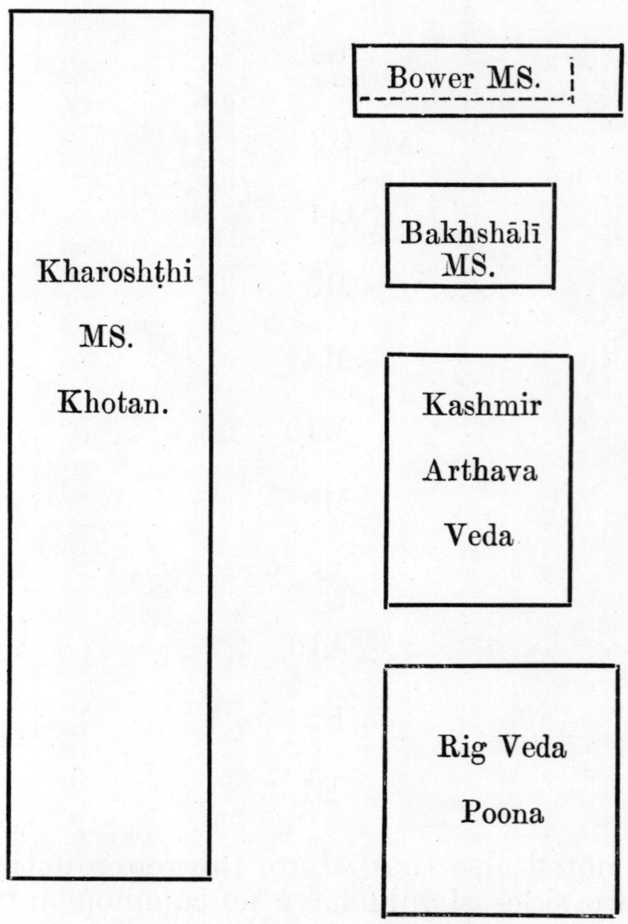

17. The format of a book is generally determined by several semi-independent factors, *e.g.*, material, economy, fashion, convenience, contents. Our first example of a birch-bark book, the Khotan manuscript, seems to have been shaped in imitation of the western papyrus roll[1]; the second example, the Bower manuscript, is probably in imitation of the palm leaf manuscripts of India; while the later examples are in imitation of paper books. The Bakhshālī manuscript appears to stand in an intermediate position and there is, of course, a possibility that its format may indicate its age within certain limits.

18. In all birch-bark manuscripts the writing is parallel to the lenticels, which on the bole of the tree are horizontal; and the reason for this direction of the writing is that the bark tends to split in the same direction. In the Bower manuscript there are one or two exceptional examples of writing inclined to the lenticels (Plates I and II) but the angle of inclination is small. It should also be borne in mind that the length of the strip of bark (parallel to the writing) is limited to the circumference of the tree, and the depth of the strip is also practically limited,[2] and that the format was, at least, partly determined by the necessity for dividing these strips economically.

19. But fashion was probably a more powerful factor than material in the determination of the format of birch-bark books, and it is possible that the method of preparing papyrus for writing upon to some extent influenced those who prepared birch-bark for the same purpose. The stem of the papyrus plant was cut into

[1] Numbers of such rolls were discovered by Sir M. A. Stein in Central Asia.

[2] See footnote on page 5.

longitudinal strips which were laid on a board side by side to the required width, and across this layer another of shorter strips was laid at right angles. After soaking, the combined sheet was hammered and dried and polished with ivory or a smooth shell. Some twenty such sheets were pasted together to form a roll. Later it was the practice to back one sheet with a second in order to obtain a more suitable paper. The depth of the rolls varied from 4 to 12 inches. The later limit for the use of papyrus as a writing material was about the 9th century A.D.

20. The palm leaf manuscripts of India are made from the leaves of *Corypha umbraculifera* or *Borassus flabellifer.* The former is indigenous in south India but the latter was probably introduced from Africa. The leaves of both of these trees are long and tapering, with central ribs. All the earlier palm leaf manuscripts are made from the leaves of Corypha. The following are examples :—

Reference.	Locality.	Length.	Depth.	Ratio.	Date.
Horiuzi	W. India . . .	11 in.	2 in.	5·5	? 520 A.D.
Bendall Add. 1049 . .	„ „ . . .	16 „	2 „	8·0	859 „
„ „ 1683 . .	Nepal . . .	21 „	2 „	10·5	1039 „
„ „ 1688 . .	Bihar . . .	22 „	2 „	11·0	1054 „
Kielhorn No. 42 . . .	W. India . . .	13 „	$2\frac{1}{4}$ „	5·8	1123 „
Peterson No. 220 . . .		$33\frac{1}{2}$ „	$2\frac{1}{4}$ „	15·0	1205 „

From measurements of a large number of Corypha manuscripts it appears that the most usual depth was about two inches and that the lengths varied from about 10 to 34 inches. The chief mode was from 10—16 inches but there was a second mode from 30—34 inches. It appears that the whole length of the prepared leaf was occasionally used, but that more often it was cut into thirds, or sometimes into halves.

About the middle of the fifteenth century the use of palm leaf as a writing material entirely ceased in western India, but in eastern India it continued to be used side by side with paper until much later.

21. The art of paper making appears to have been practised by the Chinese at a very early date. From the Chinese the Muslims learnt the process in the eighth century and they introduced it into Europe and also India about the twelfth century. Among the earliest Muslim paper manuscripts now preserved, one was written in A.D. 866, others in 974, 980, 990, etc. The earliest Indian paper manuscripts known were written in A.D. 1231 and 1343. The earlier one measures 6 by 4 inches (ratio 1·5) and the other $13\frac{1}{2}$ by 5 inches (ratio 2·7).

22. Another very common writing material in early India was copper. Bühler states that the size of the copper plate used was largely determined by the common writing material of the locality, *e.g.*, the palm leaf, birch-bark, etc. This statement, however, is not fully supported by actual evidence, for the more usual ratios of lengths to depths of copper plates vary from about 1·4 to 2·6 as against 5 to 15 for palm leaf manuscripts.[1]

[1] There are rare examples of copper plates of ratio over 4 (*e.g.*, two Taxila plates of ratio 4·7).

23. Before making definite deductions from the formats of our birch-bark manuscripts, a great deal more investigation is obviously required, and in particular the formats of the western tablets and codices, the early Chinese and Muslim paper books should be studied. Until the introduction of paper into India there does not seem to have been much consistency in the format of birch-bark manuscripts. The Khotan manuscript may have been in imitation of the papyrus rolls of the west. The width is about 8 inches (20 cm.), the length is unknown, but the total length is about 4 feet (1 m. 23) and the longest piece of the Paris portion is about 24 inches (61 cm.). The scapus was made by joining pieces of unknown depth together[1] and at each side a fine cord was threaded through the bark about one cm. from each edge. Most probably it was never intended that the manuscript should be rolled up : possibly it was to be hung on a wall. The writing is parallel to the short side, and in this matter differs entirely from the papyri, on which the writing is generally parallel to the length of the roll, and in columns of from two to three and a half inches wide. The writing is on one side only of most of the fragments of the Khotan manuscript, but on one long piece (C) both sides are written upon. This piece (C) consists of several layers of bark (it is, of course, impossible to count these layers without examining the original). The lenticels are identical on the two sides and the *verso* side is the lighter side of the bark. The manuscript appears to have been folded up to the size of 20 by 5 cm., but obviously this was not the original intention, for the folds are independent of the position of the lines of writing, and folding birch-bark manuscripts was never a wise procedure.

That the Bower manuscript format was determined by the palm leaf *pothī* is probably true, although it should be noted that 9 × 2 inches is a very uncommon size for an early palm leaf manuscript. (See Dr. Hoernle's lists in the JASB, 1900, pp. 99ff.)

24. Regarding the format of the Bakhshālī manuscript (ratio 1·6) as a criterion of age, I can come to no positive conclusion from the evidence before me. Further investigation is required and might be profitably undertaken if time allowed. Dr. Hoernle, however, writes as follows :—

> "It is noteworthy that the two oldest (Indian paper) manuscripts known to us point to their having been made in imitation of such a birch-bark prototype as the Bakhshālī manuscript."

It is not easy to accept this argument, for it would be quite as reasonable to conclude that the Bakhshālī format was determined by the paper manuscript formats, and that it is of later date than the introduction into India of paper as a writing material; and this would place the Bakhshālī manuscript about the twelfth century of our era at the earliest. However I only give this hypothesis as a set-off to Dr. Hoernle's unjustifiable deduction.

The script.

25. The Bakhshālī text is written in the Śāradā script, which flourished on the north-west borders of India from about the ninth century until within recent times. Its distribution in space is fairly definitely limited to a comparatively small area lying between longitudes 72 and 78 east of Greenwich and latitudes 32 and 36 north. Dr. Vogel distinguishes between Śāradā proper, of which the latest examples are of the early thirteenth century, and modern Śāradā.

---

[1] But I can see no evidence of this in the illustrations given in the *Journal Asiatique* (1898, p. 308).

The writing of the Bakhshālī manuscript is of the earlier period and is generally very good writing indeed. It was written by at least two scribes. In the table on p. 97 the styles of writing are indicated by the letters α and β of which β pertains wholly to the " M " section. Style α is divided into four subsections which possibly belong to the work of four separate scribes, although it is not easy to point out any fundamental differences between these styles. Folio 65 possibly exhibits the writings of two separate scribes on the two sides, which do not belong to the same original leaf. Compare also folios 29 and 17.

Possible points of differentiation between the sub-divisions α_1, α_2, α_3 and α_4 are the length of medial *i*, the length of the final up-stroke of initial *u*, the length of the *virāma* mark, the occurrence of the sickle shaped *ī* (folios 1, 35, 52, 60), of the " clubbed " *ai* and *e* (folios 16 and 60), of the " curved " medial *e* (folios 3, 18, 5), and the different methods of forming medial *e*, *ai* and *o*.

The style (β) of section " M " is distinguished by its boldness (contrast plates II and XXV), by " tails " or flourishes (including very long *virāma* marks), the methods of writing medial *ai* and *o*, the looped six, etc., etc.

In Part II the script is examined in detail and further peculiarities of the different sections are given, *e.g.*, section " M " has no examples of *ai* and contains practically all the examples of *ö* Sections L (α_4) and G (α_3) show marked differences in the methods of writing medial *e*, while section F is, in this matter, very much like section L. Section C (α_2) has no example of either the *jihvāmūlīya* or *upadhmānīya*, and so on ; but it must be borne in mind that these statistics are only of value in the mass.

The language.

26. The language of the text may be described as an irregular Sanskrit. Nearly all the words used are Sanskrit, and the rules of Sanskrit grammar and prosody are followed with some laxity. The peculiarities of spelling, sandhi, grammar, etc., that occur in the text are exceedingly common in the inscriptions of the eleventh and twelfth centuries found in the north-west of India. Dr. Hoernle, however, implies that the language is much older. He states that the text " is written in the so-called Gāthā dialect, or in that literary form of the North Western Prākṛit which preceded the employment, in secular composition, of the classical Sanskrit." He also states that this dialect " appears to have been in general use, in North Western India, for literary purposes, till about the end of the 3rd century A.D."[1]

The subject of language is discussed to some extent in Chapter viii and will be more fully dealt with in Part III of this work.

[1] *Ind. Ant.* xvii, 1888, pp. 37 & 38.

CHAPTER III.

ORDER.

27. The Bodleian Library order of the folios, which was definitely fixed by circumstances not within the control of the editor, is necessarily followed in this volume in the arrangement of the facsimiles and the first transliteration of the text. But a more satisfactory arrangement was considered desirable, and a very considerable amount of labour has been involved in the attempt to achieve an approximately correct order.

The leaves were disarranged to some extent before they reached Dr. Hoernle, but unfortunately he did not leave any proper record of the order in which the leaves reached him.[1]

28. A detailed analysis of all the leaves showed that Dr. Hoernle's rearrangement was faulty,[2] that the "find order", as far as known, was better, but not altogether reliable. Further rearrangement was necessary and had to be attempted; but leaves were missing, some were in fragments placed anyhow, not a single leaf was complete, and the connecting portions of the text were all wanting.

29. Dr. Hoernle attempted to arrange the leaves on the basis of the numbered *sūtras;* but the numbered *sūtras* were too few and too unevenly distributed to serve this purpose, and in some ways they were even misleading. The chief criterion of order is, of course, the nature of the contents of the leaves, but I have seized upon any means available that has offered any help towards the solution of the problem of order. Had all the leaves been extant, even in fragments, the problem could have been completely solved; but some leaves are completely missing and many fragments have disappeared altogether, so that the problem is only partially soluble. I have already explained how the knots in the birch-bark were of assistance, and besides this natural aid there was the accidental one of the effects of the method of storage. Possibly for some hundreds of years the bundle of leaves was subject to a certain amount of pressure, and was exposed, particularly at the edges, to chemical and other disintegrating actions. Some of the leaves stuck together, the edges of all became frayed and certain leaves became so frail as to break up into scraps when handled. On the principle that contiguous leaves would be affected approximately to the same extent, we might, if no disintegrating effects had taken place since the find, rebuild in layers the original bundle. But we know that further disintegration has taken place, and is still not altogether eliminated; nevertheless, similarity in size and shape and mechanical makings were distinctly helpful in rearranging the leaves.

30. The script was helpful in a general way: indeed it led me to differentiate the "M" section from the rest of the text. The language employed was also helpful, particularly with reference to the use of special technical terms; while such examples as the following were noted after some order had been restored:—*ghna* "multiplied by" occurs only on folios 7, 45, 46, 57 and 65, all of which belong to the same section (C); *pratyaya, yuta* and *rūpa* occur very often but never in the "M" section, etc., etc.

31. The chief criterion—that of the subject matter—could not always be utilised, owing to the absence of all the actual connecting portions and to the

---

[1] In the following notes by 'find order' I mean the order in which the leaves reached Dr. Hoernle. This order is not completely known but Dr. Hoernle marked the order of certain of the leaves.

[2] It would perhaps be more correct to say that it was incomplete. He placed leaves 10 to 18 and 60 to 63 in proper sequence.

numerous larger gaps, but occasionally it was definitely decisive : *e.g.*, one example goes from folio 65 to folio 66 and then to 64, another from 66 to 67, and another from 29 to 27, etc. We also find the same topic spread over a number of leaves, and here it was often possible to reconstruct a logical sequence that must be in some correspondence with the original.

Finally the numbered *sūtras* were useful as a check and also as indicating to some extent the relative positions of certain sections.

32. The problem of rearrangement, however, had not only to do with the order of the leaves but with the order of the pages and also the order of the fragments. In a number of cases the Bodleian arrangement places the proper first side of a leaf *in verso*, and the fragments are very often placed with wrong leaves. The illustration facing page 4 shows some of these fragments rearranged, and the table on page 14 enumerates some 13 leaves placed with the obverse *in verso*.

33. The order now given does not pretend to be the exact original order : it is merely a compromise. It brings together those portions of the remains of the text that deal with the same topics. It is not necessarily a logical order according to modern views : indeed, few mediæval mathematical works exhibit such a logical order. Neither is the order a final one, and probably it is not the best that could have been obtained ; but a working order had to be achieved even before the detailed examination of the manuscript could be completed, and now if the labour were not too great I should be tempted to revise the order once more. Further, if an attempt is made, as I trust it will be, to separate those leaves that are stuck together, then the question of order will possibly once more have to be considered.

34. The following tables show the various orders—

BODLEIAN ORDER.

(The facsimiles shown in plates ii to xlvii are in this order.)

Folio.	Revised order.	Folio.	Revised order.	Folio.	Revised order.	Folio.	Revised order.
1	A 11	21	E 4	41	J 3	61	L 2
2	A 12	22	E 5—F 1	42	M 13	62	L 3
3	A 13	23	F 2	43	M 12	63	L 4
4	B 4	24	F 3	44	M 11	64	C 6
5	C 1	25	F 4	45	C 8	65	C 4 + J 2
6	C 2	26	F 5 + ?	46	C 9 + D 1	66	E 1
7	B 3 + C 3	27	A 10	47	M 7	67	D 6
8	B 1	28	D 7	48	M 8	68	D 4
9	B 2	29	A 6—7	49	M 9	69	D 3
10	G 1	30	J 1	50	M 14	70	D 2
11	G 2	31	D 5	51	A 6 - 7		
12	G 3	32	M 3	52	A 8		
13	G 4	33	M 2	53	E 2		
14	G 5	34	M 5	54	A 4		
15	G 6	35	A 5—6	55	M 10		
16	G 7 + H 7	36	M 4	56	C 5		
17	H 2	37	M 6	57	C 7		
18	H 3	38	A 1—3	58	E 3		
19	Blank	39	A 1—3	59	K		
20	M 1	40	A 1—3	60	L 1		

REVISED ORDER, &c.

Revised order.	Folios.	Numbered Sūtras.	Find order.	Revised order.	Folio.	Numbered Sūtras.	Find order.	Revised order.	Folio.	Numbered Sūtras.	Find order.
A 1	$40^{a}+39^{a}+^{b'}+^{c}+38^{b}$		32	D 1	46 V.		9	H 1	16 V.	?	30
A 2	$39^{b}+40^{c}+38^{c}+40^{b}+39^{d}$		32	D 2	70		65	H 2	17		28
A 3	$40^{d}+39^{e}+38^{b}$		32	D 3	69		64	H 3	18	? xxviii	27
A 4	54		?	D 4	68		64				
A 5	35^{b}		?	D 5	31		63	J 1	30		32
A 6	$51^{b}+35^{a}$		37	D 6	67 V.R.		62	J 2	65 R.	?	?
A 7	51^{a}		37	D 7	28		60	J 3	41		?
A 8	52		37								
A 9	29		32	E 1	66		58	K	59	l	?
A 10	27 + 29		32	E 2	53		?				
A 11	1	ix, x	33	E 3	58		?	L 1	60	li, lii	?
A 12	2		34	E 4	21		55	L 2	61	liii	?
A 13	3	xiii, xiv	49	E 5	22 R.			L 3	62	liv, lv	?
								L 4	63	lvi, lvii	?
B 1	8	?, ?	48	F 1	22 V.	?, ?	54				
B 2	9 V.R.		43	F 2	23		52	M 1	20		26
B 3	7 V.		45	F 3	24		?	M 2	33		25
B 4	4	xv, xvi	44	F 4	25		51	M 3	32		24
				F 5	26		50	M 4	36		23
C 1	5	xviii	47					M 5	34		22
C 2	6		46	G 1	10	xxiv	42	M 6	37		21
C 3	7 R.		45	G 2	11		41	M 7	47 V.R.		20
C 4	65 V.		?	G 3	12		40	M 8	48		19
C 5	56 V.R.		16	G 4	13		39	M 9	49 V.R.		11
C 6	64		?	G 5	14		?	M 10	55		14
C 7	57 V.R.		?	G 6	15	xxv	29	M 11	44 V.R.		12
C 8	45		10	G 7	16 R.		30	M 12	43		13
C 9	46 R.		9					M 13	42		18
								M 14	50 V.R.		17

NOTE.—R. stands for *Recto*, V. for *Verso* and V.R. indicates that the facsimile should be reversed.

CHAPTER IV.

The contents of the manuscript.

35. The portions of the manuscript that have been preserved are wholly concerned with mathematics. Dr. Hoernle described the work in 1888 in the following words[1] :—

"The beginning and end of the manuscript being lost, both the name of the work and its author are unknown. The subject of the work, however, is arithmetic. It contains a great variety of problems relating to daily life. [1]The following are examples :—'In a carriage, instead of 10 horses, there are yoked 5, the distance traversed by the former was one hundred, how much will the other horses be able to accomplish?' The following is more complicated :—'A certain person travels 5 *yojanas* on the first day, and 3 more on each succeeding day; another who travels 7 *yojanas* on each day, has a start of 5 days; in what time will they meet?' The following is still more complicated :—'Of 3 merchants, the first possesses 7 horses, the second 9 ponies, the third 10 camels; each of them gives away 3 animals to be equally distributed amongst themselves. The result is that the value of their respective properties becomes equal; how much was the value of each merchant's original property, and what was the value of each animal?' The method prescribed in the rules for the solution of these problems is extremely mechanical and reduces the labour of thinking to a minimum."

36. It is necessary to emphasise the fact that this early estimate of the value of the work is inadequate and misleading. One reason for this is, that, when the estimate was made, only a comparatively small portion of the contents of the manuscript had been understood, and that was by no means the most interesting portion of the work.

37. The following is a summary list of the contents of the work as far as its present state allows of such analysis :—

	Section.
Problems involving systems of linear equations	A
Indeterminate equations of the second degree	A & K
Arithmetical progressions	B & C
Quadratic equations	C
Approximate evaluations of square-roots	C
Complex series	F
Problems of the type $x\ (1-a_1)\ (1-a_2)\ .\ .\ (1-a_n)=p$	G
The computation of the fineness of gold	H
Problems on income and expenditure, and profit and loss	L, D & E
Miscellaneous problems	M
Mensuration.	

Such is a very rough outline of the work as it now stands. Perhaps the most interesting sections are C, A and M; and of these C is the most complete and was evidently treated as of considerable importance. Section A is also of special interest as it contains examples which may be described as of the *epanthem* type. Section M is of interest principally on account of the methods of expressing the numerous measures involved and also because of its literary and social references.

38. Although the work is arithmetical in form it would be misleading to describe it as a simple arithmetical text-book. No algebraical symbolism is employed, but the solutions are often given in such a general form as to imply the complete general solution, *i.e.*, the solutions, though arithmetical in form, are really generalised arithmetic, or algebra. First of all a particular rule is given, which is

---

[1] *Ind. Ant.* XVII, 1888, p. 33.

intended to apply to the particular set of examples that follows. These rules are often expressed in language that would be impossible to interpret without the light thrown upon them by the solutions. The examples are themselves sometimes trivial, but the solutions, often expressed with what at first glance appears to be meticulous care, often redeem the examples from their apparent triviality. Proofs or verifications are often given with some elaboration and on occasions are multiplied.

39. The work may be divided roughly into algebraic, arithmetical and geometrical sections; but the boundaries of these sections are not clear, and perhaps it would be more correct to classify the problems as (*a*) academic, (*b*) commercial, (*c*) miscellaneous.

Judging by the manuscript as it now stands, the problems involving geometrical notions were comparatively very few, and we can only guess at the meanings of the remaining fragments dealing with this branch of mathematics.

One pleasing feature is the small space occupied by commercial problems. There is only one problem on interest, and a rather unobtrusive section containing problems on profit and loss.

Those problems classed as academic are concerned with particular mathematical notions that in early mediæval times had a traditional value and interest, such as the *epanthema*, the *regula virginum*, certain indeterminate equations of the second degree, and certain sets of linear equations.

The miscellaneous problems include examples where the chief interest is rather in the illustrative material than in the mathematical notions involved; *e.g.*, there are problems concerned with the abduction of Sītā by Rāvana, the prowess of Haihaya, the constitution of an army, the Sun's chariot, the daily journey of the planet Saturn, gifts to Śiva, etc., etc.

40. Such, or similar features, are, however, common to many mediæval mathematical works, and we now turn to the consideration of the distinguishing features. Characteristics that completely differentiate one such mathematical work from another of the same period are generally hard to find and generally in order to so differentiate we have to sum up a number of minor characteristics. The Bakhshālī manuscript is, however, almost unique in at least two respects of some mathematical importance. The first of these is the employment of a special sign in the form of a cross—exactly like our *plus* sign but placed after the quantity it affects—to indicate a negative quantity. The possible connexion of this with the Diophantine sign and its value as a chronological test will be dealt with in due course. The second special characteristic consists of the set of methods for indicating the change-ratios of certain measures. This characteristic also will be dealt with in some detail in due course; and it must suffice for the present to point out that neither of these peculiarities is very helpful in placing the work.

41. If the work were actually of the period to which Dr. Hoernle assigned it, then by far the most remarkable feature would be the employment of the modern place-value arithmetical notation; but Dr. Hoernle's estimate of the age of the work was wrong and the occurrence of this notation is a common-place matter. It draws our attention, however, to the skill of our author in the manipulation of numbers. Large quantities are dealt with and one particular number contains 23 digits. These large numbers do not appear to be given for mere effect for they occur quite naturally as the result of a rigorous logic. Indeed the author seems generally to prefer simple numbers.

These large quantities again lead us to note other rather special characteristics of the work, namely the apparently over-elaborated exposition of the "workings" of the solutions (to which further reference will be made), to the preservation of the generality of the solutions throughout such workings, and to the consequent necessity of preserving the quantities that occur so that the penultimate statement shall involve the whole formula. That is, although every operation is arithmetical in form, the quantities involved are not "simplified" or "cancelled" without some special justification until the final result is achieved. The (unwritten) rule followed by the author was something like this : The integrity of the method of solution must be preserved, and so long as it is preserved the calculation may be simplified—otherwise not. Indeed the numerical quantities in these problems are treated almost like algebraic symbols.[1]

42. There is no actual algebraic notation employed but the unknown quantity appears to be indicated in certain examples by the usual symbol for "nought," which symbol, however, is never the subject of operation. Where it occurs the *regula falsi* is employed—probably because of the lack of an efficient symbolism. Neither is there any symbol of operation. The negative sign (+), already alluded to, is never used as such. Operation is generally indicated by some definite *ad hoc* term but sometimes by relative position. Fractions, for example, are indicated in the modern way but without any horizontal bar between the numbers, and sometimes division proper is so indicated.

43. If we considered the Bakhshālī text to be a work of pure Indian origin, then it would be also unique in another respect, namely in the rather extensive employment of the square root rule that may be expressed by $\sqrt{A^2+b} \backsim A+b/2A$. For this rule, the early history of which is well known, was never used in the early Indian works to any extent, whereas the Bakhshālī text employs it for a comparatively large number of examples and applies the rule to second approximations in a very thorough manner.

On the same assumption as to an Indian origin the use of the sexagesimal notation in connexion with an approximation to a square-root of a non-astronomical quantity would also be unique, but it may be noted that there is only one such example preserved and this is given in a rather timid way, and the notion involved is not pursued.

44. Whether of a purely Indian origin or not, the work is Indian in form. It is written in a sort of Sanskrit and generally conforms to the Indian text-book fashion, but there are certain apparent omissions. Perhaps the most noteworthy feature of the classical Hindu texts is their treatment of indeterminate equations of the first degree, while their greatest achievement is the full solution of the so-called Pellian equation. A great part of the texts of Brahmagupta, Mahāvīra and Bhāskara are devoted to one or both of these topics, but there is no evidence of either in what remains of the Bakhshālī text; and this apparent omission is the more noticeable, inasmuch as there is evidence of considerable skill in the treatment of systems of linear equations and certain indeterminates of the second degree. Another omission to note is of a different character altogether. Every early Hindu work of this kind has a section relating to the "shadow of a gnomon," but in our text there is no evidence of such a section. We must not, however, pay too much attention to these apparent omissions. The possibility of entire sections of the manuscirpt being destroyed is not great, but negative evidence and a mutilated manuscript do not carry us very far.

---

[1] These remarks do not apply to the 'M' section.

Non-mathematical elements.

45. Many of the mathematical examples contain references to the affairs of gods and men and some of these references are of rather exceptional interest. The names of certain deities and semi-historical beings occur, and there are references to legends which may give some clue as to the origin of the work. Of interest also are the animals, crops, metals, etc., mentioned. Before indulging in conjectures as to the significance of these references it will be convenient to exhibit them in some sort of order.

46. Śiva. The name of Śiva occurs on a fragment (folio 50) that appears to exhibit a sort of colophon, in which the gift of calculation to the human race is attributed to the god. The actual name Śiva occurs nowhere else, but another name of the god—Śūlin—is given on folio 34, where the example refers to certain offerings made to him. On folio 44, in a similar example, part of the expenditure is in offerings for Śū°, and this, it is conjectured, stands for Śūlin. The term Devī occurs (folio 49) and probably indicates Śiva's consort.

The significance of these references (which all occur in section "M") is emphasised by Dr. Hoernle's remarkable claim that the work is Buddhist or Jain.[1] This appears to have been mere conjecture, for we find nothing whatever in the text that conflicts with the Śaivism indicated. Neither is the authorship of the work by a follower of Śiva in any way surprising, for, it may be noted, Kashmir Śaivism began to re-flourish about the tenth century A.D.,[2] and Kalhaṇa was a Śaivite.

Vāsudeva is a name applied to Kṛishṇa. The phrase that occurs (folio 44) is *vāsudevasya chārchanet.*

Suras and Asuras. Suras are generally classed as minor deities. In our text (folio 33) they are said to dwell on Sumeru. In the same context Asuras, demons or enemies of the gods, are contrasted with the Suras.

Rākshakas are classed as evil spirits but are not generally very clearly defined. The term occurs on folio 65, recto (part of a double leaf, and perhaps not correctly placed in section J). The problem, which is concerned with *jīva-loka,* is not understood.

Siddhas and Vidyādharas. These two terms occur together on folio 37, where the chariot of the Sun is said to be guided by the god Mahoraga among the Siddhas and Vidyādharas. The Siddhas are semi-divine beings of great holiness, who dwell in the region of the sky between the earth and the sun, while the Vidyādharas are inferior deities inhabiting the same region.

Mahoraga is Śesha or any other great serpent. The connexion between serpents and the Sun is supposed to be somewhat intimate in certain mythologies but I have not found the source of the present reference.

47. Sītā. The name Sītā is not actually preserved in our text, but on folio 32 is an example based upon her abduction by Rāvaṇa. The tale is that Sītā, the wife of Rāma, was carried off through the air by Rāvaṇa, and that, in order to attract the attention of her helpers, she tied up some jewellery in a garment and

---

[1] *Ind. Ant.* XVII, 1888, p. 38.

[2] *See* R. G. Bhandarkar, Vaishnavism, Saivism, etc., pp. 129-131.

dropped it to earth, and this circumstance forms the substance of the problem given in our text.

Rāvana is mentioned on folio 32. See Sītā.

Pārtha, irritated in a fight, shot a quiver of arrows to slay Karṇa. With exact relationship between the two terms is not certain, but the problem in which they occur relates to the destruction of an army, and the question is—how many arrows were used? A very similar problem is given twice over by Bhāskara (*Līl.* 67; *Vij. Gaṇ.* 133) as follows :—

> Pārtha, irritated in a fight, shot a quiver of arrows to slay Karṇa. With half of his arrows he parried those of his antagonist; with four times the square-root of the quiver full he killed his horses; with six arrows he slew Śalya, with three he demolished the umbrella, and with one he cut off the head of the foe. How many were the arrows which Arjuna let fly?

In Bhāskara's example Pārtha = Arjuna, the Pāṇḍava prince, but there is no connexion between him and the Haihayas. The Haihaya Arjuna was a great king but an altogether different individual. There is, however, little doubt that the examples in the two texts were connected, either directly or by some common source.[1]

Arjuna is mentioned on folio 34 in an isolated phrase......*ārjunena griddhra* ......Arjuna was connected with the Nāgas and some contest with a form of Garuḍa would not be out of place.

48. **Yudhishṭhira.** On folio 37 is the isolated fragment *rāja yudhisṭhira nāma Pāṇḍu vaṁśa*...........which implies some familiarity with the great epic of India.

Śatrudama. A similar fragment occurs on folio 47 *kaschid rāja-kumāra Śatrudama*.........If this prince could be identified the reference to him might prove a valuable clue. Compare *Śatrughna* which has the same meaning.

Sundarī. On folio 34 Sundarī, "the beautiful one," is asked to solve the problem, that is, she is addressed in exactly the same way as Līlāvatī is in Bhāskara's well-known work. Here is either imitation or a common tradition.

Chhajaka. The text (*i.e.*, section "M") is written by a Brahman, the son of Chhajaka (folio 50). Possibly this name is the same as Śajjaka, which occurs several times in the *Rājataraṅgiṇī*. Śajjaka was superintendent of the Śeda office in Kalhaṇa's time (XIIth century), but there is no real justification for connecting this individual with the author of our text.

49. Besides these names of individuals there are references to various classes of men—kings, princes, priests, learned men, bankers, servants, soldiers, tax-collectors. Merchants not only appear in money transactions but, in one case, with Brahmans and others, as recipients of propitiatory gifts. Pandits and learned men are spoken of as earning wages, and two "Rājputs" are described as servants of a king. The problems are often addressed to an individual or individuals variously described as "the best of calculators" (*gaṇakottama*), "wise" or "learned" man (*budha, prājña, paṇḍita, dharmajñaya*, etc.), or "friend."

---

[1] The mathematical prototype is common in the Greek Anthology.

50. The collection of animals is limited to elephants, horses, camels, cows, buffaloes, snakes, a worm and a vulture. There is little of significance here, but the occurrence of camels points to the north-west of India. There are, however, two classes of horses mentioned (*aśva, haya*); for example, one problem deals with certain numbers of *aśva, haya,* and *ūshtra* (camel). Dr. Hoernle translated *haya* by "yak," which would be interesting if it were justifiable; but on one occasion the two terms seem to be used synonymously (fol. 8) and the "horse" of an army are designated *haya* and *turaga* (fol. 47).

Of food stuffs and other commodities we have wheat, barley, rice and saffron (this combination being of some significance[1]); gold, iron, salt, molasses, and (?) lapis-lazuli (*ambhaloha*).

51. The mention of a boat sailing against adverse winds is not necessarily "local colouring" but we must assume that it was, at least, comprehensible. Chariots (*ratha*) are mentioned three times—twice it is the chariot of the Sun, and on the third occasion the chariots form one section of an army, which is said to consist of chariots (*ratha*), elephants (*gaja*), foot soldiers (*nara*), and "horse" (*haya*) in the ratios 1 : 1 : 5 : 3. Certain "divisions" of an army are also mentioned, namely *chamū, pṛitanā, anīkinī* and *akshauhiṇī* (fol. 47).

52. The references to religious matters are few but are significant. We have already mentioned offerings to Śiva, Devī and Vāsudeva; Brāhmans also appear to have been fed; other gifts are "for the sake of reverence" (*pūjārtha*), and for hopes for "the future world" (*paraloka*), etc. The "supreme spirit" (*paramātmana*) and (?) "creation" (*sṛishti*) are also referred to.

Is the work homogeneous?

53. In my first examination of the manuscript I noticed that the writing was not uniform, and that, in particular, certain leaves differentiated themselves from the rest by a bolder and, on the whole, a better style of writing; and I distinguished this set of leaves as the "M" section. This early differentiation was a most useful one for it marked not only a difference in style of writing but also one of matter. Indeed this "M" section proved to have so many peculiarities that the idea that it was possibly a separate work could not be ignored. But the rest of the manuscript is by no means uniform in style of writing or anything else, and I am not convinced that the "M" section is the work of a separate author, although I rather suspect that it is. It is, however, pretty certain that it was the work of a separate scribe; but, as there are slight indications that the other portions of the manuscript were *dictated*, this does not affect the question of authorship conclusively. I cannot point to any definite *evidence* of dictation that would bear examination: it is rather an impression received.

The peculiarities of the "M" section, although they may not prove heterogeneity of workmanship, call for some special mention and are here summarised.

I. The script.

(*a*) The writing is bolder and, on the whole, more uniform than that of the rest of the manuscript.

[1] "Saffron (*kuṅkuma*) has to the present day remained a famous product of Kashmir." Stein, *Rājataraṅgiṇī* II, 428. We ought also to mention birch-bark, although there is no reference to it in the text.

(*b*) Flourishes or extensions of the bottom end-strokes are common. These flourishes occur particularly at the end of ligatures, but also in the cases of the numeral figures " 5," " 7 " and " 9," and even in the case of the stop bars, and occasionally they even occur at the ends of the frame-works of the " cells " (*e.g.*, see fol. 47, etc.).

(*c*) The numerical symbols of section " M " are shown in Table IV (7) line 1, where the looped " 6 " should be noted. This is a useful but not an infallible criterion.

(*d*) The following table relating to the formation of the medial vowels *a*, *i* and *o* is taken from Part II.

	Medial ai		Medial o[1]		
	àì	*·ài*	*ŏ*	*ò·*	*·o·*
' M ' section	0%	100%	32%	32%	36%
Whole manuscript	24%	76%	75%	15%	10%

Here the indication appears to be very definite indeed, but it must be borne in mind that such criteria only apply in the mass, that the total number of *·ài* examples is only 19, and so on.[1] For a fuller discussion of these interesting statistics see Part II.

II. It is curious that all the mythological and semi-historical references (see Sections 46—48) occur in the " M " section. Indeed this section is peculiarly Hindu in contrast with the remainder of the manuscript.

III. The mathematical contents of the " M " section may be described as miscellaneous problems, which are generally solved by simple " rule of three "; but a special feature is the occurrence of numerous " measures " and a special method of exhibiting their change-ratios. But, of course, these points in no way indicate a separate work—rather otherwise—for if section " M " were an entirely different work we might expect some duplication, and there is none here.

IV. The method of exposition is somewhat different. The example is followed by a statement and the answer is then given, generally, without any detailed working, and generally there is no " proof " or verification. There are, however, exceptions.

V. There are differences in language. Certain technical terms that are extremely common in the rest of the manuscript do not occur, *e.g.*, *pratyaya*, *yuta*.

[1] The symbols used here are merely mnemonic. *See* page 95.

CHAPTER V.

EXPOSITION AND METHOD.

54. The text consists of rules (*sūtras*) and examples. There is no explanation whatever of the processes by which the rules were obtained, that is, there is no mathematical theory at all. In this the work follows the usual Indian fashion, as exhibited in all early texts. But there is a good deal of mathematical theory implied, and the rules and examples are often set forth in such a way as to convey the principles followed quite clearly to the student.

55. The rules (*sūtras*) are written in verse and are generally numbered ; and often in the solutions of the examples phrases from the *sūtras* are quoted. When Dr. Hoernle tried to re-arrange the leaves of the manuscript, he took the numbered *sūtras* as the basis of his order, but they were too few in number to lead to a satisfactory result. These *sūtras* do not represent, as might be expected, the most valuable part of the text. They are usually of particular application rather than general and are often very obscurely expressed.

56. The examples given are generally formally stated in full—without the use of notation or abbreviation of any kind ; and in most cases they are stated in verse. They are introduced by the term *udā,* an abbreviation for *udāharanaṁ* " an example." After the question sometimes comes a formal statement with numerical symbols and abbreviations, often arranged in cells. Then comes the solution or working (*karaṇa*), and here, sometimes, fragments of the *sūtras* are quoted. Finally come demonstrations—often more of the nature of verification than proof. Generally these demonstrations, by the aid of the answer found to the question, rediscover one of the original elements of the problem ; and sometimes several such demonstrations are attached to an individual problem, but sometimes the variation is merely a matter of the form of statement.

The full scheme of exposition is therefore—

Sūtram or rule.
Udāharanaṁ or " example " : indicated by *udā°*.
Sthāpanam or " statement."
Karaṇam or " solution."
Pratyayam or " verification."

The end of each *sūtra* is marked after the last example by the device  and the number of the *sūtra* is also given at the end.

57. The method of grouping sets of figures is of interest, and shows features in common with mediæval Sanskrit mathematical manuscripts, where also it is the practice to place groups of numbers in cells. In our manuscript, however, this fashion is rather more elaborated than in any Sanskrit manuscript I have examined. The mathematical possibilities of this scheme do not appear to have been realised and the student must always be careful to interpret any group of figures from the context, and not from any similarity with other groupings. However there is a certain amount of consistency in the arrangements, as the examples exhibited below will show. The real purpose of the arrangements appears to be to prevent confusion by demarcating the numerical figures from the text itself. The text is often written almost independently of the figure groups, and a word may be arbitrarily divided by the cell arrangement, which may also cut into several lines of the text not necessarily connected with it. The economic necessity of utilising

the whole of the writing surface of the birch-bark available seems to have been the determining factor. The text itself should be consulted but the following examples may be helpful :—

(*a*) Integral numbers occasionally occur without any marking off by lines or cells, but often

(*b*) each integral number has a cell to itself, *e.g.*

| 1 | | 42 | | 39 |

(*c*) Sometimes an integer is marked off by two vertical bars : thus | 14 | and invariably a series of integers is thus demarcated, *e.g.*
| 20 | 40 | 60 | 80 | evaṁ 200 ||

(*d*) Fractions and groups of fractions are placed in cells or groups of cells, *e.g.*

(i) 132 / 33

(ii) 6055040625 / 3227520000

(iii)

1	1	1	1
4	3	6	12

(iv) 1 / 1 3 / 1 4 / 1 5

(v) 40 / 1 3+ / 1 4+ / 1 5+

(vi) 21 / 20 / 21

(vii) 7 / 12 / 12

(*e*) Complete sets of operations are sometimes marked off in a similar manner. For example, (*d*) (iii) means $\frac{1}{4}+\frac{1}{3}+\frac{1}{6}+\frac{1}{12}$; (*d*) (iv) means $(1+\frac{1}{3})$ $(1+\frac{1}{4})$ $(1+\frac{1}{5})$; (*d*) (v) $40\ (1-\frac{1}{3})\ (1-\frac{1}{4})\ (1-\frac{1}{5})$.

(*f*) Series of operations may be connected together by cell arrangements, for example

2 dī° 1 2	1 di° 1 2	100000 / 947 dī°	phalaṁ	di°	60000 / 947
3 dī° 1 2	1 di° 1 3	157500 / 947 dī°	phalaṁ	di°	60000 / 947
4 dī° 1 2	1 di° 1 4	216000 / 947 dī°	phalaṁ	di°	60000 / 947

which means

$2\frac{1}{2}$ dīnāras : $1\frac{1}{2}$ days : : $\frac{100,000}{947}$ dīnāras : $\frac{60,000}{947}$ days

$3\frac{1}{2}$,, : $1\frac{1}{3}$,, : : $\frac{157,500}{947}$,, : $\frac{60,000}{947}$,,

$4\frac{1}{2}$,, : $1\frac{1}{4}$,, : : $\frac{216,000}{947}$,, : $\frac{60,000}{947}$,,

(*g*) The data of the problem and the solution may be indicated in one combined statement, *e.g.*

si° 1	16	4 a°	chhe° 2187*	phalaṁ śarā 2624400
1	1	1	1	
	1			
	4			
	1			
	3			

This is a statement of proportion where the second term means 16 $(1+\frac{1}{4})$ $(1+\frac{1}{2})$ and the number marked with an asterisk is a change-ratio.

(*h*)	0	2 1	3 3	12 4	dṛi° 300
	1	1 1	1 1	1 1	1

may be roughly expressed by means of $x+2x+3\times3x+12\times4x=300$

58. The use in our text of the sexagesimal notation in the form in which it occurs is of rather special interest, for there is, as far as I know, no other example of the kind in any of the classical Sanskrit works. The Hindus, from Āryabhaṭa onwards, were well aware of the advantages of the sexagesimal notation for astronomical purposes, but they never used it for arithmetical purposes.

Apparently there is only one purely arithmetical example of the use in the text and this example occurs, in connexion with a problem in arithmetical progression, on folio 6, verso, and 7, recto, where the fraction 178/29 is expressed as $6+8^{\text{i}}+16^{\text{ii}}+33^{\text{iii}}+6^{\text{iv}}\frac{6}{29}$. This sexagesimal fraction is actually written thus—

6	
8	
60	
16	cha°
60	
33	li°
60	
6	vi°
60	
śe° 6	
29	

The upper three figures are missing in the manuscript but the restoration is certain. Of the abbreviations *li°* stands for *liptā* (Gk. *lepté*) which in Sanskrit works ordinarily means a minute of arc, or the sixtieth part of a degree; *vi°* stands for *viliptā*, ordinarily a second of arc[1]: while *śe°* stands for *śesham* or "remainder."

The purely arithmetical (and perfectly legitimate) use of the notation here points to extra-India influence, for, although such a use is unknown in Sanskrit works, it was extremely common in mediæval Muslim works.

---

[1] On folio 37, verso; *liptā* and *viliptā* are used 'astronomically.'

It will be noticed that the term *liptā* here applies to "third parts" instead of "first parts," and *viliptā* to "fourth parts." The abbreviation *cha°* has not yet been traced to its origin.

59. The modern place-value arithmetical notation is employed throughout the text and there is not the slightest indication that, to the author, it was a new or strange invention. There is nothing in the slightest degree remarkable in the employment of this notation in the text, neither is there in the text the slightest indication of any other evidence that makes the employment of this notation in any way particularly noteworthy; but its occurrence in the text has been given an artificial importance by the arguments of Dr. Hoernle and Dr. Bühler about the age of the work. This topic will be discussed in some detail in the chapter dealing with the age of the manuscript and the age of the work.

60. As already indicated algebraic symbols are not generally employed, but a symbol for the unknown quantity is. This symbol is the arithmetical symbol • for "**nought**" or "zero," and on several occasions it is referred to by the term *śūṇya* "empty," "a cypher," and in some places by *śūṇya sthāna* or "empty place."[1]

The symbol occurs some twenty times but only in sections B, C, F, G, H, K. Its employment is illustrated in the following examples :—

(i)

ā° 1	u° 1	pa° •	labdhaṁ 10
1	1	1	1

This is a "statement" of an arithmetical progression where the first term is 1, the common difference is 1, the number of terms is unknown, and the "quotient" is 10 (where 10x = the sum of the series). Here the symbol • simply indicates that the number of terms (*pada*) is unknown, *i.e.*, that the place in the statement is empty. The symbol does not enter into any operation here or elsewhere. The giving it a denominator of unity is curious and really indicates that it is an integral number.

(ii)

ā° 5	u° 6	pa° •	dha° •
1	1	1	1
ā° 10	u° 3	pa° •	dha° •
1	1	1	1

Here are two equivalent arithmetical progressions in which the numbers of terms are equal and the sums also are the same in both cases, but both are unknown.

(iii)

•	1	1	1	1	bhā°	śe°	16
1	1	1	1	1			1
	3	3	3	3			

which means $x = 16/(1-\frac{1}{3})\ (1-\frac{1}{3})\ (1-\frac{1}{3})\ (1-\frac{1}{3})$ and since this is a deduction from a problem it is a distinct step towards a proper algebraic symbolism.

(iv)

•	2	3	4	dṛishya	200
1	1	1	1		1

which may be represented by $x+2x+3x+4x=200$. But the method of solution is by the *regula falsi*. Any number is put in the place of the • and the sum of the

[1] In Hindu works proper the dot is the symbol for negative quantity, and the zero symbol is a small circle.

series is obtained. Here the given sum multiplied by the assumed value and divided by the (false) sum is the correct solution. The connexion between this method and that of operating with unknown quantities is discussed later on.

(v)	•	5 yu	mū •	sā •	7	mū •
	1	1	1	1	1	1

which means $x+5=s^2$, $x-7=t^2$. Here the symbol 1 stands for three different unknown quantities. It simply indicates in each case an unknown number.

Negative sign.

61. The only distinct mathematical symbol employed is the sign for a negative quantity, which takes the form of a cross + placed after the number affected. This is peculiar and has given rise to discussion. In Sanskrit manuscripts a dot before the quantity affected is the usual method of indicating a negative quantity. In the transliteration of the text the original negative sign (+) is, perhaps illogically, preserved; but it is a rather special feature of the manuscript, easily printed, and leading to no ambiguity.

On this negative sign Dr. Hoernle writes[1]:

"Here, therefore, there appears to be a mark of great antiquity. As to its origin I am unable to suggest any satisfactory explanation. I have been informed by Dr. Thibaut of Benares that Diophantos in his Greek arithmetic uses the letter ψ (short for ἔλλειψις) reversed (thus ↑) to indicate the negative quantity. There is undoubtedly a slight resemblance between the two signs; but considering that the Hindus did not get their elements of the arithmetical science from the Greeks,[2] a native Indian origin of the negative sign seems more probable. It is not uncommon in Indian arithmetic to indicate a particular factum by the initial syllable of a word of that import subjoined to the terms which compose it . . . . The only plausible suggestion I can make is, that it is the abbreviation (*ka*) of the word *kanita*, diminished." (He also points out that the letter *k* in its ancient shape as used in the Asoka inscriptions is a cross and goes on to say) "Another suggestion is, that the sign represents the syllable *nū*, an abbreviation of *nyūna*, 'diminished.' The *akshara* for *nū* (or *nu*) in the Asoka characters would very closely resemble a cross (+) . . . . . However for the present, the question must be left an open one."

With this last remark I agree and I would at the same time point out the danger of attempting to trace an isolated symbolic form back through the ages. The suggested connexion with Aśoka times is only part of the special pleading for the extreme antiquity of our manuscript. The striking resemblance with the Diophantine symbol, so lightly discarded by Dr. Hoernle, is of interest, but lacks support. From the time of Diophantus to the time of the Bakhshālī manuscript is too great[3] to allow us to ignore the lack of other examples.

Abbreviations.

62. Abbreviations are employed to such an extent as to become, at times, embarrassing, and they embrace almost every type of term. None of these abbreviations is used in an algebraic sense, although, at first sight, when we find *ā*°, *u*°, *pa*° used consistently for the elements of an arithmetical progression, an algebraic symbolism does not seem very far off. But *a*°, *ha*°, *ū*°, *go*° are the abbreviations of names of animals, *ya*°, *go*° and *sa*° of plants, etc., etc. The names

---

[1] *Indian Antiquary*, xvii (1888), p. 34.

[2] They got a good deal from them, and practically all their later astronomy came from the Greeks.

[3] But Dr. Hoernle assumes that the Bakhshālī text was practically contemporaneous with Diophantus!

of measures, which are numerous, are nearly always abbreviated. Certain common terms of operation are often abbreviated, and of these the following occurs most often—

> *bhā*° for *bhāga*, placed after a term to indicate that it is a divisor.
> *śe*° for *śeshaṁ*, a remainder.
> *mū*° for *mūlaṁ*, a root, a quantity that has a root, capital.
> *pha*° for *phalaṁ*, an answer.
> etc., etc.

Indeed the writers seem to have used abbreviations whenever there was no ambiguity incurred in their own minds. We are not so fortunately placed in this matter and occasionally it has been impossible to rediscover the term implied.

Fundamental operations.

63. In the classical Sanskrit works there is generally very little formal information about mathematical principles and method. Axioms or postulates seldom or never occur and strict definition is seldom attempted. Rules are of particular application rather than general; order appears to have been a matter of convenience rather than logic; and the fundamental operations receive scanty attention. In a work that is essentially Indian in form we must not, therefore, expect much formal attention to such matters, and, indeed, in the Bakhshālī text there is little. We cannot even tell how the detailed processes of such operations as multiplication and division were actually performed. All that remains are the formal statements of terms and results, and all I can now present are examples of such statements.

64. Examples of addition

(*a*) 960 64 yutaṁ jātaṁ 1024

(*b*) $\begin{array}{|c|}10\\3\\\hline\end{array}$ sa rūpaṁ $\begin{array}{|c|}13\\3\\\hline\end{array}$

(*c*) 2 rūpa saṁyutaṁ $\begin{array}{|c|}3\\\hline\end{array}$

(*d*) 840 49 datvā jātaṁ 889

(*e*) 924 | 836 | 798 | eshāṁ yutiṁ kṛiyate jātā 2558

(*f*) 120 | 90 | 80 | 75 | 72 | eshāṁ yoga kṛite jātā 437

(*g*) $\begin{array}{|c|}120\\90\\80\\75\\12\\\hline\end{array}$ evaṁ 377

(*h*) $\begin{array}{|c|c|c|}10&30&90\\\hline\end{array}$ ekatram $\begin{array}{|c|}130\\1\\\hline\end{array}$

(*a*) 960 and 64 added: 1024 is produced. (*b*) $\frac{10}{3}$ plus unity$=\frac{13}{3}$. (*c*) 2 and unity added together=3. (*d*) 49 having been given to 840, 889 is produced. (*e*) 924, 838, 798: the sum of these may be determined: they produce 2·558. (*f*) 120, 90, 80, 75, 72: the sum of these is made: they produce 437. (*g*) 120, 90, 80, 75, 12: thus 377. (*h*) 10, 30, 90: altogether 130.

(*i*) $\begin{array}{|c|}29\\48\\58\end{array}$ 7 anena yutaṁ $\begin{array}{|c|}36\\48\\58\end{array}$

(*k*) $\begin{array}{|cc|}5 & \text{śe } 1\\1 & 16\end{array}$ $\begin{array}{|cc|}10 & \text{śe } 15\\ & 16\end{array}$ evaṁ 16

(*l*) $\begin{array}{c}45\\2\end{array}$ sārdha traya yutaṁ $\begin{array}{c}52\\2\end{array}$

65. Examples of subtraction

(*a*) | 5 | 9 | viśeshaṁ | 4 |

(*b*) $\begin{array}{|cc|}3 & 2\\2 & 1\end{array}$ viśeshaṁ $\begin{array}{|c|}1\\2\end{array}$

(*c*) | 5 | 3 | rahitaṁ jātaṁ | 2 |

(*d*) 6 | 3 śuddhi | 3 |

(*e*) | 42 | tryūnam | 39 |

(*f*) | 3 | 7 | viśoddhya | 4 |

(*g*) $\begin{array}{|c|c|}77 & 294\\11 & 11\end{array}$ pātya śeshaṁ $\begin{array}{|c|}217\\11\end{array}$

66. Examples of multiplication.

(*a*) | 2 | dvigunaṁ | 4 |

(*b*) | 30 | ashṭa gunaṁ | 240 |

(*c*) $\begin{array}{|cc|}2 & 40\\5 & 1\end{array}$ guṇita jātāṁ 16

(*d*) $\begin{array}{|cc|}4 & 1\\1 & 1\\2 & 4+\end{array}$ guṇita jātā $\begin{array}{|c|}27\\8\end{array}$

(*i*) $29\frac{48}{58}$ by this (7) increased$=36\frac{48}{58}$. (*k*) $5\frac{1}{16}+10\frac{15}{16}$ thus 16. (*l*) $\frac{45}{2}$ with three and a half added$=\frac{52}{2}$.

(*a*) 5, 9 : the difference is 4. (*b*) $\frac{3}{2}$, 2 : the difference is $\frac{1}{2}$. (*c*) 5, 3 : subtracted, 2 is produced. (*d*) 6, 3 : the difference is 3. (*e*) 42 less three is 39. (*f*) 3, 7 : having subtracted, 4. (*g*) $\frac{77}{11}$, $\frac{294}{11}$: having subtracted the difference is $2\frac{17}{11}$.

(*a*) 2 multiplied by two$=$4. (*b*) 30 multiplied by eight$=$240. (*c*) $\frac{2}{5}$ and 40 multiplied : 16 is produced. (*d*) $1-\frac{1}{4}$ and $4\frac{1}{2}$ multiplied $\frac{27}{8}$ is produced.

(*e*)	6 1	1 1 4+	anena guṇitaṁ jātaṁ	4 1 2

(*f*)	2 3	3 4	4 5	guṇita jātaṁ	2 5

(*g*)	40 1 1 3+ 1 4+ 1 5+	phalaṁ 16

(*h*)	3	4	abhyāsaṁ 12

(*i*)	8	ātma gunaṁ	64

(*j*)	880 84	964 168	guṇita jātaṁ	848320 14112

(*k*)	737 58	-----	178 29	anena guṇitaṁ jātaṁ	65593 841

(*l*)	108625 65600	pada-ghnā	6455949625 3227520000

(*m*)	405280 38724	44004 77448	saṁguṇya jātaṁ	179945781120 2999096352

67. *Division.* Fractional quantities are expressed in the usual Indian way with the numerator written above the denominator without any dividing line, and they are usually placed in cells. This indication of the operation of division often does away with the necessity for an explanatory word or phrase.

Examples of division.

(*a*)	168 4	168 6	168 7	labdhaṁ 42 | 28 | 24

(*b*)	4	vibhaktaṁ	1 4	35 guṇitaṁ	36 4

(*c*)	10	3	vibhaktvaṁ	10 3

(*e*) 6 and $1-\frac{1}{4}$: by this ($\frac{3}{4}$) multiplied, $4\frac{1}{2}$ is produced. (*f*) $\frac{2}{3}$, $\frac{3}{4}$, $\frac{4}{5}$ multiplied, $\frac{2}{5}$ is produced. (*g*) $40\ (1-\frac{1}{3})\ (1-\frac{1}{4})\ (1-\frac{1}{5}) = 16$. (*h*) 3, 4 : the product is 12. (*i*) 8 multiplied by itself=64. (*j*) $\frac{880}{84}$, $\frac{964}{168}$ multiplied ; $\frac{848320}{14112}$ is produced. (*k*) $\frac{737}{58}$, $\frac{178}{29}$, by this ($\frac{178}{29}$) multiplied, $\frac{65593}{841}$ is produced. (*l*) $\frac{108,625}{65600}$ multiplied by the number of terms ($\frac{59425}{49200}$) = $\frac{6,455,040,625}{3,227,520,000}$. (*m*) $\frac{405280}{58724}$: $\frac{44004}{77448}$: having multiplied these together $\frac{179,945,781,120}{2,999,096,352}$ is produced.

(*a*) *Labdham*=quotient. (*b*) This periphrastic exposition is common—the divisor is turned into a multiplier by being inverted. The example here means : 4 divided=$\frac{1}{4}$, which multiplied into 35=$\frac{35}{4}$. (*c*) 10, 3 : having divided=$\frac{10}{3}$.

(*d*)	$\frac{447}{29}$	dalita	$\frac{447}{58}$				
(*e*)	$\frac{132}{33}$	vartyaṁ jātaṁ	$\frac{4}{1}$				
(*f*)	$\frac{798}{1463}$	projjhya	798				
(*g*)	2558	chchheda projjhyaṁ	1095				
(*h*)	60	anena dṛishyaṁ bhājitaṁ	$\frac{1}{60}$	$\frac{300}{1}$	jātā	5	

Square-root.

68. The method of extracting square-roots exhibited in the text is of special interest. The rule is much mutilated but as it is given on three separate occasions the fragments pieced together enable us to give the complete *sūtra* :

> akṛite slishṭha kṛityūnā śesha chchhedo dvisaṁgunaḥ tad vargaḥ dala saṁslishṭha hṛiti śuddhi kṛiti kshayaḥ

This means somewhat as follows :—

> The mixed surd is lessened by the square portion and the difference divided by twice that. The difference is divided by the quantity and half that squared is the loss.

Then follows a note to the effect that " by means of this rule an approximation (*anaya*) to the proper root of a mixed quantity is found . . . ."

The rule as it stands is cryptic and hardly translatable, but fortunately there are examples given in some detail, and these show that the rule was extended to " second approximations."

The rule means that the first approximation to $\sqrt{Q}=\sqrt{A^2+b}$ is $A+b/2A$ or q_1; but $q_1^2-Q=(b/2A)^2=e_1$.

No rule for 'second approximations' is preserved but there are several examples; and, of course, no fresh rule is really required,[1] for $\sqrt{Q}=\sqrt{q_1^2-e_1}\backsim q_1-e_1/2q_1=A+\frac{4A^2b+b^2}{8A^3+4Ab}$, and the 'second error' may be indicated by $e_2=(e_1/2q_1)^2=\left(\frac{b}{2A}\right)^4/4\left(A+\frac{b}{2A}\right)^2$.

(*d*) $\frac{447}{29}$ halved $=\frac{447}{58}$. (*e*) $\frac{132}{33}$ reduced gives 4. (*f*) Having discarded the denominator $\frac{798}{1463}$ becomes 798. (*g*) 2558—1463= 1095 where 1463 is the denominator of the fraction. (*h*) By this (60) the known quantity (300) is divided and $\frac{1}{60}$ of 300=5.

[1] Note that if $\sqrt{A^2+b}=r$ then $r=A+\frac{b}{2A+}\,\frac{b}{2A+}\,\frac{b}{2A+}\cdots\cdots$ The text uses $r'=A+\frac{b}{2A}$, $r'''=A+\cfrac{b}{2A+\cfrac{b}{2A+\cfrac{b}{2A}}}=A+\frac{4A^2b+b^2}{8A^3+4Ab}$, but not $r''=A+\cfrac{b}{2A+\cfrac{b}{2A}}=A+\frac{2Ab}{4A^2+b}$.

All the examples preserved belong to section C and appear to be merely subsidiary to the solution of certain quadratic equations arising out of problems **in arithmetical progressions.** These problems will be found fully worked out in § 86 and I give here merely a summary of the square-root evaluations.

(i) $\sqrt{41}=\sqrt{36+5}$ and $q_1=6\frac{5}{12}$ while $e_1=\frac{45}{144}$ and $q_2=6\frac{745}{1848}$.

(ii) $\sqrt{105}=\sqrt{100+5}$ and $q_1=10\frac{1}{4}$; $e_1=\frac{1}{16}$; $q_2=10\frac{1}{4}-\dfrac{\frac{1}{16}}{2\times 10\frac{1}{4}}$

$$=10\frac{81}{328};\ e_2=\left(\dfrac{\frac{1}{16}}{2\times 10\frac{1}{4}}\right)^2=\frac{1^*}{107584}$$

(iii) $\sqrt{481}=\sqrt{21^2+40\dagger}$; $q_1=21\frac{40}{42}$; $\frac{e_1}{8}=\frac{1}{8}\left(\frac{40}{42}\right)^2=\frac{1600}{14112}$;

$$q_2=21\frac{20}{21}-\left(\frac{20}{21}\right)^2/2\times 21\frac{20}{21}=21\frac{9020}{9681};\ \frac{e_2}{8}=\frac{1}{8}\left(\frac{40}{42}\right)^4/\left(2\times 21\frac{40}{42}\right)^2=\frac{160{,}000}{2{,}999{,}096{,}352}.$$

(iv) $\sqrt{889}=\sqrt{29^2+48}$; $q_1=29\frac{48}{58}$; $\frac{e_1}{24}=\frac{1}{24}\left(\frac{48}{58}\right)^2=\frac{24}{841}$.

(v) $\sqrt{336009}=\sqrt{579^2+768}$; $q_1=579\frac{384}{579}$; $q_2=579\frac{515{,}225{,}088}{777{,}307{,}500}$;

$$\frac{e_2}{8d}=\frac{21{,}743{,}271{,}936}{7{,}250{,}483{,}394{,}675{,}000{,}000}.$$

69. There is not much doubt about the exegesis of the rule. It was neither connected with continued fractions nor with the so-called Pellian equation.[1] Brahmagupta gave the converse of the rule, namely $(A+x)^2 \sim A^2+2Ax$ from which the square-root rule given in our text is immediately deducible. But, as already pointed out, the square-root rule itself was not used by the Hindus and was not even noticed by them until the sixteenth century.[2] Indeed the Hindus had a very good practical rule of their own, which was given by Śrīdhara (*Triśatika,* 46) and Bhaskara (*Līlāvatī,* 138), namely—" Multiply the quantity whose square-root cannot be found by any large square number, take the square-root of the product —leaving out of account the remainder—and divide by the square-root of the multiplier." For example $\sqrt{41}=\sqrt{41\times 1000000}\div 1000 \sim 6{\cdot}403+$

This rule was used in conjunction with the usual modern scholastic square-root rule, which dates at least from the time of Heron.

The examples in the text may be further conveniently summarised thus :—

Q	q_1	q_2	q (approximately correct).
41	6·41667	6·40313	6·403124
481	21·9524	21·9307 . .	21·9317122
889	29·828 . .		29·8161030
336009	579·664	579·6615	579·65 . . . .

* In the text we find $\frac{e}{8d}$ and not *e* by itself, for reasons explained in §85.

† The form $\sqrt{481}=\sqrt{22^2-3}\sim 22-\frac{3}{44}=21\frac{37}{44}$ suggests itself; but the negative sign does not appear to have been used in first approximations.

Note.—The Appendix to Chapter vi (page 53) may be helpful to those who are interested in the details of the calculations.

[1] For a full discussion of the topic see P. Tannery *Mémoires scientifiques,* I, pp. 189ff. ; and II, pp. 157ff.

[2] A short historical note is given on page 45.

"*Rule of three.*"

70. The term *trairāśika* "relating to three quantities" occurs about a dozen times—always employed in the sense of arithmetical proportion. Most often it occurs in the phrase *pratyaya trai-rāśikena* or "proof by the rule of three" and the proportion is generally set out in the following manner—

1	1	4	phalaṁ	18
3	1	1		1
	2			

which means $\frac{1}{3} : 1\frac{1}{2} :: 4 : 18$.

The term *trairāśika* is orthodox but the term *phalaṁ* is used throughout our text quite appropriately as equivalent to "answer", and is not applied to the second term of a proportion as in the *Līlāvatī*.

There are very many examples of the "rule of three" used either as the direct method of solution or as a proof. There are also some very much damaged examples of what appear to be problems in so-called compound proportion. There is nothing of the nature of a theory of proportion discussed but several examples of the following principle occur—

If $\frac{a_1}{b_1}=\frac{a_2}{b_2}\ldots\ldots=\ldots=\frac{a_n}{b_n}=\frac{A}{B_1}$, then $\frac{\Sigma a}{\Sigma b}=\frac{A}{B}$.

Regula falsi.

71. The rule of "false position" or "supposition" is used in two ways in our text[1] in solving linear equations.

(i) $f(x)=p$ is solved by assuming a value e for x. This gives $f(e)=p'$ and the correct value of x is ep/p'

An example is

> The amount received by the first is not known. The second receives twice as much as the first, the third thrice as much as the first two and the fourth four times as much as all the others. Altogether they receive 300. How much did the first receive?

Suppose the first receives one, then the second receives 2, the third 9, and the fourth 48; or altogether they receive 60. Actually, therefore, the first received $300/60=5$.

(ii) If $f(x)=p$ reduces to $bx+c=p$ and $f(e)=p'$ reduces to $be+c=p'$ then $x=(p-p')/b+e$, which appears to have been considered useful when it was desired to keep b and c unchanged.

---

[1] No rule is preserved in the text but Bhāskara gives the following (*Līlāvatī*, 50).

"Any number assumed at pleasure is treated as specified in the particular question, being multiplied and divided, raised and diminished by fractions; then the given quantity, being multiplied by the assumed number and divided by the result yields the number sought."

For example if $x_1+x_2=a_1$, $x_2+x_3=a_2$, $x_3+x_1=a_3$ we have $2x_1+(a_2-a_1)=a_3$; and if in place of x we put e then $2e+(a_2-a_1)=a'_3$ and the correct value of x_1 is $(a_3-a'_3)/2+e$

72. This rule of "false position" is interesting as being, in a way, a precursor of algebraic symbolism. It connotes the idea of an unknown quantity and even of a symbol for that quantity (*e.g.*, as in our text) but it does not embrace the notion of such a quantity being subject to operation or being isolated. As soon as we introduce algebraic symbols the rule, as it were, disappears. Algebraically the rule solves $bx=p$ by $x=ep/p'$ where $p'=be$, which shows that the transformation from $bx=p$ to $x=p/b$ was not conceived.

There is no algebraic symbolism in our text but it may be noted that Bhāskara gives both the *regula falsi* and also an early form of algebraic symbolism. Neither Āryabhaṭa, Brahmagupta nor Śrīdhara gives this rule or makes use of the principle. Its occurrence in the *Līlāvatī* therefore seems to indicate that it was introduced into northern India after the time of Śrīdrara (XIth cent.). Mahāvīra (IXth cent.), however, uses the method in rather a special way in connexion with a geometrical problem.[1]

The rūpoṇa method.

73. There are several references to the *rūpoṇa* method, and the phrases *rūpoṇā karaneṇa* or *pratyaya rūpoṇā karaneṇa* occur. In all the cases the application is to the summation of a series of terms in arithmetical progression according to the rule

$$s=[(t-1)d/2+a]t$$

The recurrence of the phrase *rūpoṇā karaneṇa* seems to imply that the rule in question began with the term *rūpoṇā* which corresponds to the $(t-1)$ of the formula.[2] The rule is not preserved in our text but we find the following in the *Ganita Sāra-sangraha* of Mahāvīra (ii, 63)

rūpenono gachchho dalī kṛitaḥ pra chayatādi to miśraḥ |
prabhavena padābhyas tas saṅkalitaṁ bhavati sarveshām ||

"The number of terms is diminished by one, halved and multiplied by the increment. This when combined with the first term of the series and multiplied by the number of terms becomes the sum of all."

The rule is exemplified in two ways in our text of which the following are particular cases :—

(i)	ā°	1	u°	1	pa°	19	rūpoṇā karaneṇa phalaṁ	190
		1		1		1		1

which means : "a=1, d=1, t=19 ; by the rūpona method the answer is s=190."

Similar bare summaries of the "rūpona" method are found in section "B" mainly.

---

[1] *Gaṇita Sāra-saṅgraha* vii, 112.

[2] Is this a faint echo of Pythagorean style ? Anatolius writes : "The Pythagoreans state that their master, in connexion with numbers that form a right-angled triangle, showed how such could be composed by means of unity." *i.e.*, $[\frac{1}{2}(a^2-1)]^2+a^2=\frac{1}{2}(a^2+1)^2$. See P. TANNERY *Mémoires scientifiques*, iii, 14.

(ii) In section "C" the application of the *rūpoṇa* method is always worked out step by step. For example when a=5, d=3, t= 178/29; then s=((t−1)$\frac{3}{2}$ +5) t; and the *rūpoṇa* method is applied as follows :—

$$t-1=\frac{149}{29};\ 3\times\frac{149}{29}=\frac{447}{29};\ \frac{447}{29}\times\frac{1}{2}=\frac{447}{58};\ \frac{447}{58}+5=\frac{737}{58};\ \frac{737}{58}\times\frac{178}{29}=\frac{65593}{841}=77\frac{836}{841}.$$

Proofs.

74. To many solutions are attached proofs, which are generally introduced by the term *pratyayaṁ* "proof" or "verification." This is sometimes amplified into *pratyaya-trai-rāśikena* "proof by the rule of three," or *pratyaya-rūponā-karanena*, "proof by the *rūpona* method."

Sometimes the "proof" seems to be merely a matter of rewriting a statement in another form, but generally the answer is utilised and one of the original terms of the problem is rediscovered. Occasionally the "proof" consists of a solution of the original problem in another way, *e.g.*, by steps; but sometimes it deals with a subsidiary aspect of the original problem. In certan problems approximations are employed and then the "proof" may be described as a process of reconciliation. These "reconciliations" entail a comparatively high degree of mathematical skill. Occasionally several different proofs are attached to a problem. The following are specimens.

(1) *Problem.* $\frac{5}{3}t-7=\frac{6}{5}t+7$

Solution. $t=2\times 7/\left(\frac{5}{3}-\frac{6}{5}\right)=30$

Proof. 3: 5 : : 30: 50, 5:6 : : 30: 36 and 50 − 7 = 36 + 7

(2) *Problem.* A gives $2\frac{1}{2}$ in $1\frac{1}{2}$ days, B gives $3\frac{1}{2}$ in $1\frac{1}{3}$ days, C gives $4\frac{1}{2}$ in $1\frac{1}{4}$ days. In what time will they have given 500 dīnāras altogether?

Solution. $t=\dfrac{500}{\frac{2\frac{1}{2}}{1\frac{1}{2}}+\frac{3\frac{1}{2}}{1\frac{1}{3}}+\frac{4\frac{1}{2}}{1\frac{1}{4}}}=\dfrac{60,000}{947}$ days;

and A gives 100,000/947, B gives 157,000/947, C gives 216,000/947 dīnāras.

Proof. $2\frac{1}{2}:1\frac{1}{2}::100,000/947:60,000/947$

$3\frac{1}{2}:1\frac{1}{3}::157,000/947:60,000/947$

$4\frac{1}{2}:1\frac{1}{4}::216,000/947:60,000/947$

(3) *Problem.* One earns *e* and spends *f* daily. How long will a capital of C last?

Solution. t = C/(f − e)

Proof. 1:f: :t:F (the total expenditure)

1:e: :t:E (the total earnings) and F − E = C.

(4) *Problem.* If 7 are bought for 2 and 6 sold for 3, and the capital is 24, what will be the profit?

Solution. $p = C\,(c/s - 1) = 24\,(\frac{7}{2} \div \frac{6}{3} - 1) = 18.$

Proof. 2 : 7 : : 24 : 84 (the number of articles)

and 6 : 3 : : 84 : 42 (the total proceeds), and $42 - 24 = 18$.

or—1 : c : : C : n, s : 1 : : n : C+p and $C + p - C = p$.

(5) *Problem.* $x\,(1 - \frac{1}{2})\,(1 - \frac{1}{4})\,(1 - \frac{1}{5}) = x - 280.$

Solution. $x = 280/(1 - \frac{3}{10}) = 400.$

Proof. $400/2 = 200$, $400 - 200 = 200$,

$200/4 = 50$, $200 - 50 = 150$,

$150/5 = 30$, $150 - 30 = 120$, and $400 - 120 = 280$.

(6) *Problem.* $1 + \frac{1}{2}\,(2 + \frac{1}{2}\,(3 + \frac{2}{3}(4 + \frac{3}{2}\,(5 + \frac{1}{2})))).$

Solution. $= \frac{93}{16}.$

Proof. $(((((\frac{93}{16} - 1)\;2 - 2)\;2 - 3) - 4)\,\frac{2}{3} - 5 - \frac{1}{2}) = 0.$

(7) *Problem.* Solve $x + 5 = s^2$, $x - 7 = t^2$.

Solution. $x = (\frac{1}{2}(\frac{5+7}{2} - 2))^2 + 7 = 11.$

Proof. $11 + 5 = 4^2$ $11 - 7 = 2^2$

(8) *Problem.* $F = \dfrac{\frac{1}{2}(1) + \frac{1}{3}(2) + \frac{1}{4}(3) + \frac{1}{5}(4)}{1+2+3+4}.$

Solution. $F = \dfrac{162 \div 60}{10} = \dfrac{163}{600}.$

Proof. 10 : 163/60 : : 1 : 163/600

10 : 163/60 : : 2 : 163/300

10 : 163/60 : : 3 : 163/200

10 : 163/60 : : 4 : 163/150,

or, since $F = \dfrac{\Sigma f \times w}{\Sigma w}$, $\Sigma w : \Sigma f \times w : : w_r : w_r F.$

(9) *Problem.* $Dt = ((t - 1)\,d/2 + a)\,t.$

Solution. $t = 2\,(D - a)/d + 1.$

Proof by the *rūpona* method: $s = (\,(t - 1)\,\frac{d}{2} + a)t$ and $Dt = s$

(10) *Problem.* $s=((t-1)d/2+a)\ t.$

Solution. $t=\{\sqrt{(2a-d)^2+8s}-(2a-d)\}\div 2d.$

Generally the solution is an approximation t_1 or t_2 depending on the method of evaluation of the surd quantity; and the *rūpona* method gives s_1 or s_2, neither of which is the same as the original *s*. A process of reconciliation is therefore called for, and this is given by $\acute{s}-s=e/8d$ where $\acute{s}$ is the approximation to s given by the proof and e may be termed the square-root error.

Problem. $a=1,\ d=1,\ s=60.$

First solution. $t_1=\{\sqrt{(2\cdot1-1)^2+480}-1\}\div 2\cdot 1=(21\frac{40}{42}-1)\div 2=\frac{880}{84}.$

First proof. $s_1=t_1\ (t_1+1)\div 2=\frac{880}{84}\times\frac{964}{168}=\frac{848,320}{14112}$

$$e_1=\left(\frac{40}{42}\right)^2 \text{ and } s=s_1-\frac{e_1}{8a}=\frac{848,320-1600}{14112}=60.$$

Second solution. $t_2=\left(\frac{424,642}{19360}-1\right)\div 2=\frac{405,280}{38724}$

Second proof. $s_2=\frac{405,280}{38724}\times\frac{444,004}{77448}=\frac{179,945,941,120}{2,999,096,352}.$ Now $e_2=\left(\frac{40}{42}\right)^4\div 4\ \left(21+\frac{40}{42}\right)^2$, and $\frac{e_2}{8d}=\frac{160,000}{2,999,096,352}$;

$$\text{and } s=s_2-e_2/8d=\frac{179,945,781,120}{2,999,096,352}=60.$$

(11) The problems in section "G" are characterised by numerous proofs—as many as five separate proofs being attached to one example. The following is fairly typical:—

Problem. $x=500(1-\frac{1}{4})(1-\frac{1}{4})(1-\frac{1}{4})(1-\frac{1}{4}).$

Solution. $x=158\frac{13}{64}$

Proof (i). $x'(1-\frac{1}{4})\ (1-\frac{1}{4})\ (1-\frac{1}{4})\ (1-\frac{1}{4})=158\frac{13}{64}$, whence $x'=500$. This is written horizontally.

Proof (ii). $x'(1-\frac{1}{4})(1-\frac{1}{4})(1-\frac{1}{4})(1-\frac{1}{4})=158\frac{13}{64}$, whence $x'=500$. This is written vertically.

Proof (iii). $x'=\frac{158\frac{13}{64}}{(1-\frac{1}{4})\ (1-\frac{1}{4})\ (1-\frac{1}{4})\ (1-\frac{1}{4})}=500.$

Proof (iv). By steps.

$500\div 4=125,\ 500-125=375;$

$375\div 4=93\frac{3}{4},\ 375-93\frac{3}{4}=281\frac{1}{4};$

$281\frac{1}{4}\div 4=70\frac{5}{16},\ 281\frac{1}{4}-70\frac{5}{16}=210\frac{15}{16};$

$210\frac{15}{16}\div 4=52\frac{47}{64},\ 210\frac{15}{16}-52\frac{47}{64}=158\frac{13}{64}.$

Solutions.

75. The solutions are sometimes very detailed, proceeding most carefully step by step, so that they become expositions in general terms of the processes involved. Also they often give actual quotations from the rules—to such an extent sometimes that the original wording of the rule can be reconstructed. Unfortunately, however, such helpful suggestions are more often missing than not: in many cases the

particular portions of the manuscript are lacking or damaged beyond repair. In some sections, *e.g.*, section " M ", nothing but the bare answer is generally given, and in others the outline only of the working is given.

The following examples illustrate these remarks : 1. The problem is $DT+Dt = ((t-1)+a)\, t$ where *a* and *d* are respectively the first term and common difference of an arithmetical progression, and *t*, the number of terms, to be found. The rule is

$$t = \{ 2\,(D-a)+d+\sqrt{(2\,(D-a)+d)^2+8dDT}\,\} \div 2d$$

and the particular example gives $D=5$, $T=6$, $a=3$, $d=4$. The actual working of the solution, translated as literally as possible, is as follows :—

> *The daily rate diminished by the first term ;* the daily rate is five yojanas, 5 ; the first term is 3 ; their difference is 2 ; this doubled is 4 : *this increased by the common difference* is 8 ; and squared is 64, which is *known as the *kshepa* quantity.* *Multiplied by eight* ; the fixed term (30) multiplied by eight is 240 ; and multiplied by the common difference—multiplied 960. *The quantity known as *kshepa* is added.* Now the quantity known as *kshepa* is 64, which added gives 1024. The root of this is 32 ; the quantity set aside is 8 and this added gives 40. *Divided by twice the common difference :* twice the common difference is 8—divided 5.

The phrases placed between asterisks are quotations from the rule. A perfectly literal rendering would not be very intelligible to the ordinary reader and one or two gaps have been filled in ; but the translation is a perfectly fair representation of the original.

ii. The problem may be represented by $x\ (1-\frac{1}{3})(1-\frac{1}{4})(1-\frac{1}{5})=x-24$

and the solution is given as follows :—

> *Having calculated the loss on unity* the terms become $\frac{2}{3}$, $\frac{3}{4}$, $\frac{4}{5}$ and these multiplied together give $\frac{2}{5}$. This subtracted from unity gives $\frac{3}{5}$, which, inverted and multiplied† by the given amount 24, is $\frac{5}{3}$ of $24=40$.

iii. Something travels 3 *yavas* a day. How long will it take to go 5 yojanas ?

Here the solution is indicated by the following proportion only :

3 *ya°* : $\frac{1}{360}$ years : : $5\times 4{,}608{,}000$ *ya°* : 21,333 years 4 months.

---

* The phrases marked off by asterisks are quotations from the *sūtra* or rule.

† Literally "divided and multiplied" into. This is generally used to indicate division by a fraction.

CHAPTER VI.

AN ANALYSIS OF THE CONTENTS.

76. The present chapter contains an analysis of the mathematical contents of the text. To those who are chiefly concerned with the history of mathematics this chapter will possibly be of rather special interest. Indeed, taken with chapter I of Part III (which will appear in the second volume) it forms the real corpus of the present study, for the latter gives the re-arranged text, while the present chapter is an attempt at a fairly complete explanation of the intelligible parts of that text. The analyses of the different sections are not exhibited on a uniform plan, some are given in great detail, while others are mere summaries; some are arithmetical in form and some algebraical; but these differences correspond to some extent to the spirit of the text, and it is hoped that the analysis as a whole gives a proper representation of the original.

77. In order to avoid innumerable cross references the following table is given to show the connexion between the paragraphs of the present chapter and the original and re-arranged texts.

Chapter VI.		Illustrations (Bodleian order).		New order.
§78		Folio(s)	29, 27	$A_{11, 12}$.
§79		,,	1, 2, 3	$A_{13, 15}$.
§80		,,	38[a], 39[e], 40[c, d], 58	A_7.
§81		,,	59	K.
§82		,,	27 recto	A_{12}.
§83	(i)	,,	4 recto	B_1.
	(ii)	,,	4 verso	B_1.
	(iii)	,,	8 recto, 9 recto, 7 verso	$B_{2, 3, 4}$.
§84	(i)	,,	8 verso	B_2.
	(ii)	,,	9 verso	B_3.
§85		Section 'C' generally		....
§86	(i)	Folio(s)	5	C_1.
	(ii)	,,	6, 7 recto	$C_{2, 3}$.
	(iii)	,,	65 verso, 56 verso, 56 recto, 64 recto	$C_{4, 5, 6}$.
	(iv)	,,	64 verso, 57	$C_{6, 7}$.
	(v)	,,	45 recto	C_8.
	(vi)	,,	45 recto, et verso, 46 recto	$C_{8, 9}$.
§87		,,	23, 24, 25, 26	$F_{2, 3, 4, 5}$.
§88		,,	51	A_9.
§89		,,	10, 11, 12, 13, 14, 15, 16 recto	G_{1-6}.

Chapter VI.	Illustrations (Bodleian order).	New order.
§90	Folio(s) 16 verso, 17, 18	$H_{1, 2, 3.}$
§91 (i)	,, 60 verso	$L_{1.}$
(ii)	,, 60 recto, 61	$L_{1, 2.}$
(iii)	,, 21 verso, 22 recto	E_4, $F_{1.}$
§92 (i)	,, 61	$L_{2.}$
(ii)	,, 62	$L_{3.}$
(iii)	,, 62 verso	$L_{3.}$
(iv)	,, 63	$L_{4.}$
§93	Section 'M'	M.
§94	Folio(s) 47 recto	$M_{7.}$
§95	,, 44 recto, 43 verso	$M_{12.}$
§96	,, 44 verso	$M_{11.}$
§97	,, 36 verso	$M_{4.}$
§98	,, 20 verso	$M_{1.}$
§99	,, 32	$M_{3.}$
§100 (i)	,, 37 recto	$M_{6.}$
(ii)	,, 37 recto	$M_{6.}$
(iii)	,, 37 verso	$M_{6.}$
§101 (i)	,, 55 recto	$M_{10.}$
(ii)	,, 55 verso	$M_{10.}$
(iii)	,, 49 recto	$M_{9.}$
(iv)	,, 49	$M_{9.}$

Systems of linear equations.

78. Examples of the following type occur

$x_1 + x_2 = a_1, \quad x_2 + x_3 = a_2, \ldots\ldots, x_n + x_1 = a_n$ where *n* is always odd.

We have $a_n = (a_2 - a_1) + \ldots. + (a_{n-1} - a_{n-2}) + 2x_1$; and if we assume $x_1 = p$

then $a'_n = (a_2 - a_1) + \ldots\ldots + (a_{n-1} - a_{n-2}) + 2p$.

Subtracting we get $x_1 = p + (a_n - a'_n)/2$, which is the solution employed in the text.

The following examples occur

(i) $x_1+x_2=13,\quad x_2+x_3=14,\quad x_3+x_1=15.$

The value of x_1 is assumed to be 5, then by 'subtraction in order' $x_2=8$, $x_3'=6$ and $x_3'+x_1=11$. The correct values are therefore $x_1=5+(15-11)\div 2=7$, $x_2=6$ and $x_3=8$.

(ii) $x_1+x_2=16,\ x_2+x_3=17,\ x_3+x_4=18,\ x_4+x_5=19,\ x_5+x_1=20.$

Here the value of x_1 is assumed to be 7 and $x_5'+x_1'=16$, therefore the correct values are $x_1=9$, $x_2=7$, $x_3=10$, $x_4=8$, $x_5=11$.[1]

The following are implied

(iii) $x_1+x_2=9,\ x_2+x_3=5,\ x_3+x_1=8.$

(iv) $x_1+x_2=70,\ x_2+x_3=52,\ x_3+x_1=66.$

(v) $x_1+x_2=1860,\ x_2+x_3=1634,\ x_3+x_1=1722.$

(vi)[2] $x_2+x_3+x_4+x_5=317$

$x_1+x_3+x_4+x_5=347$

$x_1+x_2+x_4+x_5=357$

$x_1+x_2+x_3+x_4=365$

and the following, which, however, is too mutilated to be sure of

(vii) $x_1+x_2=36,\ x_2+x_3=42,\ x_3+x_4=48,\ x_4+x_5=54,\ x_5+x_1=60.$

The directions seem to indicate that we should cancel by six. We then get

$a_1'=6,\quad a_2'=7,\quad a_3'=8,\quad a_4'=9,\quad a_5'=10$ of which the solution is $x_1'=4,\ x_2'=2$, etc.; whence $x_1=24$, $x_2=12$, $x_3=30$, etc.

79. The next set of examples can be represented in the form $\Sigma x-x_1=c-d_1x_1$, $\Sigma x-x_2=c-d_2x_2\ \ldots\ldots\ \Sigma x-x_n=c-d_nx_n$.

If we set $a=1-d$ these become

$\Sigma x-a_1x_1=\Sigma x-a_2x_2=\ldots\ldots=\Sigma x-a_nx_n=c$, and $a_1x_1=a_2x_2=\ldots\ldots=a_nx_n=\Sigma x-c=k$.

---

[1] Since in these two particular examples the values of a_1, a_2, etc., are in arithmetical progression a simpler solution would be $x_1=\frac{m}{2}$ where *m* is the mean of the series a_1, a_2, etc.

[2] This can be arranged in the form $y_1+y_2=365$, $y_2+y_3=347$, $y_3+y_4=362$, $y_4+y_5=317$, $y_5+y_1=357$. Solving this we get $x_1+x_2=210$, $x_2+x_3=170$, $x_3+x_4=155$, $x_4+x_5=147$, $x_5+x_1=192$, which solved gives $x_1=120$, $x_2=90$, $x_3=80$, $x_4=75$, $x_5=72$. (See folio 1.)
Also note that most of these six examples can be expressed in the form

$$\Sigma x-x_1=a_1,\ \Sigma x-x_2=a_2,\ \ldots\ldots\ldots\ \Sigma x-x_n=a_n.$$

The examples given in the text are undoubtedly akin to the 'Ephanthema', usually attributed to Thymaridas, which may be expressed by $x_0=\frac{\Sigma a-\Sigma x}{n-1}$, where $x_0+x_1=a_1$, $x_0+a_2=x_2$, $\ldots\ldots x_0+x_n=a_n$. In Āryabhaṭa's *Gaṇita* is a similar rule $\Sigma x=\frac{\Sigma d}{n-1}$ where $\Sigma x-x_1=d_1$, $\Sigma x-x_2=d_2\ldots\ldots\Sigma x-x_n=d_n$, which Cantor (*Vorlesungen über der Geschichte der Mathematik* i, 624) considers to be a modification of the rule given by Thymaridas. See also P. TANNERY *Mémoires scientifiques*, Tome ii, pp. 192-195.

Therefore $\Sigma x=(\frac{1}{a_1}+\frac{1}{a_2}+\dots\frac{1}{a_n})k=\frac{p}{q}k.$

A solution is $x_1=\frac{q}{a_1}$, $x_2=\frac{q}{a_2}$ ...... $x_n=\frac{q}{a_n}$ and $c-p=q$[1]

There are four examples illustrating this process which may be tabulated thus—

	d_1	d_2	d_3	d_4	d_5	$p \div q$	c	x_1	x_2	x_3	x_4	x_5
(i)	$\frac{1}{2}$	$\frac{1}{3}$	$\frac{1}{4}$	$\frac{1}{5}$	$\frac{1}{6}$	$\frac{437}{60}$	377	120	90	80	75	72
(ii)	$-\frac{7}{12}$	$-\frac{3}{4}$	$-\frac{11}{6}$			$\frac{2558}{1463}$	1095	924	836	798		
(iii)	5	7	8				262	42	28	24		
(iv)	2	3	4				17	6	3	2		

Of these only fragments remain. For example, of the first we have :[2]

$$\frac{1}{2} \quad \frac{1}{3} \quad \frac{1}{4} \quad \frac{1}{5} \quad \frac{1}{6} \quad \frac{120}{60}+\frac{90}{60}+\frac{80}{60}+\frac{75}{60}+\frac{72}{60}=\frac{437}{60}$$

$$90+80+75+72=317, \qquad 317+\frac{120}{2}=377.$$

$$120+80+75+72=347, \qquad 347+\frac{90}{3}=377.$$

$$120+90+75+72=357, \qquad 357+\frac{80}{4}=377.$$

$$120+90+80+72=362, \qquad 362+\frac{75}{5}=377.$$

$$120+90+80+75=365 \qquad 365+\frac{72}{6}=377.$$

The second of these examples may be expressed in the form

$$x_1+x_2-(\tfrac{1}{3}+\tfrac{1}{4})\ x_3=x_2+x_3-(\tfrac{1}{4}+\tfrac{1}{2})\ x_1=x_3+x_1-(\tfrac{1}{2}+\tfrac{1}{3})\ x_2.$$

From this $\frac{19}{12}x_1=\frac{7}{4}x_2=\frac{11}{6}x_3$, and $\Sigma x/(\Sigma x-c)=\frac{12}{19}+\frac{4}{7}+\frac{6}{11}=\frac{2558}{1463}$.

The solution given is $\Sigma x=2558$ and $c=2558-1463=1095$, whence $x_1=924$, $x_2=836$, $x_3=798$.

Of the third there is sufficient of the formal question preserved to enable it to be restored.

One possesses seven horses, another nine mules (?), and a third ten camels. Each gives one of his animals to each of the others and then their possessions become of equal value.

---

[1] Compare with this the treatment of the *epanthem* by Jamblichus (HEATH, *Greek Mathematics*, i, 93-94).

[2] The nearest approach to this that I have come across in an Indian work is given by Mahāvīra in his *Gaṇita-Sāra-Saṅgraha* (vi, 239-240).

Five merchants saw a purse of money. They said one after another, by obtaining $\frac{1}{6}$, $\frac{1}{7}$, $\frac{1}{8}$, $\frac{1}{9}$ or $\frac{1}{10}$ of the contents of the purse I shall become three times as rich as all of you.

Let p be the value of the purse and x_1, x_2, etc., the original capitals; then $3(\Sigma x-x_1)=\frac{p}{6}+x_1$, $3(\Sigma x-x_2)=\frac{p}{7}+x_2$, etc.; whence $11\Sigma x=(\frac{1}{6}+\frac{1}{7}+\frac{1}{8}+\frac{1}{9}+\frac{1}{10})\ p=\frac{6508p}{10080}$, or $\Sigma x=\frac{6508p}{110880}$. If $p=110880$ then $x_1=261$.

But al-Karkhī (xith cent.) gave (iii, 6) the following :—A certain sum is divided among three people, one-half being given to the first, one-third to the second and one-sixth to the third. But then one-half the share of the first, one-third the share of the second and one-sixth that of the third is pooled and shared equally by the three.

In A.D. 1225 this problem became famous as it was one of those propounded to Leonardo of Pisa and solved by him.

If x_1, x_2, x_3 be the shares and 3c be the amount pooled, then $x_1-\frac{x_1}{2}+c=\frac{\Sigma x}{2}$, $x_2-\frac{x_2}{3}+c=\frac{\Sigma x}{3}$, $x_3-\frac{x_3}{6}+c=\frac{\Sigma x}{6}$ and $\Sigma x-x_1=2c$, $\Sigma x-2x_2=3c$, $\Sigma x-5x_3=6c$, whence $\Sigma x=\frac{47c}{7}$. Make $c=7$ and then $x_1=33$, $x_2=13$, $x_3=1$.

The fourth example is of similar form.[1]

80. The following occurs on some very mutilated scraps that have been pieced together. It is now impossible to say how it originated or what its context was.

$$x(x+y+z)=60$$
$$y(x+y+z)=75$$
$$z(x+y+z)=90$$

We have $(x+y+z)^2=225$ whence $x+y+z=15$, and $x=4$, $y=5$, $z=6$.

80. (*a*). There is one nearly complete statement and solution of the following pair of equations :—

$$x+y+z=20$$
$$3\,x+\tfrac{3}{2}\,y+\tfrac{1}{2}\,z=20$$

of which the only solution in positive integers is $x=2$, $y=5$, $z=13$.

This type of problem was a favourite in Europe and Asia in early mediæval times. Later it was known as the 'Regula Virginum,' 'Regula Potatorum,' etc. It was given by Chang-ch'iu-chien (sixth century A.D.),[2] by Alcuin the Englishman (eighth century) and others. About 900 A.D. it was pretty fully treated by Abū Kāmil al-Misrī who gives some six problems, varying from three to five terms and attempts to find *all* the positive integral solutions.[3]

In the earlier Indian works problems of this type do not occur, but *exactly* the same example is given by both Mahāvīra[4] (? ninth century) and Bhāskara[5] (twelfth century).

Quadratic indeterminate equations.

81. There are two types of quadratic indeterminates preserved.

$$\text{(i)}\ x+a=s^2, \qquad x-b=t^2$$

The solution given may be represented by

$$x=\left(\tfrac{1}{2}\left(\tfrac{a+b}{c}-c\right)\right)^2+b$$

which makes both $x+a$ and $x-b$ perfect squares. In the actual example preserved $a=5$ and $b=7$ and the solution is $x=11$, which is the only possible integral (positive) solution obtainable from the formula. No general rule is preserved but the solution itself indicates the rule. It proceeds by steps thus : $5+7=12$, $12\div2=6$, $6-2=4$, $4\div2=2$, $2^2=4$, $4+7=11$. The value of this type of detailed exposition is here

---

[1] This set of examples is introduced by a rule which simply means : change the fraction $\frac{A}{B}$ into $\frac{B}{B-A}$. This illustrates the fact that the *sūtras* or rules often make no pretence of indicating in any way the general theory : they are merely intended to be helpful.

[2] Yoshio Mikami.

[3] H. Suter, *Das Buch der Seltenheit*, etc. Bib. Math. xi (1910-11), pp. 100-120. See also Suter's *Die Mathematiker und Astronomen der Araber und Ihrer Werke*, p. 43.

E.g. of $x+y+z+w=100$, $2x+\frac{y}{2}+\frac{z}{3}+w=100$ there are said to be 304 solutions and twenty are given. See also L. E. Dickson's *History of the Theory of Numbers*, vol. ii, pp. 77, *et seq.*

[4] *Gaṇita-Sāra-Saṅgraha* vi, 152.

[5] *Vīja Gaṇita*, §158.

self-evident. It seems to have been almost necessitated by the absence of a suitable algebraic symbolism.[1]

82. The second type is

$$\text{(ii)}\quad xy-ax-by-c=o$$

The solutions which appear to be followed in the text are $x=(ab+c)\div m+b$, $y=a+m$; or $y=(ab+c)\div m+a$, $x=b+m$

where m is any assumed number. The only example that is preserved[2] is $xy-3x-4y\pm1=0$ of which the solutions given are 15 and 4, and 16 and 5, *i.e.*,

$$(3{\cdot}4-1)\div1+4=15,\ 3+1=4;\ (3{\cdot}4-1)\div1+3=16,\ 4+1=5.$$

Simple 'motion' problems.

83. From the mathematical point of view this section is not of particular interest but it was possibly intended as a sort of introduction to the section (C) that follows directly after. The examples can be classified into three types :

(i) *A* travels at a certain rate r_1 for a number of days *T*, and then *B* starts at a daily rate r_2. When will *A* and *B* have travelled equal distances ?

Since $r_1T+r_1t=r_2t$, $t=r_1T/(r_2-r_1)$ which is expressed by a *sūtra* and illustrated by two examples.

(ii) *A* travels a distance a_1 the first day, a_1+d_1 the second day and so on ; *B* travels a_2 the first day, a_2+d_2 the second day and so on. When will they have travelled equal distances ?

Since $((t-1)d_1\div2+a_1)t=((t-1)d_2\div2+a_2)t$, $t=2(a_1-a_2)/(d_2-d_1)+1$

(iii) One goes at a fixed rate *A* and another goes *a* the first day, $a+d$ the second day and so on. In what time will they have travelled equal distances ?

Since $tA=t((t-1)d\div2+a)$, we have $A=(t-1)d\div2+a$, and $t=(2A-2a)\div d+1$

---

[1]This type of equation occurs in many mediæval works from the time of Diophantus onwards (*e.g.*, see DIOPHANTUS ii, 11ff. ; BRAHMAGUPTA, xviii, 84 ; AL-KARKHĪ, p. 63 ; etc.) and has here no very special interest beyond the indication it gives that the Bakhshālī text followed the fashion. Dr. Hoernle, however, thought that it indicated a 'peculiar' connexion between the Bakhshālī MS. and Brahmagupta's work ; and from this deduced that our text 'may have been one of the sources from whence the later astronomers took their arithmetical information.' (*Ind. Ant.*, xvii, 1888, p. 37.)

[2] Unfortunately the text (folio 27, recto) is so mutilated that the correctness of the interpretation here given cannot be guaranteed. But the equation was well known to mediæval mathematicians and has historical interest. Brahmagupta gave a general solution (xviii, 61) which, however, he appears to have thought unnecessary. Bhāskara also gives a general solution (*Vīja Gaṇita*, 212-214) together with demonstrations in both algebraical and geometrical forms ; but he also on another occasion (ib. 208-209) gives an arbitrary solution. The general solution is also given by al-Karkhī. Mahāvīra gives (vi, 284) the equation in a different form, namely $(a+\frac{1}{x})(b+\frac{1}{y})=ab+A$ which reduces to $xy-\frac{a}{A}x-\frac{b}{A}y-\frac{1}{A}=o$ of which the solution is $x=\frac{b}{A}+\frac{1}{m}(\frac{ab+A}{A^2})$, $y=\frac{a}{A}+m$; or $y=\frac{a}{A}+\frac{1}{m}(\frac{ab+A}{A^2})$, $x=\frac{b}{A}+m$. If we make $m=\frac{ab}{A}+1$ then $x=\frac{b+1}{A}$, $y=\frac{a+ab+A}{A}$ which is the solution given by Mahāvīra.

His examples are : The product of 3 and 5 is 15, and the required product is 18 or 14. What are the quantities to be added or subtracted ? *i.e.* (i) $(3+\frac{1}{x})(5+\frac{1}{y})=15+3$, (ii) $(3-\frac{1}{x_1})(5-\frac{1}{y_1})=15-1$. His answers are (i) 2 and 7, (ii) 6 and 17.

84. There are two examples in this section of different types—

(i) A chariot is drawn by A horses a at a time. How many stages should there be in a distance l, and how many stages should each horse do?

The argument seems to be : the 'mileage' done is la therefore each horse does la/A miles ; the number of stages is A consisting of 1/A miles each and each horse does a stages.[1]

(ii) A and B start on a journey together. A goes at the rate of r_1 and meets B, whose rate is r_2, on the return journey. If the distance between the two places is l, when will they meet?

Let x be the distance gone by the slower traveller then the other goes $2l-x$ and $tr_1=2l-x$, $tr_2=x$ whence $t=2l/(r_1+r_2)$. A proof is given in the form $l : r_1 :: t : tr_1$ and $l : r_2 :: t : tr_2$

Quadratic equations and square-root approximations.[2]

85. Ostensibly this section deals with certain problems in arithmetical progression that were common property in comparatively early times; but it is of special interest because it elaborates a square-root rule and gives a number of examples, some of which involve large numbers. It also employs, although to a very limited extent, the sexagesimal notation. Of particular interest are the methods of reconciliation necessitated by employing approximations. There are two types of arithmetical progression exhibited—

(i) $DT+Dt=((t-1)d\div 2+a)t$, where T is a given period and t is to be determined. The solution is $t=\{2(D-a)+d\pm\sqrt{(2(D-a)+d)^2+8DT}\}\div 2d$.

The second type is the ordinary one of finding the number of terms of an arithmetical progression when the other elements are given, the solution being

(ii) $t=\{-(2a-d)\pm\sqrt{(2a-d)^2+8ds}\}\div 2d$.

These two types may conveniently be reduced to one by the formula $t=(p\pm\sqrt{q})\div 2d$ or $(p\pm q)\div 2d$, where $q=\sqrt{Q}$ and $p=2D-(2a-d)$ or $-(2a-d)$. The negative sign for the root quantity is not used so $t=(p+q)\div 2d$ always. Unless Q is a square number the solutions given depend upon approximations, and the most interesting results exhibited in the text are those connected with the proofs or verifications of the approximate answers obtained.

In type (i) let $s'=DT+Dt$ and $s''=((t-1)d\div 2+a)t$. If approximate values of t are used such that $s_1'=DT+Dt_1$, and $s_1''=((t_1-1)d\div 2+a)t_1$, it is evident that s_1' does not equal s_1, and the problem of reconciliation arises.

Now $s_1'=DT+Dt_1=DT+(Dp+Dq_1)\div 2d=(8dDT+4Dp+4Dq_1)\div 8d$, $s_1''=((t_1-1)d/2+a)t_1=(4Dp+4Dq_1-p^2+q_1^2)\div 8d$, and $s_1''-s_1'=q_1^2-(p^2+8dDT)\div 8d$; but $p^2+8DdT=Q=q^2$, therefore $s_1''-s_1'=(q_1^2-q^2)\div 8d=e/8d$ (see below).

---

1 This example is evidently traditional and we find it in Mahāvīra (vi, 157-158) in the following form :—

"It is well known that the horses belonging to the Sun's chariot are seven. Four horses being harnessed to the yoke to draw it. They have to do a journey of seventy yojanas. How many times are they yoked and unyoked in four ?"

His rule is the same as that given above.

2 Approximations are here indicated by subscripts: thus s_1 represents a first approximation to the value of s, and s_2 represents an approximation of the second order, etc.

This form of reconciliation occurs only once, but that about to be explained occurs a number of times. In type (ii) $s=((t-1)d/2+a)t=t^2d/2+(2a-d)t/2=(t^2d-pt)/2$; and since $2dt-p=q$, we have $4d^2t^2-4dtp=q^2-p^2$, and $8ds=4d^2t^2-4dtp=q^2-p^2$. Therefore $s_1-s=(q_1{}^2-p^2)/8d-(q^2-p^2)/8d=(q_1{}^2-q^2)/8d$ $=e/8d$ as before.

The square-root rule comes in incidentally when a surd quantity occurs. In this way it is actually given three times (see §68) and this seems to indicate that it was not treated separately and was considered as subsidiary to the arithmetical progression problem. To the student of the history of mathematics, however, it will be probably considered the most interesting part of the work. The rule may be expressed by $\sqrt{A^2+b}=A+b/2A$ approximately.[1] But since $(A+b/2A)^2=A^2+b+(b/2A)^2$, the error may be shown by $e_1=(b/2A)^2$. Again $\sqrt{A^2+b}=\sqrt{r_1{}^2-e_1}=r_1-e_1/2r_1$ approximately, where $r_1=A+b/2A$ and the second error may be represented by $e_2=(e_1/2r_1)^2=b^4/64A^4(A+b/2A)^2$. Thus for a first approximation $s_1-s=e_1/8d$, and for a second approximation $s_2-s=e_2/8d$.

86. The examples of this section are fragmentary but in most cases they are quite unambiguous and can be interpreted with certainty. I now give an accurate representation of these examples; but it should be noted that those portions that I have enclosed in angular crotchets do not occur in the text as it now stands.

Rule. $t=\{2(D-a)+d+\sqrt{(2(D-a)+d)^2+8DTd}\}\div 2d$

(i) *Example.* $D=5,\ T=6,\ a=3,\ d=4$

. . . . . $DT=30,\quad D-a=5-3=2,\quad 2(D-a)=4,$

$2(D-a)+d=8,\quad (2(D-a)+d)^2=64.\quad DT=30,$

$8DT=240,\quad 8DTd=960,\quad \{2(D-a)+d\}^2+8DTd=960+64$

$=1024,\quad \sqrt{1024}=32,\quad 2(D-a)+d=8,\quad 8+32=40,$

$<2d=8$ and $40\div 8=5=t>$.

Proof by the rule of three; $1:5::5:25$ and $25+30=55$.

(ii) *Example.* $D=7,\ T=5,\ a=5,\ d=3$.

$DT=35,\ D=7,\ a=5,\quad <D-a=2,\ (2(D-a)+d)^2=49>$

$8DTd=840,\ 840+49=889,\quad <\sqrt{889}=\sqrt{29^2+48}>\ =29\frac{48}{58}.$

$2\,(D-a)+d=7,\ 29\frac{48}{58}+7=36\frac{48}{58}=\frac{2136}{58},\quad <2d=6$ and $\frac{2136}{58}\div 6=\frac{178}{29}=t_1.>\frac{178}{29}=6+8^{i}+16^{ii}+3^{iii}+6\frac{6}{29}^{iv}.$

$t_1-1=\frac{178}{29}-1=\frac{149}{29},\ (t_1-1)\,d=\frac{149}{29}\times 3=\frac{447}{29},\ (t_1-1)d/2=\frac{447}{28},\ (t_1-1)d/2+a=\frac{447}{58}+5=\frac{737}{58}.$

---

[1] The history of this rule is of considerable interest. It was given by Heron (Heath, Greek Math. ii, 324) who indicates the second approximation; and it was known to Planudes (xiiith century), Barlaam and Nicholas Rhabdas (xivth cent.) and to al-Qalasādī (xvth cent. See Woepcke, *Jour. Asiatique*, 1854, p. 384). It does not appear in any early Indian work but it is given by Sūryadāsa (xvith cent.), who attributes it to his father, Jnyāna Rāja. Colebrooke's translation (*Algebra, etc., from the Sanscrit*, p. 155) of this Indian version is as follows:—"The root of a near square, with the quotient of the proposed square divided by that approximate root, being halved, the moiety is an approximate root; and repeating the operation as often as necessary, the nearly exact root is found. Example 5. This divided by two which is first put for the root gives $\frac{5}{2}$ for the quotient, which added to the assumed root 2, makes $\frac{9}{2}$; and this divided by 2, yields $\frac{9}{4}$ for the approximate root." Sūryadāsa's approximation may thus be represented by $\frac{A+\frac{A^2+b}{A}}{2}$. Jnyāna Rāja's work on astronomy is entitled *Siddhānta Sundara*. See the reference to Sundarī on folio 34 of our text.

$s_1'' = ((t_1 - 1)d/2 + a)t_1 = \frac{737}{58} \times \frac{178}{29} = \frac{65,593}{841}.$

$s_1' = DT + Dt_1 = 7\ (5 + \frac{178}{29}) = \frac{65569}{841}$; and $s_1'' - s_1' = \frac{24}{841} = < \frac{48^2}{4 \times 841 \times 24} = \frac{b^2}{4A^2.8d}. >$

Proof by the 'rule of three', $1:7::\frac{178}{29} : 42\frac{28}{29} <$ and $42\frac{28}{29} + 35 = 77\frac{812}{841}. >$

(iii) *Rule.* $t = \{-(2a - d) + \sqrt{(2a-d)^2 + 8ds}\} \div 2d.$

Example. $a = 1,\ d = 1,\ s = 60.$

Solution. $8ds = 480,\ 2a - d = 1,\ 480 + 1 = 481, \sqrt{481} \sim 21\frac{40}{42} = \frac{882 + 40}{42} = \frac{922}{42}$; $<$ and $t_1 = \frac{1}{2}(\frac{922}{42} - 1) > = \frac{880}{84}.$

$<$ Now $s_1 = t_1(t_1 + 1)/2 > = \frac{880}{84} \times \frac{964}{168} = \frac{848,320}{14112}.$

$<$ But $s_1 - s = e_1/8d = \frac{1}{8}(\frac{40}{2.21})^2 > = \frac{1600}{14112}$, and $\frac{848,320}{14,112} - \frac{1600}{14112} = \frac{846,720}{14112} = 60.$

$<$ Again $\sqrt{481} = \sqrt{(21\frac{20}{21})^2 - (\frac{20}{21})^2} > \sim 21\frac{20}{21} - \frac{(20/21)^1}{2 \times 21\frac{20}{21}} = \frac{425,042 - 400}{19,362} = \frac{424,642}{19,362}$, $<$ and $t^2 = (\frac{424,642}{19,362} - 1) \div 2$

$= \frac{405,280}{38,724}$; $<$ and $s_2 = ((t_2 - 1)\ d \div 2 + a)t_2 = t_2(t_2 + 1)/2\ > = \frac{405,280}{38,724} \times \frac{444,004}{77,448} = \frac{179,945,941,120}{2,999,096,352}.$

$<$ But $s_2 - s = e_2/8d = \frac{40^4}{8^3 \times 21^4 \times (21\frac{20}{21})^2} > = \frac{160,000}{2,999,096,352}$ and $\frac{179,945,941,120 - 160,000}{2,999,096,352} = < \frac{179,945,781,120}{2,999,096,352} = 60. >$

(iv) ***Example.*** $< a = 1,\ d = 1,\ s = 5$; therefore $t = (-1 + \sqrt{41})/2; > \sqrt{41} \sim 6\frac{5}{12} = \frac{77}{12}$, and $t_1 = \frac{65}{24}.$

But $s_1 = ((\frac{65}{24} - 1)\frac{1}{2} + 1)\frac{65}{24} = (\frac{1}{2}.\frac{41}{24} + 1)\frac{65}{24} = (\frac{41}{48} + 1)\frac{65}{64} = \frac{89}{48} \times \frac{65}{24} = < \frac{5785}{1152} = 5\frac{25}{1152} = 5 + \frac{e_1}{8d} >.$

Again a second approximation to $\sqrt{41}$ is given by $6\frac{5}{12} - \frac{1}{2}(\frac{5}{12})^2/6\frac{5}{12} = \frac{11858}{1848} - \frac{25}{1848} = \frac{11833}{1848}$; and t_2

$= \frac{1}{2}(\frac{11833}{1848} - 1) = \frac{1}{2}(\frac{9985}{1848}) = \frac{9985}{3696}\ldots\ldots$

(v) ***Example.*** $< a = 1\frac{1}{2},\ d = 1\frac{1}{2},\ s = 2$; therefore $t = \{-\frac{3}{2} + \sqrt{(\frac{3}{2})^2 + 8.\frac{3}{2}.2}\} \div 3 = (-3 + \sqrt{105}) \div 6.$

Now $\sqrt{105} = 10\frac{1}{4}$ approximately, and a second approximation is given by $10\frac{1}{4} - \frac{1}{16}/2(10\frac{1}{4}) = 10\frac{81}{328}.$

With this value $t_2 = (-3 + 10\frac{81}{328})/6 = \frac{59425}{49200}$, and $s_2 = ((t_2 - 1)\frac{d}{2} + a)t_2 = ((\frac{59425}{49200} - 1)\frac{3}{4} + \frac{3}{2})t_2 >$

$= (\frac{10225}{32800} \times \frac{1}{2} + \frac{3}{2})t_2 = (\frac{10225}{65600} + \frac{3}{2})t_2 = \frac{108625}{65600} \times \frac{59425}{49200} = \frac{6,455,040,625}{3,227,520,000}.$ Now $s_2 - s = e_2/8d^* = \frac{625}{3,227,520,000}$

and $s = \frac{6,455,040,625 - 625}{3,227,520,000} = \frac{6,455,040,000}{3,227,520,000} = 2.$

(vi) *Example.* $a = 1\frac{1}{2},\ d = 1\frac{1}{2},\ s = 7000$; therefore $< t = (-3 + \sqrt{33609})/6$, and since $579^2 = 335,241 =$ $33,609 - 768$, we have as a first approximation to the root quantity $579\frac{384}{579}$, while the second approximation is $579\frac{384}{579} - (\frac{384}{579})^2 \div 2(579\frac{384}{579}) > = 579\frac{384}{579} - (\frac{384}{579})^2 \times \frac{1158}{671,250} = 579\frac{768}{1158} - \frac{294,912}{777,307,500} = 579\frac{515,520,000 - 294,912}{777,307,500}$

$= 579\frac{515,225,088}{777,307,500} = \frac{450,576,267,588}{777,307,500}.$ $<$ Therefore $t_2 = (\frac{450,576,267,588}{777,307,500} - 3) \div 6 > = \frac{448,244,345,088}{4,663,845,000}.$

---

[1] The Appendix to this Chapter (page 53) may be useful to those who are interested in the details of the calculations.

* Here instead of $d = 1\frac{1}{2}$ we must use $4.d = 6$, for we have really given *a* and *d* new values by cancelling, namely $a = 3,\ d = 3$ and e is four times greater than it should have been (since $e = q_2^2 - q^2$) ; otherwise since $q^2 = 26\frac{1}{4},\ q_2 = \frac{3361}{2.328}$, $e = q_2^2 - q^2 = (\frac{3361}{16.41})^2 - 26\frac{1}{4}$ $= \frac{1}{16^2.41^2}$; and $\frac{e}{8.\frac{3}{2}} = \frac{1}{3.4.16.^2 41^2} = \frac{25^2}{3.4.^3 41.^2 (4.25)^2} = \frac{625}{3,227,520,000}.$

Now $s_2=((t_2-1)\frac{3}{4}+\frac{3}{2})t_2$ and $t_2-1=\frac{443,580,500,088}{4,663,845,000}$; $(t_2-1)d=\frac{221,790,250,044}{1,554,615,000}$; $(t_2-1)\ \frac{d}{2}=\frac{110,895,125,022}{1,554,615,000}$;

$(t_2-1)d/2+a=\frac{113,227,047,522}{1,554,615,000}$; and finally $s_2=\frac{50,753,383,762,746,743,271,936}{7,250,483,394,675,000,000}$.

But $s_2-s=e_2/8d=\frac{768^4}{6.8^3.579^4.(579\frac{768}{1158})^2}=\frac{21,743,271,936}{7,250,483,394,675,000,000}$, which given $s=\frac{50,753,383,762,725,000,000,000}{7,250,483,394,675,000,000}=7{,}000$.

Series.

87. There is a set of series that may be represented as follows, where T_1, T_2, T_3, etc., represent the successive terms of any of the series irrespective of their form.

	T_1	+ T_2	+ T_3	+ T_4	$= S$
	x	$+\ 2T_1$	$+\ 3T_1$	$+\ 4T_1$	$= 200$
	$x\ (1+\frac{3}{2})$	$+\ 2T_1+\frac{5}{2}x$	$+\ 3T_1+\frac{7}{2}x$	$+\ 4T_1+\frac{9}{2}x$	$= \frac{71}{2}$
	$x\ (1+\frac{3}{2})$	$+\ 2T_1-\frac{5}{2}x$	$+\ 3T_1-\frac{7}{2}x$	$+\ 4T_1-\frac{9}{2}x$	$= \frac{29}{2}$
	x	$+\ 2T_1$	$+\ 3T_2$	$+\ 4T_2$	$= 132$
	$x\ (1+\frac{3}{2})$	$+\ 2T_1+\frac{5}{2}x$	$+\ 3T_2+\frac{7}{2}x$	$+\ 4T_2+\frac{9}{2}x$	$= 144\frac{1}{2}$
Missing	$<x\ (1+\frac{3}{2})$	$+\ 2T_1-\frac{5}{2}x$	$+\ 3T_2-\frac{7}{2}x$	$+\ 4T_3-\frac{9}{2}x$	$= \frac{41}{2}>$
	x	$+\ 2T_1$	$+\ 3(T_1+T_2)$	$+\ 4\ (T_1+T_2+T_3)$	$= 300$
	$x\ (1+\frac{3}{2})$	$+\ 2T_1+\frac{5}{2}x$	$+\ 3\ (T_1+T_2)+\frac{7}{2}x$	$+\ 4\ (T_1+T_2+T_3)+\frac{9}{2}x$	$= 222$
	$x\ (1+\frac{3}{2})$	$+\ 2T_1-\frac{5}{2}x$	$+\ 3\ (T_1+T_2)-\frac{7}{2}x$	$+\ 4\ (T_1+T_2+T_3)-\frac{9}{2}x$	$= 78$

These are obviously built up on a definite plan. There are three fundamental series, namely

$$a_1+2a_1+3a_1+\ldots.+na_1$$

$$a_1+2a_1+3a_2+\ldots.+na_{n-1}$$

$$a_1+2a_1+3\ (a_1+a_2)+\ldots\ldots+n\ (a_1+a_2+\ldots\ldots a_{n-1})$$

and a subsidiary series which may be represented by

$$0+e+e(1+d)+e(1+2d)+\ldots\ldots\ldots \text{etc.}$$

There is no attempt at general summation, which, indeed, is not generally possible, and I am afraid we can credit the author with little of mathematical value here. It may be noted that the sums of the series may be given as $10T_1\pm3(e+d)$, $33T_1\pm(22e\pm7d)$, $60T_1\pm(26e+7d)$, and that in all cases $e=T_1=\frac{5}{2}$, and $d=\frac{2}{5}T_1$. The answers are therefore $25\pm\frac{21}{2}=\frac{71}{2}$ or $\frac{29}{2}$, $\frac{156}{2}\pm\frac{124}{2}=\frac{289}{2}$ or $\frac{41}{2}$, and $\frac{300}{2}\pm\frac{144}{2}=222$ or 78. Of these $\frac{41}{2}$ is missing.

88. There are other series of which the most notable is also a double series. It may be represented by

$$\left\{\begin{matrix} a_1+3a_1+3^2a_1+3^3a_1+3^4a_1 \\ 0+\frac{3}{4}a_1+\frac{3}{4}(a_1+a_2)+\frac{3}{4}(a_1+a_2+a_3)+\frac{3}{4}(a_1+a_2+a_3+a_4) \end{matrix}\right\}=329,$$ whence the value of a_1 is 2.

There are other simple series on certain fragments, that, as they now stand, are not of particular interest.

89. There is a set of some 17 examples that may be represented by

$$C\ (1-a_1)(1-a_2)\ .\ .\ .\ .\ .\ .\ .\ (1-a_n)=R=C-L$$

where R is the 'remainder' and L the 'loss' after successive deductions from C. The values of *a* are generally simple fractions with unity as numerator, and if $(1-a_1)\ .\ .\ .\ .\ (1-a_n)=p/q$ then $Cp/q=R=C-L$ whence $C=\frac{Rq}{p}$ and $C=L/(1-\frac{p}{q})$. The following are specimens

$$4(1-\tfrac{1}{4})(1-\tfrac{1}{4})(1-\tfrac{1}{4})(1-\tfrac{1}{4})=x$$

$$8(1-\tfrac{1}{3})(1-\tfrac{1}{3})(1-\tfrac{1}{3})=x$$

$$60(1-\tfrac{1}{2})(1-\tfrac{1}{3})(1-\tfrac{1}{4})(1-\tfrac{1}{5})=x$$

$$x(1-\tfrac{1}{3})(1-\tfrac{1}{4})(1-\tfrac{1}{5})=x-24$$

$$x(1-\tfrac{1}{2})(1-\tfrac{1}{4})(1-\tfrac{1}{5})=x-280$$

$$x(1-\tfrac{1}{3})(1-\tfrac{1}{4})(1-\tfrac{1}{5})=2x-32$$

The following questions are somewhat restored.

(i) A traveller takes on a journey a bottle containing four *prasthas* of wine. At the end of each stage he drinks one *prastha* and then fills up his bottle with water. How much wine and how much water will there be in the bottle after four stages?

The solution is first given by

$4(1-\tfrac{1}{4})(1-\tfrac{1}{4})(1-\tfrac{1}{4})(1-\tfrac{1}{4})=1\tfrac{17}{64}$ *prasthas*, the amount of wine, and $2\tfrac{47}{64}$ *prasthas* as the amount of water. Secondly it is worked out by steps.

(ii) A tax is paid at the rate of one-third, one-fourth and one-fifth on three separate occasions and the total amount paid in taxes is twenty-four. What was the original amount.[1]

The solution given is
$x\ (1-\tfrac{1}{3})\ (1-\tfrac{1}{4})\ (1-\tfrac{1}{5})=x-24$ and $x=40$.

This section is interesting[2] on account of the numerous proofs or verifications of the calculations, which have been referred to in the previous Chapter (§ 74).

Computation of gold.

90. This topic is treated in most mediæval mathematical works and our text follows the usual treatment but with some slight irregularity due, possibly, to a slight misconception of the author, who apparently looked upon the 'quality' or 'touch' factor as a negative quantity.

---

[1] See BHĀSKARA'S *Līlāvatī*, 73.

[2] It is also interesting because of the occurrence of a similar section in a papyrus of Akhmīm (HEATH, *Greek Math.* ii, 544; but Heath's reference is misleading and so is Cantor's: the papyrus of Akhmīm was published by J. BAILLET, *Mémoires du Caire*, IX, 1892, pp. 1-88).

The fundamental theorem is, of course, the usual one of averages, namely :—

$$\text{(i)}\quad x = \frac{f_1w_1+f_2w_2+\ldots\ldots f_nw_n}{w_1+w_2+\ldots w_n}$$

where w_r is the mass of a piece of gold, and f_r is its quality or 'touch.' That is f is the amount of pure gold per unit in some unknown measures. There is some indication in the text that f is expressed in *māshakas* and w in *suvarṇas*, where 12 *māshakas*=1 *suvarṇa* but this is not supported by Bhāskara or Mahāvīra, and indeed the money measures in our text are nowhere well defined.

Modifications of (i) that are indicated are

$$\text{(ii)}\quad \frac{f_1w_1+f_2w_2+\ldots\ldots x.w_n}{w_1+w_2+\ldots\ldots w_n}=F$$

$$\text{(iii)}\quad \frac{f_1w_1+f_2w_2+\ldots\ldots f_nw_n}{w_1+w_2+\ldots\ldots x}=F$$

$$\text{(iv)}\quad \frac{(f_1w_1+\ldots\ldots f_nw_n)+f'_1w'_1+\ldots\ldots f'_nw'_n}{w_1+w_2+\ldots\ldots w_n}=x$$ which is not understood.

Earning and spending.

91. (i) A earns e_1 in d_1 days ; B earns e_2 in d_2 days. If A gives g to B when will they have equal amounts ?

$$t = 2g \div \left(\frac{e_1}{d_1}-\frac{e_2}{d_2}\right)$$

(ii) A earns e in d_1 days and spends f in d_2 days. How long will his capital C last ?

$$t = c/\left(\frac{f}{d_1}-\frac{e}{d_2}\right)$$

(iii) One gives a_1 *dīnāras* in d_1 days ; a second gives a_2 *dīnāras* in d_2 days ; and so on. In what time will they have given b *dīnāras* altogether ?

In one day the total gift is $\frac{a_1}{d_1}+\frac{a_2}{d_2}+\ldots\ldots+\frac{a_n}{d_n}=\frac{P}{Q}$ and therefore b *dīnāras* are given in $\frac{b\,Q}{P}$ days and the respective gifts are $\frac{a_1\,b\,Q}{d_1\,P}$, $\frac{a_2\,b\,Q}{d_2\,P}$, $\frac{a_3\,b\,Q}{d_3\,P}$, &c.

The following occurs—

In $1\frac{1}{2}$ days one gives $2\frac{1}{2}$ *dīnāras*, in $1\frac{1}{3}$ days a second gives $3\frac{1}{2}$ *dīnāras*, and in $1\frac{1}{4}$ days a third gives $4\frac{1}{2}$ *dīnāras*. In what time will they have given 500 *dīnāras* altogether ?

Since $\frac{2\frac{1}{2}}{1\frac{1}{2}}+\frac{3\frac{1}{2}}{1\frac{1}{3}}+\frac{4\frac{1}{2}}{1\frac{1}{4}}=\frac{947}{120}$ *dīnāras* are given in 1 day, 500 *dīnāras* will be given in $500/\frac{947}{120}=\frac{60{,}000}{947}=63\frac{339}{947}$ days.

Profit and Loss.

92. Let C be the capital, p the profit, M=C+p, n=Cc the number of articles, c the cost rate and s the sale rate [c=a/b where a is the number purchased for b

***drammas*, say, but the money measure is not generally stated; and $s = \frac{e}{d}$ where *c* is the number sold for *d*.]**

The following rules are given with illustrative examples.

(i) $C = \frac{p}{c/s - 1}$ (ii) $p = \frac{Cc}{s} - C$

(iii) $C = \frac{M}{c/s} = \frac{C+p}{c/s}$ (iv) $C = \frac{-p}{1 - c/s}$

The 'M' Section.

93. The examples that belong to the so-called 'M section' may be described as miscellaneous, but most of them are solved by 'the rule of three terms.' Their chief interest is hardly mathematical although certain of them exhibit in very interesting forms the change-ratios of certain measures; it lies rather in what may be termed the social nature of the formal questions (see §§ 45-52). The translations of the questions here offered are 'informal' and sometimes very much 'restored'.

94. An army consists of chariots, elephants, horse and foot in the ratios 1 : 1 : 5 : 3 : If a complete army contains 10×3^7 of these how many of each kind are there?

The answers are

$10 \cdot 3^7 = 21{,}870$ chariots.

$10 \cdot 3^7 = 21{,}870$ elephants.

$10 \cdot 3^7 \cdot 5 = 109{,}350$ foot.

$10 \cdot 3^7 \cdot 3 = 65{,}610$ horse.

$(10^2 \cdot 3^7) = 218{,}700$ (1 akshauhiṇī).

This is the traditional subdivision of a Hindu army. There was probably more information given in the text for the terms *akshauhiṇī, anīkinī, chamū* and *pṛitanā* occur. According to the dictionaries 1 *akshauhiṇī* consists of 10 *anīkinīs* or 218,700 in all; a *chamū* consists of 129 chariots, 129 elephants, 2,187 horse, 3,645 foot, or 6,090 (*i.e.*, $3 \times 43 + 3 \times 43 + 3^7 + 3^6 \times 5 = 10 \times 3 \times 7 \times 29$); while a *pṛitanā* is said to consist of $243 + 243 + 729 + 1215 = 2430$ (or $3^5 + 3^5 + 3^6 + 5 \times 3^5 = 10 \times 3^5$). See also Albīrūnī's *India*, chap. xlviii.

95. One produces ten and a half in two and one-third days. For the sake of religion he gives thirteen and one-third in three and one-eighth days. He offers to VĀSUDEVA one quarter less than thirteen in eight and a half days. Desiring reward in a future world he gives to Brāhmans for food one and one-third in three and one-fifth days (. . . $2\frac{1}{4}$ in 5 days . . .) and also twelve and a half in thirty-three and one-third days for the best wine for the consumption of the merchants. In the treasure house is stored twelve hundred. Say, O Pandit, how long can this expenditure continue?

The daily income is $\frac{10\frac{1}{2}}{2\frac{1}{3}} = \frac{9}{2}$ and the daily expenditure is $\frac{13\frac{1}{2}}{3\frac{1}{8}} + \frac{12\frac{3}{4}}{8\frac{1}{2}} + \frac{1\frac{1}{3}}{3\frac{1}{5}} + \frac{\frac{1}{3}}{1\frac{1}{2}} + \frac{1}{3\frac{1}{3}} + \frac{2\frac{1}{4}}{5} + \frac{12\frac{1}{2}}{33\frac{1}{3}} = \frac{1807}{240}$. The daily loss is therefore $\frac{1807}{240} - \frac{9}{2} = \frac{727}{240}$ and $\frac{727}{240} : 1 :: 1200 : x$ gives the period, and $x = \frac{1200}{727/240} \times \frac{1}{360}$ years $= \frac{800}{727}$ years.

A proof is given in the following way $2\frac{1}{3} : 10\frac{1}{2} :: \frac{800}{727} \times 360$ days: x, and $x = 1782\frac{486}{727}$ which is the total amount earned. Also 1 day: $\frac{1807}{240} :: \frac{800}{727} \times 360 : x^1$, and $x^1 = 2982\frac{486}{727}$, which is the total amount spent; and $1782\frac{486}{727} + 1200 = 2982\frac{486}{727}$.

96. A certain person earns one and a half in one and one-third days and gives away eight in five and one-third days for Bha(vanī), one in thirty-two days for pa(raloka), and as an offering to ŚŪLIN one quarter of two and a half in thirty-six days. If he already possesses seven hundred when will it all be consumed?[1]

The daily income is $1\frac{1}{2} \div 1\frac{1}{3} = \frac{9}{8}$, and the gross daily expenditure is $\frac{8}{5\frac{1}{3}} + \frac{1}{32} + \frac{2\frac{1}{2}}{4 \times 36} = \frac{223}{144}$. The net expenditure is therefore $\frac{233}{144} - \frac{9}{8} = \frac{61}{44}$, and $\frac{61}{144}$: 1 day : : 700 : x, whence $x = \frac{700 \times 144}{61 \times 360}$ years $= \frac{280}{61}$ years = 4 years 7 months $2\frac{8}{61}$ days.

A proof is given thus:—1 day : $\frac{223}{144}$: : $\frac{280 \times 360}{61}$: $2559\frac{1}{61}$ and this is the gross expenditure. Again: 1 day : $\frac{9}{8}$: : $\frac{280 \times 360}{61}$: $1859\frac{1}{61}$, which is the total income; and $700 + 1859\frac{1}{61} = 2559\frac{1}{61}$. Further $\frac{280}{61}$ years : $2559\frac{1}{61}$: : $\frac{1}{360}$: $\frac{223}{144}$, which is the daily expenditure.

97. A boat[2] goes one-half of a third of a *yojana* plus one-third less one quarter in one-half of one-third of a day, but then it is driven back by the wind one-half of one-fifth of a *yojana* in one-eighth of three days. In what time will it travel one hundred and eight *yojanas?*

The details of the question and the solution are not clear.

98. A snake eighteen *hastas* long enters its hole at the rate of one-half plus one-ninth of that *aṅgulas* less one twenty-first part daily. In what time will it have completely entered the hole?

Since 24 *aṅgulas* = 1 *hasta* we have $\frac{1}{2} + \frac{1}{9}$ of $\frac{1}{2} - \frac{1}{21}$: $\frac{1}{360}$: : 18×24 : x, and $x = \frac{189}{80} = 2$ years 4 months $10\frac{1}{2}$ days.

99. A snake gets out of its skin as quickly as possible at the rate of half an *aṅgula* in one day. Its length is 100 *yojana*, 6 *krośa*, 3 *hasta* plus 5 *aṅgulas*.[3] In what time will it be free?

$\frac{1}{2}$ aṅgula : 1 day : : 100 *yo.* + 6 *kro.* + 3 *ha.* + 5 *aṅg.* : x or $\frac{1}{2}$ *aṅgula* : 1 day : : 77,376,077 *aṅg.* : x, and x = 154,752,154 days = 429867 years, 1 month and 4 days.

100. On folio 37 are three very simple astronomical problems:

(i) Bhanu (The Sun) travels 500,000,000 *yojanas*[4] in one day. State with certainty the amount of progress in one *ghaṭika?*

Since 60 *ghaṭika* = 1 day (of 24 hours) we have

60 *gha.* : 500,000,000 *yo.* : : 1 *gha.* : x and x = $8,333,333\frac{1}{3}$ *yojana.*

(ii) The Sun's chariot is guided by the god MAHORAGA among the Siddhas and Vidyādharas. The clever scientist says that, according to the general rule, it travels half a hundred *koti*[5] in a day and night. Tell me, O best of calculators, what it will go in one *muhūrta?*

---

[1] The wording of the example is 'restored' on the basis of certain indications given in the solution preserved, *e.g.*, *Sū°* may stand for ŚŪLIN. The numerical portions are certain.

[2] MAHĀVĪRA (V, 23-31) gives a small section dealing with problems of 'forward and backward movement', which he illustrates by a boat problem and a snake problem amongst others. His boat problem is:—

In the course of $\frac{3}{7}$ of a day a boat traverses $\frac{1}{5}$ of a *krośa* of the ocean, but owing to adverse winds it loses $\frac{1}{9}$ of a *krośa*. State in what time it will have advanced $99\frac{3}{5}$ *yojana*, thou who hast powerful arms in crossing easily the ocean of numbers. (*Answer*: 5 years 117 days.)

The snake problem is:—

A powerful, unvanquished, excellent, black snake, 32 *hastas* long enters a hole at the rate of $7\frac{1}{2}$ *aṅgulas* in $\frac{5}{14}$ of a day; and in the course of $\frac{1}{4}$ of a day its tail grows by $2\frac{3}{4}$ of an *aṅgula*. O ornament of arithmeticians, tell me by what time it enters fully into the hole. (*Answer*: $76\frac{4}{5}$ days.)

[3] The snake is here nearly 1,000 miles long! This may have been a modest estimate of the size of the serpent Śesha, who supports the world, causes earthquakes whenever he yawns, and occasionally destroys the universe. (See the *Vishṇu Purāṇa*, ii, v, etc.) But Mr. Hargreaves suggests that the example refers to the Nāga Elāpattra, whose head was at Benares and whose tail reached to Taxila—a distance of just about 100 *yojanas*.

[4] An orbit of 5.10^8 *yojanas* or a mean radius of nearly 80 million *yojanas* or, say, 727 million miles. What is the origin of this?

[5] A *koṭi* = 10^7.

Since 30 muhūrta = 1 day (of 24 hours) we have

30 *mu.* : 500,000,000 : : 1 *mu.* : x and x = 16,666,666$\frac{2}{3}$ *yojana*.

(iii) If Bhanujā (Saturn) traverses one sign in two and a half years, **tell me, thou who art thoroughly learned, how far he will go in one solar[1] day.**

Since 1 sign = 30° = 108000″ we have

2$\frac{1}{2}$ years : 108,000″ : : $\frac{1}{360}$: x and x = 120″ or 2 minutes of arc.[2]

101. The following examples[3] seem to form a set but the text is mutilated and only the bare skeletons of the problems remain:

(i) 1 *tola*: 5$\frac{1}{3}$ years : : 1 *to* + 1 *dhā* + 1 *aṁ* + 1 *ya* + 1 *si* + 1 *ka* + 1 *mū* : x, or 1 to : 5$\frac{1}{3}$ years : : $\frac{21681}{19200}$ **to : x and x = 6 years** 8$\frac{1}{10}$ **days.**

(ii) 1 ***to*** : **35 *drammas*** : : **1$\frac{1}{2}$** *to* + 1$\frac{1}{2}$ *mā* + **1$\frac{1}{2}$ *ani*** + 1$\frac{1}{2}$ ***ya*** : x, **or 1 : 35 : :** $\frac{319\frac{1}{3}}{192}$: x, **and x = 58** $\frac{31}{128}$ ***drammas.***

(iii) 1 *to* : 400 *dīnāra* : : 1 *dhā* + 1 *aṁ* + 1 *ya* + 1 *ka* + 1 *pā* + 1 *mū* : x, or 12 *dhā* : 400 *dī* : : $\frac{2441}{1600}$ *dhā* : x, **and x** = 50$\frac{41}{48}$ *dīnāras* = 50 *dī* + 10 *dhā* + 1 *ani*.

(iv) 1 day : (3 to + 2 mā + 3 ani + 3 ya + 1 ka + 1 pā + 1 mū) − **(4 ra + 4 si) : : 25 years 5 months 20 days : x, or 1 day : :** $\frac{62321}{19200}$ tola − $\frac{1440}{19200}$ tola **: : 9170 days : x, and x =** $\frac{558,278,770}{19200}$ tola = 1 *bhā* + 1634 *pā* + 5 *to* + 0 *mā* + 0 *ani* + 3 *ya* + 4 *ka* + 1 *pā* + 2 *mū*.

---

[1] See my *Hindu Astronomy*, p. 57.

[2] It should be about 122″.

[3] The importance of these examples lies in the use of numerous measures and in the methods of expressing change-ratios. The subject of measures is discussed in chapter VII.

APPENDIX TO CHAPTER VI.

(Aids to the calculations.)

FACTORS.

$384=2^7.3$

$441=3^2.7^2$

$481=13.37$

$492=2^2.3.41$

$576=2^6.3^2$

$579=3.193$

$621=3^3.23$

$656=2^4.41$

$737=11.67$

$768=2^8.3$

$841=29^2$

$889=7.127$

$1024=2^{10}$

$1158=2.3.193$

$1681=41^2$

$3696=2^4.3.7.11$

$5785=5.13.89$

$9681=3.7.461$

$9985=5.1997$

$10225=5^2.409$

$11858=2.7^2.11^2$

$14112=2^5.3^2.7^2$

$19362=2.3.7.461$

$38724=2^2.3.7.461$

$40528=2^4.17.149$

$59425=5^2.2377$

$65593=11.67.89$

$67125=3.5^3.179$

$84672=2^6.3^3.7^2$

$84832=2^5.11.241$

$107\ 584=2^6.41^2$

$108\ 625=5^3.11.79$

$112\ 003=31.3613$

$322\ 752=3.2^6.41^2$

$335\ 241=3^2.193^2$

$336\ 009=3.31.3613$

$425\ 042=2.461^2$

$444\ 004=2^2.11.10091$

$645\ 504=2^7.3.41^2$

$1\ 554\ 615=3^2.5.179.193$

$4\ 663\ 845=3^3.5.179.193$

$7\ 773\ 075=3^2.5^2.179.193$

$515\ 225\ 088=2^9.3^2.7.15973$

$2\ 999\ 096\ 352=2^5.3^2.7^2.461^2$

$6\ 455\ 042\ 625=5^5.11.79.2377$

$17\ 994\ 594\ 112=2^6.11.17.149.10091$

$21\ 743\ 271\ 936=2^{28}.3^4=768^4/2^4$

$113\ 227\ 047\ 522=2.3^2\ (6\ 290\ 391\ 529)=?$

$221\ 790\ 250\ 044=2^2.3^2.11.560,076,389$

$347\ 892\ 350\ 976=2^{32}.3^4=768^4$

$7\ 250\ 483\ 394\ 675=3^5.5^2.179^2.193^2$

$50\ 753\ 383\ 762\ 725=3^5.5^2.7.179^2.193^2$

$448\ 244\ 345\ 088=2^8.3^2\ (194\ 550\ 497)$

$50\ 753\ 383\ 762\ 746\ 743\ 271\ 936=113\ 227\ 047\ 522\times 448\ 244\ 345\ 088$
$=2^2.3^4.\ (6,290,391,529)\ (194,550,497)$
$=2^9.3^4\ (2^{19}+3.5^{11}.7.179^2.193^2)$

Unresolved 6 290 391 529 and 194 550 497

SQUARES.

$21^2=441$

$24^2=576$

$29^2=841$

$41^2=1681$

$77^2=5929$

$179^2=32041$

$193^2=37249$

$461^2=212521$

$579^2=335241$

$768^2=589824$

$3361^2=11296321$

Powers of 2, 3 and 5

$2^4=16$ $2^5=32$ $2^6=64$ $2^7=128$ $2^8=256$ $2^9=512$ $2^{10}=1024$ $2^{28}=268\ 435\ 456$ $2^{30}=1073\ 741\ 824$ $2^{32}=4294\ 967\ 296$

$3^4=81$ $3^5=243$ $3^6=729$ $3^7=2187$; $5^4=625$ $5^5=3125$ $5^6=15625$

Prime numbers greater than 100

127 149 179 193 241 409 461 1997 2377 3361 3613 10091 15973

CHAPTER VII.

Measures.

102. The measures exhibited in the manuscript are of rather special interest. As a whole they are Indian and the terminology is Sanskrit; but there are some Sanskritised western terms such as *liptā, dramma, dīnāra, satera* employed. Most of the terms are well defined but the values of some are doubtful. Money measures, however, are, as in most early Indian works, very ill defined and hardly show any differentiation from measures of weight.

103. *Change ratios.** The change ratios are often given with considerable care and elaboration, and are expressed in several different ways. The change ratio appears to be considered as a divisor for it is most frequently marked by the term *chhedam*† which indicates the operation of division.

Examples are

chhe° 80 *rakti*°-*su*° i.e. 80 *raktikā* = 1 *suvaana*¶

chhe° 24 *aṁ*°-*ha*° i.e. 24 *aṅgula* = 1 *hasta*

chhe° 2 *gha*°-*mu*° i.e. 2 *ghaṭika* = 1 *muhūrta*

chhe° 4608000 *ya*°-*yo*° i.e. 4,608,000 *yava* = 1 *yojana*

ūrdha‡ *chchhe*° 768000 *a*°-yo°, i.e. 768,000 *aṅgula* = 1 *yojana*, and in this particular example the operation of multiplication is to be performed.

ūrdha chchhedaṁ 108000 *viliptānām rāśi*, i.e. 108000 viliptā = 1 *rāśi* and multiplication is indicated.

adha chchhedaṁ 2000 *pa*°-*bhā*°, i.e. 2000 pala = 1 *bhāra* and division is indicated.

Another form is illustrated in the following examples

tolenāsti dhāne 12, i.e. 1 *tola* = 12 *dhāna*.

dhāṇenāsti aṁ° 4, i.e. 1 *dhāna* = *aṇḍikā*

dīnāranāsti dhāṇe 12, i.e. 1 *dīnāra* = 12 *dhāna*

---

* By a *change ratio* I mean a number by which one denomination is multiplied or divided to change it to another. Thus, since 20 shillings = 1 pound, 20 is the *change ratio* between pounds and shillings. It is a *multiplier* or a *divisor* according to the direction of change.

† Often abbreviated *chhe.*

¶ Or 'division by 80 changes *raktikas* to *suvarnas*.'

‡ Dict. *ūrdhva*.

104. In tabulated examples the methods of expressing the change ratios are modified. In two cases sets of cumulative ratios are given thus

to° 1	dhā° 1 1*	aṁ° 1 1*	ra° 1 1*	ya° 1 1*	si° 1 1*
1	1 12	1 48	1 60	1 192	1 480

ka° 1 1*	pā° 1 1*
1 1200	1 4800

which means primarily

$1\ to^{\circ} + 1\ dh\bar{a}^{\circ} + 1\ a\dot{m}^{\circ} + 1\ ra^{\circ} + 1\ ya^{\circ} + 1\ si^{\circ} + 1\ ka^{\circ} + 1\ pa^{\circ}$

while the other numbers are simply explanatory and mean that $1\ dh\bar{a}^{\circ} = \frac{1}{12}\ to^{\circ}$, $1\ a\dot{m}^{\circ} = \frac{1}{48}\ to^{\circ}, \ldots\ldots.\ 1\ p\bar{a}^{\circ} = \frac{1}{4800}\ to.$

A similar table is given without naming the terms in the form

1	1	1*	1	1*	1	1*
1	1	12	1	48	1	192
2	2		2		2	

which means

$1\frac{1}{2}\ to^{\circ} + 1\frac{1}{2}\ dh\bar{a}^{\circ} + 1\frac{1}{2}\ a\dot{m}^{\circ} + 1\frac{1}{2}\ ya^{\circ}$ and that $1\ dh\bar{a}^{\circ} = \frac{1}{12}\ to^{\circ}$, $1\ a\dot{m}^{\circ} = \frac{1}{48}\ to^{\circ}$, $1\ ya^{\circ} = \frac{1}{192}\ to^{\circ}$

Both of the sets given above are also tabulated in a different manner, e.g.—

to	1	
dhā	1 12*	
aṁ	1 4	
ra	1	1* bhā 1 4
ya	1	3* bhā 1 5
si	1	2* bhā 1 2
ka	1	2* bhā 1 2
pā	1	

This again primarily means $1\ to^{\circ} + 1\ dh\bar{a} + 1\ a\dot{m} + 1\ ra + 1\ ya + 1\ si + 1\ ka + 1\ p\bar{a}$ and the other numbers are simply change ratios and indicate that 12 dhā = 1 to°, 4 aṁ = 1 dhā, $1\frac{1}{4}$ ra = 1 aṁ, $3\frac{1}{5}$ ya = 1 ra, $2\frac{1}{2}$ si = 1 ya, $2\frac{1}{2}$ ka = 1 si.

Note that where the change ratios are fractional they are written at the side, obviously to prevent confusion.

Examples of the type just given occur on folios 7 recto, 48 recto 48 verso, 49 verso.

105. A variation of the above scheme in which the charge ratios are *halved* occurs thus

1 1 2		to
1 1 2	1* 6	mā
1 1 2	1* 2	aṁ
1 1 2	1* 2	ya

which means $1\frac{1}{2}$ *to*° + $1\frac{1}{2}$ *mā*° + $1\frac{1}{2}$ *aṁ*° + $1\frac{1}{2}$ *ya*°. The notion of halving the change ratios seems to be like this : $\frac{3}{2}$ *to*° multiplied by $\frac{1}{6} = \frac{3}{12}$ *to*° = 3 mā and so on. Another example of this practice occurs on folio 20 recto and gave considerable trouble. It is written thus

1 3			
1 2	1 2	chhe	1* 2
1 4		chhe	1* 4

and means $\frac{1}{3}\,a + \frac{1}{2}\cdot\frac{1}{2}\,b + \frac{1}{4}\,c$. Until the change ratios between a, b and c could be found the problem was insoluble. The real change ratios are given by $\frac{1}{2}\cdot\frac{1}{3}\,\mathbf{a} = \mathrm{b}$, $\frac{1}{4}\cdot\frac{1}{2}\cdot\frac{1}{2}\,\mathrm{b} = \mathrm{c}$.

106. Before proceeding to discuss the several kinds of measures that occur a summary table of them is given for reference. The names of the measures are nearly always abbreviated in the manner indicated : the full terms are

Aḍh = Ādhaka	Guṁ = Guñja	Pra = Prastha
Aṁ = Aṁśa	Ha = Hasta	Pra = Prasṛiti
Aṁ = Aṇḍikā	Kā = Kākini	Ra = Raktikā
Aṁ = Aṅgula	Ka = Kalā	Rā = Rāśi
Bhā = Bhāra	Khā = Khāri	Sa = Satera
Dhā = Dhānakā	Kro = Krosa	Si = Siddhārtha
Dha = Dhanus	Ku = Kuḍava	Su = Suvarṇa
Di = Dina	Li = Lipta	To = Tola
Dī = Dīnāra	Mā = Māsa	Va = Varsha
Dram = Dramma	Mā = Māshaka	Vi = Viliptā
Dra = Draṅkshana	Mū = Mūdṛikā	Ya = Yava
Dro = Drona	Mu = Muhūrta	Yo = Yojana
Gav = Gavyūti	Pā = Pāda	
Gha = Ghaṭika	Pa = Pala	

The Bakhshālī measures.

Time.

	gha	mu	di	mā	va	
gha	1					=24 min.
mu	2	1				=48 „
di	60	30	1			=24 hr.
mā			30	1		
va			360	12	1	

Arc.

	vi	li	aṃ	rā	
vi	1				=1″
li	60	1			=1′
aṃ	3,600	60	1		=1°
rā	108,000	1,800	30	1	=1 sign.

Money.

	ra	dhā*	dra	dī	su
ra	1				
dhā	5	1			
dra	30	6	1		
dī	60	12	2	1	
su	80	16	$2\frac{2}{3}$	$1\frac{1}{3}$	1

* Also 4 *aṁ*=1 *dhā* (49r.)

Weight.

	mū	pā	ka	si	ya	ra	aṅ	dhā	dra	to	pa	bhā
mū	1											
pā	4	1										
ka	16	4	1									
si	40	10	$2\frac{1}{2}$	1								
ya	100	25	$6\frac{1}{4}$	$2\frac{1}{2}$	1							
ra	320	80	20	8	$3\frac{1}{5}$	1						
aṅ	400	100	25	10	4	$1\frac{1}{4}$	1					
dhā	1,600	400	100	40	16	5	4	1				
dra					96			6	1			
to	19,200	4,800	1,200	480	192	60	48	12	2	1		
pa	153,600	38,400	9,600	3,840	1,536	480	384	96	16	8	1	
bhā											2,000	1

Length.

	ya	aṅ	ha	dha	kro	ga	yo	
ya	1							= ·12 in.
aṅ	6	1						= ·75 in.
ha	144	24	1					=18·00 in.
dha	576	96	4	1				6 feet.
kro	576,000	96,000	4,000	1,000	1			1·14 m.
ga	2,304,000	384,000	16,000	4,000	8	1		4·52 m.
yo	4,608,000	768,000	32,000	8,000	8	2	1	$9\frac{1}{11}$ m.

Capacity.

	pa	pra	ku	pra	āḍ	dro	khā
pa	1						
pra	2	1					
ku	4	2	1				
pra	32	8	4	1			
āḍ	128	32	16	4	1		
dro	256	128	64	16	4	1	
khā	4,096	2,048	1,024	256	64	16	1

Measures of Time.

107. The measures of time used in the text are the usual practical measures employed in India and there is nothing in any way remarkable about them. The measures found in early Hindu texts are given below.[1] All the authorities agree on one point, namely in dividing the day (24 hours) into 30 *muhūrtas*. The half-*muhūrta* was possibly introduced for astronomical purposes, as was the *vinadi* of the *Sūrya Siddhānta*. Also it is noteworthy that the tables give neither hours nor weeks, tlthough the 'hour' was used for astrological purposes and the 'week' came into general use in the early centuries of the present era. Neither of these measures occurs in our manuscript but in this matter the text is quite orthodox.

The measures of time employed therefore call for little comment, but Dr. Hoernle suggested[2] that the year of 360 days might be an indication of the age of the work! He could hardly have been aware that it was the common practice to give this value in Hindu arithmetical text-books.[3] Śrīdhara, indeed, remarks: "Time is calculated according to this rule in all mathematical works."

Measures of Arc.

The measures of arc, which with the early Hindus were lengths rather than angles,[4] are such as occur in all mediæval Hindu astronomies. The term *liptā* is a Sanskritised form of λεπτη.

---

[1] In the sections that follow I have used with considerable freedom the chapter on 'Weights and Measures' given in L. D. Barnett's *Antiquities of India,* which is the most useful contribution to the subject (of those at my command) since that given in ALBĪRŪNĪ'S *India.*

[2] Indian Antiquary, xvii, 1888, p. 37.

[3] See Mahāvīra and Śrīdhara.

[4] See my Hindu Astronomy, p. 54.

Hindu Measures of Time.

Manu.

Nimesha	1					·2 sec.
Kāshṭhā	18	1				3·2 ,,
Kalā		30	1			1·6 min.
Muhūrta		900	30	1		48 ,,
Aho-rātra		27,000	900	30	1	24 hrs.

Vishṇu Purāṇa.

Nimesha	1					·21 sec.
Kāshṭhā	15	1				3·2 ,,
Kalā		30	1			1·6 min.
Muhūrta		900	30	1		48 ,,
Day		27,000	900	30	1	24 hrs.

The Kauṭilīya.

Truṭis	1								·06 sec.
Lava	2	1							·12 ,,
Nimesha		2	1						·24 ,,
Kāshṭhā			5	1					1·2 ,,
Kalā				30	1				36 ,,
Nālikā					40	1			24 min.
Muhūrta						2	1		48 ,,
Day						60	30	1	24 hrs.

Sūrya Siddhānta.

Gurvakshara	1					·4 sec.
Prāṇa	10	1				4 ,,
Vināḍī	60	6	1			24 ,,
Nāḍī	3,600	360	60	1		24 min.
Day			3,600	60	1	24 hrs.

Mahāvīra.

Uchchhvāsa	1											= ·75 sec.
Stoka	7	1										= 5·3 ,,
Lava		7	1									= 37½ ,,
Ghaṭī			38½	1								= 24 min.
Muhūrta				2	1							= 48 ,,
Day					30	1						= 24 hrs.
Paksha						15	1					= 15 days.
Month						30	2	1				= 30 ,,
Ṛitu						60	4	2	1			= 60 ,,
Ayana						180	12	6	3	1		= 180 ,,
Year						360	24	12	6	2	1	= 360 ,,

Śrīdhara.

Ghaṭī	1			
Day	60	1		
Month	1,800	30	1	
Year	21,600	360	12	1

Bhāskara.

Truṭis	1								·00003 sec.
Tatpara	100	1							·003 ,,
Nimesha	3,000	30	1						·1 ,,
Kāshṭhā	54,000		18	1					1·6 ,,
Kalā	$162 \cdot 10^4$			30	1				48 ,,
Ghaṭika	$486 \cdot 10^5$				30	1			24 min.
Kshaṇa	$972 \cdot 10^5$					2	1		48 ,,
Day	$2{,}916 \cdot 10^6$					60	30	1	24 hrs.

Bakhshālī.

	Gha	Mu	Di	Mā	Va
Gha	1				
Mu	2	1			
Di	60	30	1		
Mā			30	1	
Va			360	12	1

Measures of Length.

108. (*a*) Although individual measures of length occur pretty often there are few examples in which the change ratios are exhibited. The following (fol. 32, verso) is the most complete of such examples :

100	urdha chchhe	768000*	a° – yo°
6		1	
8*			
3			
4000*			
5			
24*			

which means 100 *yojanas* + 6 (*krośas*) + 3 *hastas* + 5 *aṅgulas*, the numbers marked with asterisks being change ratios. These numbers inform us that 24 *aṅgulas* = 1 *hasta*, 4000 *hastas* = 1 *krośa*, 8 *krośas* = 1 *yojana*, and that to reduce *yojanas* to *aṅgulas* the former must be multiplied by 768,000. The term *krośa* does not actually occur, but it, or an equivalent, must have been mentioned in the question to which the table was attached.

In another statement (fol. 32, recto) the equation 8,000 *dhanus* = 1 *yojana* is given, hence 1,000 *dhanus* = 1 *krośa*. This equation occurs in Hindu works but most of the texts give 2,000 *dhanus* = 1 *krośa*.

In another example (fol. 36) the statement

chhe 4,608,000 *ya° – yo°*

indicates that 4,608,000 *yavas* = 1 *yojana*, whereas Mahāvīra and Bhāskara give 6,144,000 *yavas* = 1 *yojana*. The difference in the change ratios is noteworthy.

The measures *gadyūti* and *yojana* are connected (fol. 12), but unfortunately the actual change ratio is missing. The term *gavyūti* is not common in Hindu works but it occurs in the *Mārkaṇḍeya Purāṇa, the Mahābhārata*, etc. Its value is variously given as 4000, or 2000, or 200 *dhanus*.

Of the Indian tables given below the last is what the *Sūrya Siddhānta* designates *a-mūrta* or 'unreal.' It was never used seriously but it is interesting as exhibiting the constant change ratio of 'eight.'

108. (*b*) Possibly a measure of area is referred to on folio 39° verso where a change ratio of 64 is mentioned ; and possibly *bhu* on fol. 26 also indicates an area.

Indian Measures of Length.

Mārkaṇḍeya Purāṇa.

	Ya.	Aṅ.	Pa.	Vi.	Ha.	Da.	Nā.	Ga.	Yo.
Yava* .	1								
Aṅgula .	8	1							
Pada .		6	1						
Vitasti .			2	1					
Hasta .				2	1				
Daṇḍa .					4	1			
Nāḍī . .						2	1		
Gavyūti .						4,000	2,000	1	
Yojana .						16,000	8,000	4	1

Kauṭilīya Artha Sāstra.

	Ya.	Aṅ.	Vi.	Ara.	Da.	Go.	Yo.
Yava* .	1						
Aṅgula .	8	1					
Vitasti .	96	12	1				
Aratni .		24	2	1			
Daṇḍa .		96	8	4	1		
Goruta .				4,000	1,000	1	
Yojana .				16,000	4,000	4	1

Śrīdhara.

	Aṅ.	Ha.	Da.	Kro.	Yo.
Aṅgula .	1				
Hasta .	24	1			
Daṇḍa .	96	4	1		
Krośa .	192,000	8,000	2,000	1	
Yojana .	768,000	32,000	8,000	4	1

Mahāvīra.

	Se.	Ya.	Aṅ.	Pa.	Vi.	Ha.	Da.	Kro.	Yo.
Sesamum* .	1								
Yava .	8	1							
Aṅgula .		8	1						
Pada . .		48	6	1					
Vitasti .		96		2	1				
Hasta .		192	24	4	2	1			
Daṇḍa .		768	96	16	8	4	1		
Krośa .		1,536,000	192,000	32,000	16,000	8,000	2,000	1	
Yojana .		6,144,000	768,000	128,000	64,000	32,000	8,000	4	1

Bhāskara.

	Ya.	Aṅ.	Ha.	Da.	Kro.	Yo.
Yava .	1					
Aṅgula .	8	1				
Hasta .	192	24	1			
Daṇḍa .	768	96	4	1		
Krośa .	1,536,000	192,000	8,000	2,000	1	
Yojana .	6,144,000	768,000	32,000	8,000	4	1

Amūrta (un-real) Table.

Paramāṇus .	1							
Parasūkhsmas	8	1						
Trasareṇu .	64	8	1					
Reṇu . .	512	64	8	1				
Vālāgra . .	4,096	512	64	8	1			
Likshā . .	32,768	4,096	512	64	8	1		
Yūka . .	8^6	32,768	4,096	512	64	8	1	
Yava . .	8^7	8^6	32,768	4,096	512	64	8	1

Bakhshālī.

	Ya	Aṅ	Ha	Dha	Kro	Ga	Yo
Ya	1						
Aṅ	6	1					
Ha	144	24	1				
Dha	576	96	4	1			
Kro	576,000	96,000	4,000	1,000	1		
Ga	2,304,000	384,000	16,000	4,000	8	1	
Yo	4,608,000	768,000	32,000	8,000	2	2	1

* These tables are accompanied by the *Amūrta* table.

Measures of Capacity.

109. The following is the most complete table given (fol. 13).

ā°	2	
	4*	dro°
pra°	0	
	4*	ā° pra°
ku°	2	
	4*	prasthi

which means 2 *āḍhakas* + 0 *prasthas* + 2 *kuḍavas*. The figures here marked with asterisks are change-ratios and indicate that 4 āḍhakas = 1 droṇa

4 prasthas = 1 āḍhaka

4 kuḍavas = 1 prastha.

The terms *ghaṭaka* and *pala* are mentioned together (fol. 15), where possibly *ghaṭaka* may be equivalent to *droṇa*; and in another place (fol. 53) *ghaṭaka* is obviously used as a capacity term.

A comparison with the Hindu measures shows that the Bakhshālī measures of capacity are strictly orthodox; and it may be noted that the Hindu measures of capacity are generally more consistent in the matter of change-ratios than any other Hindu measures.

Hindu Measures of Capacity.

Atharva Veda.

	Kri.	Mā.	Pa.	Pra.	Āḍh.	Dro.
Krishṇala .	1					
Māshaka .	5	1				
Pala . .		64	1			
Prastha .			32	1		
Āḍhaka .			128	4	1	
Droṇa .			512	16	4	1

Kauṭilīya Artha Śāstra.

	Ku.	Pra.	Āḍh.	Dro.	Khā.	Kum.	Va.
Kuḍumba .	1						
Prastha .	4	1					
Āḍhaka .	16	4	1				
Droṇa .	64	16	4	1			
Khārī .	1,024	256	64	16	1		
Kumbha .	2,048	5,120	1,280	320	20	1	
Vaha .	20,480	51,200	12,800	3,200	200	10	1

Varāha Mihira—Dry Measure.

	Pa.	Ku.	Pra.	Āḍh.
Pala . .	1			
Kuḍava .	4	1		
Prastha .	16	4	1	
Āḍhaka .	64	16	4	1

Varāha Mihira—Liquid Measure.

	Pa.	Ku.	Pra.	Āḍh.
Pala . .	1			
Kuḍava .	8	1		
Prastha .	32	4	1	
Āḍhaka .	128	16	4	1

Mahāvīra.

	Sho.	Ku.	Pra.	Āḍh.	Dro.	Mā.	Khā.	Pra.	Ku.
Shoḍasika .	1								
Kuḍaha .	4	1							
Prastha .	16	4	1						
Āḍhaka .	64	16	4	1					
Droṇa .	256	64	16	4	1				
Mānī . .		256	64	16	4	1			
Khārī .		1,024	256	64	16	4	1		
Pravartikā .							5	1	
Kumbha .								5	1

Bakhshālī.

	Pa.	Pra.	Ku.	Pra.	Āḍh.	Dro.	Khā.
Pa.	1						
Pra.	2	1					
Ku.	4	2	1				
Pra.	32	8	4	1			
Āḍh.	128	32	16	4	1		
Dro.	256	128	64	16	4	1	
Khā.	4,096	2,048	1,024	256	64	16	1

Money measures.

110. Although the terms *dramma* and *dīnāra* occur pretty often their relationship with each other is nowhere indicated; and, indeed, there is very little definite information in our manuscript about money beyond the mere money names. In early times in India there were no special measures for money beyond the weight measures for different metals, and sometimes the difference in these is more apparent than real. Thus Mahāvīra's gold and silver tables given below have much in common.

In our manuscript we have only the following scraps of information—

(*a*)[1] *chhe* 80 *rakti*° – *su*°

(*b*) *chhe* 9 *gu*° – *va*°

(*c*)

1	1	1	1	1	1	1	108	*pha*°	*dī*° 1 *dhā*° 8 *aṁ*° 1.
1	2	2	2	2	2	2	1		

Of these examples (*a*) is orthodox, for according to Manu 80 *raktikās* or *guñjas* of copper or gold are equal to one *suvarṇa*. Example (*b*) is isolated and cannot be interpreted with any certainty. Possibly *gu*° stands for *guñja*, and possibly *va*° for *valla*. Example (*c*) means $1 : (\frac{1}{2})^6 : : 108 : 1$ *dī*° + 8 *dhā* + 1 *aṁ*°, where *dī*° = *dīnāra*, *dhā* = *dhānakā*, and *aṁ* = *aṁśa*; and it is implied that 4 *aṁśas* = 1 *dhānakā*, and 12 *dhānakās* = 1 *dīnāra* (fol. 33) and the same relationship is given on folio 49 recto. The term *kākini* (cowrie) occurs once; *satera* also occurs (fol. 34) but with doubtful import.

The Guptas adopted the term *dīnāra*[2] from the Kushāns together with the coin of weight from 118 to 122 grains; and in a number of Gupta inscriptions certain gold coins are termed *dīnāras*.[3] But the *dīnāra* was not invariably a gold coin[4] and in the Bakhshālī text it probably is a copper coin, for a day's wages is stated to be from $1\frac{1}{2}$ to 3 *dīnāras* (fol. 60); and also according to Mahāvīra (vi, 231) wages work out about 18 *dīnāras* a day per coolie.[5] The term occasionally occurs in Sanskrit works of a more literary character and has provoked some discussion.[6]

The term *dramma* (δραχμή) also occurs in various Indian inscriptions[7] and in Indian mathematical works. Bhāskara makes it one-sixteenth of a *nishka*[8] and in our text wages are about 1 *dramma* a day.

Such are the facts but Dr. Hoernle writes[9]—

> "The way in which the two terms are used in the Bakhshālī arithmetic seems to indicate that the gold *dīnāra* and the silver *dramma* formed the ordinary currency of the day. This circumstance again points to some time within the first three centuries of the Christian era as the date of its composition."

The subject of monetary measures is thus treated in a vague and unsatisfactory manner in what remains of the manuscript; but this is by no means a peculiarity: it is almost a characteristic of Sanskrit texts of mediæval times. The connexion between measures of money and weight was then fairly intimate and they can be hardly considered apart.

---

1 (*a*) 80 *raktikās* = 1 *suvarna*. (*b*) 9 *gu*° = 1 *va*°. (*c*) 108/64 *dīnāra* = 1 *dīnāra* + 8 *dhānakās* + 1 *aṁ*°.

2 Gk. δηνάριον, Lat. *denarius*.

3 J. Allan, *Catalogue of the Coins of the Gupta Dynasty, etc.*, p. cxxxiv; J. F. Fleet, *CII*, iii, nos. 5, 7, 8, 9, 62, 64.

4 Compare our penny and read Sir A. Stein's note on the Kashmir *dīnāra* in his *Kalhaṇa's Rājataraṅgiṇī* ii, 308ff. See also sec. 133 below.

5 We must not, however, place very much reliance on these text-book wages.

6 *JRAS*, 1907, pp. 408 and 681, etc.

7 *Epigraphia Indica*, I, 167; P. Vogel, *Chambā Inscriptions*, 204, etc.

8 *Līlāvatī*, sec. 2.

9 *Ind. Ant.* xvii, 1888, p. 37.

Hindu Measures of Money.

Mahāvīra—Gold.

	Ga.	Gu.	Paṇ.	Dha.	Ka.	Pa.
Gaṇḍaka .	1					
Guñjā .	4	1				
Paṇa . .	20	5	1			
Dharaṇa .	160	40	8	1		
Karsha .	320	80	16	2	1	
Pala . .	1,280	320	64	8	4	1

Mahāvīra—Silver.

	Gr.	Gu.	Mā.	Dha.	Ka.	Pa.
Grain .	1					
Guñjā .	2	1				
Māsha .	4	2	1			
Dharaṇa .	64	32	16	1		
Karsha .	160	80	40	2½	1	
Pala . .	640	320	160	10	4	1

Śrīdhara.

	Va.	Kā.	Pa.	Pu.
Varāṭaka .	1			
Kākiṇī .	20	1		
Pana . .	80	4	1	
Puraṇa .	1,280	64	16	1

Bhāskara.

	Va.	Kā.	Pa.	Dra.	Ni.
Varāṭaka .	1				
Kākiṇi .	20	1			
Pana . .	80	4	1		
Dramma .	1,280	64	16	1	
Nishka .	20,480	1,024	256	16	1

Albīrūnī.

	Mdr.	Pā.	Ka.	Ya.	Aṇḍ.	Mā.	Su.
Mdrī . .	1						
Pāda . .	4	1					
Kalā . .	16	4	1				
Yava .	100	25	6¼	1			
Aṇḍi . .	400	100	25	4	1		
Māsha .	1,600	400	100	16	4	1	
Suvarṇa .	25,600	6,400	1,600	256	64	16	1

Bakhshālī.

	Ra.	Dhā.	Dra.	Dī.	Su.
Ra.	1				
Dhā.	5	1			
Dra.	30	6	1		
Dī.	60	12	2	1	
Su.	80	16	2¾	1½	1

Measures of Weight.

111. Besides the examples already given (sec. 104), the following are noteworthy :—

(a)

dhā°	1		
aṁ°	1		
	4*		
ra°	1	$1\frac{1}{4}$*	bhā°
ya°	1	3 *	bhā°
ka°	1	$6\frac{1}{4}$*	bhā°

(b)

to°	3	
mā°	2	
	12*	
aṁ°	3	
	4*	
ya°	3	
	4*	
ka°	1	$6\frac{1}{4}$*
pā°	1	
	4*	

(c)

ra°	1		
ya°	1	$3\frac{1}{3}$ *	bhā°
ka°	1	$6\frac{1}{4}$*	bhā°
pā°	1		
	4*		
śe°	$\frac{1}{3}$		

(d)

chhe°	8*		
dhā°	2		
chhe°	12*		
guṁ°	3		
chhe°	5*		
ya°	2	$3\frac{1}{3}$*	bhā°

(e)

216	bhā°
270	pa°
2000*	chhe°
6	to°
8*	chhe°
8	dhā°
12*	chhe°

These measures are combined in the following table.

	Mū°.	Pā.	Ka°.	Si°.	Ya°.	Ra°.	Aṇ°.	Dhā°.	To°.	Pa°.
Mūdṛikā .	1									
Pāda . .	4	1								
Kalā . .	16	4	1							
Si° . .	40	10	$2\frac{1}{2}$	1						
Yava .	100	25	$6\frac{1}{4}$	$2\frac{1}{2}$	1					
Raktikā .	320	80	20	8	$3\frac{1}{5}$	1				
Aṇḍika .	400	100	25	10	4	$1\frac{1}{4}$	1			
Dhānaka .	1,600	400	100	40	16	5	4	1		
Tola . .	19,200	4,800	1,200	480	192	60	48	12	1	
Pala . .	153,600	38,400	9,600	3,840	1,536	480	384	96	8	1

(*a*) In all cases the change-ratios are marked with asterisks (not, of course, in the original). The table means 1 *dhā°* +1 *aṁ°*+ 1 *ra°*+1 *ya°*.

(*b*) 3 *to°*+2 *ma°*+3 *aṁ°*+3 *ya°*+1 *ka°*+1 *pā°*.

(*c*) 1 *ra°* +1 *ya°*+1 *ka°*+$1\frac{1}{3}$ *pā°*.

(*d*) ....2 *dhā°*+3 *gum°*+2 *ya°*.

(*e*) 216 *bhā°*+270 *pa°*+6 *to°*+8 *dhā°*

To this we should add 12 *māshas* = 1 *tola*, 5 *tolas* = 1 *suvarṇa* and 2000 *pala* = 1 *bhāra*.

This agrees with the Hindu tables generally but more particularly with that of Varāhamihira as given by Albīrūnī, which, however, contains neither *si*° nor *raktikā* – both interpolations here, as their fractional change ratios show. Obviously Albīrūnī's *mdṛi* and our *mūdṛika* are identical : the term occurs in no other Hindu work known to me. The remainder of the table—*dhānakas* to *bhāras*—does not seem so orthodox. Compare with Mahāvīra. As stated before there is some uncertainty as to monetary measures in our text and it is possible that some of the measures included in the above table should be treated as such.

The combination that occur in the text are as follows :—

mā, su, to	fol. 11
ya, ra, am, dhā	„ 49
mū, pā, ka, ya, ra, am, mā, to	„ 49
dhā, to, pa, bhā	„ 48
dhā, gum, ya	„ 48
mū, pā, ka, si, ya, ra, am, dhā, to	„ 55
ya, am, mā, to	„ 55

Possibly the term *dra*° should be included in the above table. It occurs only once (fol. 20) in the phrase *chhedaṁ* 6 *dhā*° – *dra*° which means 6 *dhā* = 1 *dra*°, and this must stand for 6 *dhānakās* = 1 *draṅkshana*; for Albīrūnī gives 6 *māsha* = 1 *draṅkshana*; and Mahāvīra gives 6 *bhāga* = 1 *draṅkshuna*, and *dhānaka*, *māsha* and *bhāga* are synonymous terms as applied to measures of weight.

The term *si*° occurs thrice and the values given are 40 *si*° = 1 *mā*°, $2\frac{1}{2}$ *si*° = 1 *ya*° and 480 *si*° = 1 *to*° – all of which are in agreement. The abbreviation *si*° probably stands for *siddhārtha*. Connected with it there is another measure *ku*° also unidentified. The statement in which both these occur is as follows :—

| ku° $1\frac{1}{2}$ chhe° 128 mā° – ku° $\frac{1}{2}$ mā° chhe° 40 si° – mā° | sa° 55 |

Here we have 128 *mā*° = 1 *ku*° which so far has puzzled me. *Sa*° is an abbreviation for *satera*; but its connexion with other measures is not clear. Mahāvīra makes 2 *dīnāra* = 1 *satera*. The equation 8 *suvarṇa* = 1 *ku*° can be deduced, but is not helpful. *Śāna* (Dict. = 4 māshakas) occurs.

The measures are generally expressed by the abbreviations *mū*, *pā*, *ka*, *si*, *ya*, *ra* or *guṁ*, *aṁ*, *dhā* or *mā*, *to*, *pa*, *bhā*; but occasionally the full terms are employed and in one place (fol. 49) a fairly complete set of terms is given, namely—

dhānakā aṁdika raktikā yavā kalā pāda mūdṛikā

Hindu Measures of Weight.

1. Manu and Yājñavalkya, 2nd—4th Century A.D.

					Ya.	Ra.	Mā.	Ka.	Pa.	Dha.
Trasa-reṇu .	1									
Likshā . .	8	1								
Rāja-sarshapa .	24	3	1							
Gaura-sarshapa	72	9	3	1						
Yava . .				6	1					
Raktikā . .				18	3	1				
Māsha . .					15	5	1			
Karsha . .					240	80	16	1		
Pala . .					960	320	64	4	1	
Dharaṇa . .					9,600	3,200	640	40	10·	1

(*a*) The Bṛihaspati Smṛiti gives 4 karshas=1 ḍhānaka and 12 dhānakas=1 dīnāra.

2. Nārada, 5th—6th Century, A.D.

	Kā.	Mā.	Aṇ.	Dhā.	Dī.
Kākaṇī .	1				
Māsha .	4	1			
Aṇḍikā .	80	20	1		
Dhānaka .	320	80	14	1	
Dīnāra .	3,840	960	48	12	1

Kauṭilīya.

	Mā.	Guñ.	Mā.	Su.	Pa.
Māsha .	1				
Guñja .	2	1			
Māshaka .	10	5	1		
Suvarṇa .	160	80	16	1	
Pala . .	640	320	64	4	1

3. Varāha-mihira, 6th Century. (After Albīrūnī.)

	Md.	Pā.	Ka.	Ya.	Aṇ.	Mā.	Su.
Mdrī . .	1						
Pāda . .	4	1					
Kalā . .	16	4	1				
Yava .	100	25	$6\frac{1}{4}$	1			
Aṇḍī . .	400	100	25	4	1		
Māsha .	1,600	400	100	16	4	1	
Suvarṇa .	25,600	6,400	1,600	256	64	16	1

4. Mahāvīra, 9th Century.

	Pā.	Ka.	Ya.	Aṁ.	Bhā.	Dra.	Dī.	Sa.
Pāda . .	1							
Kalā . .	4	1						
Yava .	25	$6\frac{1}{4}$	1					
Aṁśa .	100	25	4	1				
Bhāga .	400	100	16	4	1			
Drakshuṇa	2,400	600	96	24	6	1		
Dīnāra .	4,800	1,200	192	48	12	2	1	
Satera .	9,600	2,400	384	96	24	4	2	1

4 (a). Mahāvīra.

	Pa.	Pra.	Tu.	Bhā.
Pala .	1			
Prastha .	12	1		
Tulā .	200	16	1	
Bhāra .	2,000	160	10	1

5. Śrīdhara, 11th Century.

	Gu.	Mā.	Ka.	Pa.
Guñja .	1			
Māsha .	5	1		
Karsha .	80	16	1	
Pala .	320	64	4	1

6. Bhāskara, 12th Century.

	Ya.	Ra.	Va.	Dha.	Ga.
Yava .	1				
Raktikā .	2	1			
Valla .	6	3	1		
Dharaṇa .	48	24	8	1	
Gadyāṇaka	96	48	16	2	1

Bakhshālī.

	mū.	pā.	ka.	si.	ya.	ra.	aṇ.	dhā.	dra.	to.	pa.	bhā.
mū	1											
pā	4	1										
ka	16	4	1									
si	40	10	$2\frac{1}{2}$	1								
ya	100	25	$6\frac{1}{4}$	$2\frac{1}{2}$	1							
ra	320	80	20	8	$3\frac{1}{5}$	1						
an	400	100	25	10	4	$1\frac{1}{4}$	1					
dhā	1,600	400	100	40	16	5	4	1				
dra								6	1			
to	19,200	4,800	1,200	480	192	60	48	12	2	1		
pa	153,600	38,400	9,600	3,840	1,536	480	384	96	16	8	1	
bhā											2,000	1

CHAPTER VIII.

THE SOURCES.

112. In considering the question of exotic influence the period of composition of the work is of importance. We must forget for the time being Dr. Hoernle's suggestion that the work was written in the early centuries of our era and bear in mind that it possibly belongs to a much later period,[1] by which time western mathematics, although almost all traceable to Greek sources, had assumed new forms and had included some new notions. On the whole western mediæval mathematics tended to become less rigorous and more mixed than the mathematics of classical Greece. Practical calculation (logistic) altogether supplanted the earlier pure arithmetic, mensuration took the place of pure geometry, and algebra slightly developed. Such changes are, indeed, indicated in the later Alexandrian works, and what we sometimes term degeneration had already set in there. The introduction of a place-value arithmetical notation made calculation easier and more popular, and more intricate arithmetical problems than those, for example, exhibited in the Greek Anthology, appeared in the later mediæval text-books. The general body of popular mathematical knowledge became more diffused. (Identical problems occur in Chinese, Indian, Arabic and European text-books of a comparatively early period). Indeed the mediæval mathematical works of Asia and Europe had so much in common that at first it seems almost impossible to pick out that which is definitely western or eastern in origin. In this connexion it should be remembered that the early Hindu astronomers (Āryabhaṭa and Varāha Mihira) were among the first to exploit Greek mathematical learning; that later the Arabs, after sampling Indian works, turned to those of Greece; and that it was from the Arabs that Europe received once more the learning it had previously rejected.

113. Such facts indicate to some extent the difficulties of making a direct comparison of, say, a twelfth century work with those of the classical Greek period. A knowledge of the development or degradation of mathematics during the intervening period is obviously demanded, and without such knowledge sound judgment is impossible. That knowledge must be sought elsewhere, but I give here a summary chronological table of the period, that may serve to recall the chief mathematical writers and their works.

			A.D.
5th century	Hypatia		d. 415
	Proclus		410-485
	Boethius		b. 470
	Āryabhaṭa		b. 476
6th century			
	Eutocius		
	Damascius	visited Persia	529
	Simplicius		
	Dominus		
	Chang ch' iu-chien		550
	Varāhamihira		d. 587
	Isidore of Seville		570-636

[1] Specific reasons for arriving at the later date are given in the next chapter.

7th century	*Brahmagupta*	b. 598
	Fall of Alexandria	640/1
	Papyrus of Akhmin	
	Bede	b. 735
8th century	Muhammad b. Mûsa	
	Alcuin	d. 804
9th century	*Mahāvīra*	
	Tabit b. Qorra	836-901
10th century	al-Battānī	
	Avicenna	
	Pope Sylvester ii (Gerbert)	d. 1003
11th century	Albīrūnī visited India	1017-1030
	Śrīdhara	b. 991
	al-Kārkhī	
	Psellus	
	Omar Khayyām	b. 1046
12th century	Adeland at Cordova	1120
	Bhāskara	b. 1114
	Leonardo	b. 1175

114. Although the manuscript was found at Bakhshālī it cannot be assumed that it originated there. The evidence of the script, however, limits the area of origin to the neighbourhood of Gandhāra. Roughly the Śārada script area is limited by longitudes 72 and 78 east of Greenwich and north latitudes 32 and 36. This area includes, besides Gandhāra, Kashmir, Kaṅgra and territories near by. Vogel enumerates six inscriptions from Gandhāra, some of which belong to the Swat valley. The evidence therefore does not militate against Bakhshālī itself as the actual place of origin.

115. The position is peculiarly interesting with reference to the routes of transmission of knowledge and the consequent liability to outside influence. Between the fifth century B.C. and the fifth century A.D. the country round Bakhshālī was actually under the dominion of seven different nations : the Persians, the Macedonians, the Mauryas, the Bactrian Greeks, the Scythians, the Parthians, the Kushans : " and " writes Sir John Marshall,[1] " it may be taken for granted that, with the exception of the Macedonians whose conquest was merely transitory, each of these nations in turn left some impress on the arts and culture of the country." From the sixth to the tenth centuries of our era Gandhāra was more or less subject to the Guptas and their successors and then came the Muhammadan invaders. The following are some notable dates relating to the country with which we are concerned.

B.C.	326	Alexander receives the submission of Āmbhi, king of Taxila.
	190	Demetrius of Bactria conquers the Punjab.
	85-50	Maues, the Scythian, conquers Taxila.

[1] *A Guide to Taxila*, p. 23.

A.D.	20 circa	**Gondopharnes the Parthian ruling over Kabul, Taxila and Arachosia.**
	60	**Hermæus and Kujula Kadphises annex Gandhāra.**
	120	Kanishka the Kushān, king of Gandhāra.
	319	The Gupta era begins.
	400	Fa-hien entered India by way of Gandhāra.
	400-500	Invasion of India by the Epthalities or White Huns.
	500	Sung Yun in Gandhāra.
	630 circa	Hiuen Thsang in Gandhāra.
	720	Kashmir subject to China.
	753	Ou-K'ong visited Gandhāra.
	1001	Mahmud of Ghaznī defeats Jaipal near Peshawar.
	1175	Muhammad Ghorī attacks Multan.

116. The achievements of the Greeks in mathematics and art form the most wonderful chapters in the history of civilisation, and these achievements are the admiration of western scholars.[1] It is therefore natural that the western investigator into the history of knowledge should seek for traces of Greek influence in later manifestations of art and mathematics in particular. The position of Bakhshālī in the heart of Gandhāra, and the political history of that country are such, that not only warrant the search for traces of Greek influence, but make it practically imperative. Indeed the neglect of such an enquiry would stamp any investigation of this kind as incomplete.

Evidence of Greek influence in the realm of art has been discovered in profusion in Gandhāra and the surrounding country. Sir John Marshall writes[2] "The monuments and antiquities that have recently been recovered from the soil at Taxila and other places, all consistently bear witness to the strong hold which Hellenistic art took upon this part of India. This hold was so strong, that long after the Greek kingdoms of the Punjab had passed away, even after the Scythians and Parthians, who overthrew the Greeks, had themselves been supplanted by the Kushans, Greek art still remained paramount in the North West, and continued to exercise considerable influence until the fifth century of our era, although it was growing more and more decadent year by year."

It would not therefore be unreasonable to imagine that in mathematics also there was at least a possibility of Greek influence in the same country and at the same period; and Sir John himself supplies a sort of connecting link between art and mathematics. He continues[3] "This persistence and this slow decadence of Greek ideas is best illustrated by the coins, the stylistic history of which is singularly lucid and coherent. In the earliest examples every feature is Hellenistic. The standard weight of the coins is the standard established by Athens : the legends are in Greek . . . . Later on, when the Greek power in India became consolidated, the old Attic standard gave place to one, possibly based on Persian coinage . . bilingual legends were substituted for the Greek ; and little by little the other Greek qualities gradually faded . . ."

---

[1] The modern Indian does not seem to be attracted in this way.

[2] *A Guide to Taxila*, p. 26.

[3] *Ib*, p. 27.

The case for mathematics is almost exactly parallel, allowance, of course, being made for the more abstract nature of the subject; and the history of mathematics gives abundant illustration of the same types of change as Sir John traces in the Indo-Greek coinage.

Enough has been said to show that it would be stupid to exclude the possibility of Greek influence in the realm of mathematics on general grounds; and if such influence is to be negatived finally it must be for special reasons to be discovered by subjecting our manuscript to a detailed examination.

117. It is unfortunate that the question of origin of the Bakhshālī Manuscript was discussed and judged upon before the work was thoroughly examined. The announcement made by Weber, on the authority of Bühler, at the fifth International Oriental Congress was based upon no real knowledge at all; and Dr. Hoernle himself only examined in detail about one-third of the manuscript. We do not, therefore, feel bound to treat the pronouncements of these eminent scholars on this particular subject with the reverence that is usually due to them. Also we know that Weber was misled by Bühler and Dr. Hoernle told me that he was prepared to modify his earlier views to some extent. With reference to the last point the quotations given below must therefore not be taken as expressing Dr. Hoernle's final views. However, they are the basis of all that has been subsequently written and it is on them that the position of the Bakhshālī Manuscript has been determined by the historians of mathematics. Dr. Hoernle summarised his views in the following words :—

"I believe that it is generally admitted that Indian arithmetic and algebra, at least, are of entirely native origin. While Siddhānta writers, like Brahmagupta and his predecessor Āryabhaṭa, might have borrowed their astronomical elements from the Greeks or from books founded themselves on Greek science, they took their arithmetic from native Indian sources. Of the Jains it is well known that they possess astronomical books of a very ancient type, showing no traces of western or Greek influence.[1] In India arithmetic and algebra are usually treated as portions of works on astronomy.[2] In any case it is impossible that the Jains should not have possessed their own treatises on arithmetic, when they possessed such on astronomy. The early Buddhists too, are known to have been proficients in mathematics.[3] The prevalence of Buddhism in North-Western India, in the early centuries of our era, is a well known fact. That in early times there were also large Jain communities in those regions, is testified by the remnants of Jain sculpture found near Mathurā and elsewhere. From the fact of the general use of the North-Western Prakrit (or the 'Gāthā dialect') for literary purposes among the early Buddhists it may reasonably be concluded that its use prevailed also among the Jains, between whom and the Buddhists there was so much similarity of manners and customs. There is also a diffusedness in the mode of composition of the Bakhshāli work[4] which reminds one of the similar characteristics observed in Buddhist and Jain literature. All these circumstances put together seem to render it probable that in the Bakhshālī manuscript there has been preserved to us a fragment of an early Buddhist or Jain work on arithmetic (perhaps a portion of a larger work on astronomy) which may have been one of the sources from which the later Indian astronomers took their arithmetical information."

118. There is not the slightest evidence in the manuscript itself of its being connected either with the Jains or Buddhists. It is Hindu (Śaivite). The author was a Brahman (fol. 50); to Śiva is attributed the gift of calculation to the human race (fol. 50); offerings to Śiva are mentioned on more than one occasion (fols. 34, 44); references are made to certain incidents recorded and persons named in the Hindu epics (fol. 32, etc.);[5] and there is not a single reference that could be construed as indicating any connexion with Buddhism or Jainism.

---

[1] The reference is probably to the *Sūryaprajñapti*. See my *Hindu Astronomy*, pp. 19-21.

[2] Not before the advent of western astronomy.

[3] Is Dr. Hoernle here thinking of the incident related in the *Lalitavistara* and popularised by Arnold? The legend, according to Weber, "carries no weight whatever."

[4] I should not myself have noted this as a characteristic of the Bakhshālī work.

[5] And these references occur in the portions of the MS that were not critically examined by Dr. Hoernle.

119. Further there are indications of a connexion with Muslim mathematicians (§ 120) and there is other internal evidence that points to a much later date than Dr. Hoernle's thesis allows. It is rather curious that the connexion between arithmetical and astronomical works that Dr. Hoernle points out only holds for those later works which distinctly show western influence. For the major assumption " that Indian arithmetic and algebra are of entirely native origin " Dr. Hoernle was not himself responsible. It was a common opinion, that still obtains to some extent. One point appears at first sight to be in favour of Dr. Hoernle's argument (but of which he could not have been aware), namely that in some matters of detail[1] the Bakhshālī work more closely resembles the *Ganita-Sāra-saṅgraha* of Mahāvīra than any other Indian work on mathematics.

120. Although the mediæval works of East and West have so much in common yet there are differences even in the later treatment of topics and notions that had originally a common origin. For example the Indians, although they adopted the western sexagesimal notation for astronomical purposes did not utilise this notation for purposes of arithmetical calculation. On the other hand the Muslim mathematicians commonly employed this notation to express ordinary fractional quantities. Now in the Bakhshālī Manuscript is an example of the transformation of a simple fraction expressed in the ordinary way to the sexagesimal notation. This transformation may be represented by

$$\frac{178}{29} = 6 + 8' + 16'' + 33''' 6^{iv} \frac{6}{29}$$

No such example occurs in any early Hindu work and there is not the slightest doubt that it indicates direct western influence. Indeed our author could have hardly provided us with a more conclusive piece of evidence.

Again our manuscript exhibits a method for finding approximate roots of surd quantities that is not Indian. The method may be represented by $\sqrt{A^2+b} = A + b/2A$ approximately, and closer approximations may be achieved by continuing the process. The *sūtra* embodying this method is given three times and a number of examples of first and second approximate evaluations is given. Indeed this square-root method is one of the most prominent topics of the work. Its history is quite well known (See §69). It occurs in many western works from the time of Heron onwards but it occurs in no Indian work earlier than the twelfth century : indeed the earliest record of this method in an Indian work (other than the Bakhshālī Manuscript) known to me is of the 16th century !

There is an interesting similarity between part of our text and the arithmetical papyrus of Akhmin (See §89). There are problems of the type of the Epanthem which can also be traced to a definite Greek source (See p. 40) ; but in this and other cases it is possible that the problems reached the Bakhshālī Manuscript by way of other Indian works.

121. But, of course, this evidence of western influence does not mean that the work was not Indian. It is, indeed, almost as Indian as any other mathematical work of the period. It contains references to Hindu mythology and to Hindu deities : and the language is Indian of a sort : the script is an off-shoot of the classical script of northern India ; the form of presentation is Indian ; and the material of most of the examples is Indian.

The general conclusion is that the work is mainly Indian, but that, as was to be expected, it shows signs of outside influence, and it gives rather special prominence to the non-Indian material used.

---

[1]But these have not the slightest Jaina significance.

CHAPTER IX.

The Age of the Manuscript and the Age of the Work.

122. Dr. Hoernle held that the mathematical treatise which is written out in the so-called Bakhshālī Manuscript was considerably older than the manuscript itself. Indeed he thought that the work was composed about six centuries earlier than the copy we are considering. He excluded the possibility of our manuscript being a translation for he largely based his estimate of the age of the work on the antiquity of the language employed in the manuscript. If Dr. Hoernle were right in his differentiation between the age of the work and of the Bakhshālī copy then we should have to consider the possibility of other copies being preserved and the probability of the work being known to other mathematical writers of the intervening centuries. Dr. Hoernle, indeed, did suggest that the work was one of the sources from which the early Hindu mathematicians drew inspiration, but without any justification. Unfortunately Dr. Hoernle's reasons for his views as to the ages of the manuscript and the work are not satisfactory and we are compelled to reject his conclusions altogether. Of course it will be impossible to say definitely that the manuscript is the original and only copy of the work but we shall be able to show that there is no good reason for estimating the age of the work as different from the age of the manuscript to any considerable degree.[1]

There are certain general causes for Dr. Hoernle's mistaken conclusions. He examined in detail only a comparatively small portion of the manuscript ; the history of the Śāradā script was not well known thirty years ago ; there has since that time been light thrown upon the type of language used in the manuscript, and the knowledge of mediæval mathematics has been extended.

It is proposed therefore to re-examine in detail the whole question of age and in the course of the re-examination Dr. Hoernle's arguments will receive due consideration.

The age of the manuscript.

(*a*) *The circumstances of the find.*

123. The circumstances of the find have already been described in detail (§3). They led Bühler and Weber to suggest that the manuscript might prove to be of the age of Kanishka, *i.e.*, of the second century of our era. We need not labour this point. The suggestion was based upon a misunderstanding and there is not the slightest evidence to support it. It was discarded by Dr. Hoernle, who, however, follows much the same line of thought in arguing for a slightly later date. He writes "The country in which Bakhshālī lies and which formed part of the Hindu kingdom of Kabul, was early lost to Hindu civilisation through the conquests of the Muhammadan rulers of Ghaznī, and especially through the celebrated expeditions of Mahmūd towards the end of the tenth and the beginning of the eleventh centuries A.D. In those troublous times it was a common practice of the learned Hindus to bury their manuscript treasures. Possibly the Bakhshālī manuscript may be one of these. In any case it cannot well be placed much later than the tenth century A.D. It is quite possible that it may be somewhat older."

Dr. Hoernle assumes that the manuscript could not well have been written after the time of Mahmūd. Regarding the alleged burial custom I can say nothing.

---

[1] There is evidence that the MS is not a copy at all. It is not the work of a single scribe : there are cross references to leaves of the manuscript ; there is a case of wrongly numbering a *sūtra* and the mistake is noted in another hand-writing.

It may have been prevalent among 'learned Hindus'; but there is not the slightest evidence to show that the Bakhshālī manuscript was *deliberately* buried.

(*b*) *The material.*

124. The material on which the work is written is birch-bark, which was the common writing material for a considerable period of time in Kashmir and its neighbourhood. Unfortunately we have very few birch-bark manuscripts earlier than the fifteenth century preserved, so it would be rash to fix definitely the earlier limit of its use. However the earliest known birch-bark manuscript belongs to about the second century of our era and we know that this material was in common use in Kashmir until about the seventeenth century. There will be little danger in placing our manuscript within the limits here indicated.

But there was fashion in birch-bark manuscripts, and the process of preparation of the material developed. The strips of bark from which the leaves for writing upon are obtained can generally be split up into a number of laminæ. Some scribes were content with comparatively coarse material, obtained by dividing the original strip into two similar strips; but sometimes the sub-division was carried on much further and very thin strips were obtained, two of which cut to the required size were pasted together to form a writing leaf. Portions of the Bower manuscript and all of the Bakhshālī manuscript are of the cruder form, while the Kashmerian Artharva Veda consists of the more elaborately prepared and finer writing material. There might appear to be little doubt that the use of the cruder material denotes an earlier period, but there is a good deal of doubt really; and the scarcity of known specimens reduces the value of any criterion based upon this fashion.

(*c*) *Format.*

125. The shape and size of the birch-bark leaf might also be expected to give some indication of age. The format of the Kharoshṭhī Dhammapada from Khotan was probably due to early western influence while the Bower manuscript format was probably due to the Indian palm-leaf pothī. The Bakhshālī format differs considerably from both of these and is certainly of a later date. Dr. Hoernle thought the Bakhshālī manuscript was the prototype of the early Indian paper book, but it might have been the other way round (See §16).

The Script.

126. The script will be described in some detail later on and the chronological evidence it gives will be indicated. The earlier orientalists had rather inaccurate notions about its chronology. Bühler stated that the oldest Śāradā inscription was that of Baijnāth, which he dated A. D. 804 instead of A. D. 1204, while Dr. Hoernle took the Śāradā script back to A. D. 500. The main facts relating to the chronology of the Śāradā script are as follows : The earliest known examples are of the ninth century and are found on certain coins of the Varma dynasty of Kashmir. There are at least two inscriptions of the tenth century, namely the Sarāhan inscription and an inscription of the reign of Queen Diddā. Many dated inscriptions of the eleventh and twelfth centuries have been preserved. At the beginning of the thirteenth century is placed the Baijnāth inscription, with which, according to Vogel, the history of the Śāradā proper comes to an end.

Of the script of the Bakhshālī manuscript Dr. Hoernle first wrote[1]—"Some of the forms which very frequently occur in the manuscript, especially of vowels, very closely resemble the forms used in the Aśoka and early Gupta inscriptions." The implication here made he slightly modified later[2] by writing as follows :—"The Śāradā characters used in it exhibit in several respects a rather archaic type, and afford some ground for thinking that the manuscript may go back to the 8th or 9th century." "But," he wisely continues, "in the present state of our epigraphical knowledge,[3] arguments of this kind are always somewhat hazardous."

Every letter of the manuscript has now been examined and the script has been compared with other available examples and the following age criteria have been applied :—

(*i*) Vogel thinks that the form of the letter *n* is a fairly reliable test of age. Table I, Part II exhibits three distinct types. The Sarāhaṇ example has a horizontal connecting stroke in the middle of the letter, the Baijnāth example is without this horizontal stroke but has a tail turning inwards from the left. The Bakhshālī manuscript shows no examples of either the horizontal middle stroke or of the left hand tail, and thus seems to place itself between the Sarāhaṇ inscription (tenth century) and the Baijnāth *praśastis* (early thirteenth century).

(*ii*) The common method of forming medial *ā* is to add a knob to the top right hand of the mātṛikā and of this method there are hundreds of examples in our manuscript ; but in the case of *jā* modification of the mātṛikā was the rule. There is, however, an isolated example (folio 16, recto, the sixth akshara of the third line from the bottom) where the commoner method is applied for *jā*. According to Vogel this method of writing *jā* came into fashion about A. D. 1200. This test, if reliable, is rather intriguing, as is also the next one.

(*iii*) Medial *i* and *ī* are generally formed as in Devanāgarī, but there was an older practice of forming them by sickle-shaped curves above the mātṛikās and of this older type there are two or three in our manuscript (folios 1 verso, 52, recto et verso, 60 recto). According to Vogel these superscribed short and long *ī*s dropped out of use about A. D. 1200.

(*iv*) The slanting superscribed medial *e* tends to become horizontal in later Śāradā, *e.g.*, in the Baijnāth inscription. In our manuscript the stroke is nearly always slanting but there are a few examples where it is horizontal. Also another method of expressing medial *e* is by a stroke behind the mātṛikā and this method is exemplified 269 times in our manuscript. The anecdote related by Vogel (p. 96) shows that this back stroke method was not in general use in the fifteenth century.

(*v*) Medial *ai* is formed in two ways and according to Vogel the change took place about A. D. 1100. In the Bakhshālī Manuscript there are 19 examples of the older method as against 61 of the more modern method.

---

[1] *Indian Antiquary* xii, 1883, p. 89.

[2] Ib. xvii, 1888, p. 36.

[3] Our epigraphical knowledge of the Śāradā script has been largely extended by the researches of Dr. Vogel, whose volume on *The Antiquities of Chambā* has been of the greatest help in the present enquiry.

(*vi*) Medial *o* is expressed in three ways of which the most modern method largely predominates in the text.

The following statistics are interesting. (See pages 95—98.)

MEDIAL E.

Older [·e]	Newer [è]		
70%	30%	Sarāhaṇ inscp.	xth cent.
57%	43%	Chambā, No. 25.	xith ,,
44%	56%	Bakhshālī manuscript.	
0%	100%	Ant. of Chambā, p. 63.	xiiith ,,

MEDIAL O.

Old [·o·]	Middle [ò]	New [ŏ]		
58%	12%	30%	Sarāhaṇ inscp.	xth cent.
10%	20%	70%	Chambā, No. 25.	xith ,,
13%	10%	77%	Bakhshālī manuscript.	
0%	10%	90%	Ant. of Chambā, pp. 64, 65.	xiiith ,,

The second table may be combined thus

MEDIAL O.

Older	Newer		
70%	30%	Sarāhaṇ inscp.	xth cent.
30%	70%	Chambā No. 25.	xith ,,
23%	77%	Bakhshālī manuscript.	
10%	90%	Ant. of Chambā, pp. 64, 65.	xiiith ,,

This scriptual evidence is rather remarkable and is as convincing as such evidence can well be; but I must add a word of caution. Fashion in writing is not altogether a matter of chronology : it is also largely a matter of locality. Indeed no single criterion based upon script is infallible, but it is significant that *all* the evidence that the Bakhshālī script gives points to some time about the twelfth century, and there is not a single item of evidence of any type against this conclusion.

Symbols and notation.

127. Not unconnected with the scriptual questions are the forms of symbols, and particularly those connected with the arithmetical notation.

The negative symbol, or minus sign, is a cross +, the use of which for this particular purpose is unique. The nearest approach to this use is the inverted ψ (⫛) employed by Diophantus to indicate a negative quantity. Dr. Hoernle rejected the Diophantine origin of the Bakhshālī symbol on the ground that "the Hindus did not get their elements of the arithmetical science from the Greeks."[1] That such a sweeping assumption as this is altogether unjustifiable has been already shown, and the implication that the work is wholly Hindu in origin has never been proved. Dr. Hoernle attempts to trace the sign back to the Aśoka *ka* but, as he confesses, not with much success. However he suggests that the Bakhshālī symbol is "a mark of great antiquity." See § 61 for some further discussion. All that we can now say is that the use of this symbol cannot be traced to an Indian source.

The arithmetical notation employed will be referred to again (see § 130). Here we are concerned with the forms of the symbols only. In my paper published by the Asiatic Society of Bengal in 1912, I attempted to place these symbols epigraphically. Unfortunately we have very few examples of numerical symbols written in early Śāradā, but on the whole the Bakhshālī symbols resemble most closely those of the Devī-rī-koṭhī Fountain inscription[2] of A. D. 1159. The Śāradā numerical symbols are fairly consistent in form but there are rather peculiar variations of the "four" and "six." The Bakhshālī "four" most closely resembles the example in No. 15 Chambā which probably belongs to the eleventh century.[3] The evidence is really too scanty for us to form any definite conclusion.

Language.

128. Dr. Hoernle writes[4]:—

"The Bakhshālī arithmetic is written in that peculiar language which used to be called the Gāthā dialect, but which is rather the literary form of the ancient North-Western Prākṛit (or Pāli). It exhibits a strange mixture of what we should now call Sanskṛit and Prākṛit forms. As shown by the inscriptions (*e.g.*, of the Indo-Scythian kings in Mathurā) of that period, it appears to have been in general use, in North-Western India, for literary purposes till about the end of the third century A.D., when the proper Sanskṛit, hitherto the language of the Brāhmaṇic schools came into general use also for secular compositions . . . . . Its use, therefore, in the Bakhshālī arithmetic points to a date not later than the 3rd or 4th century A.D. for the composition of that work."

It would be presumptuous for me to contradict Dr. Hoernle on linguistic matters, and his opinion here carries great weight. But from the evidence before me I am compelled to disagree here, as elsewhere, when he is speaking of the age of the work. He gives a number of examples of the "peculiar characteristics" of the language of the Bakhshālī work, on which he appears to have based his views. Now every single one of these "peculiar characteristics" is common in Śāradā inscriptions of the eleventh and twelfth centuries. Indeed their very occurrence helps to

---

[1] *Ind. Ant.* xvii, 1888, p. 34.

[2] *Ant. of Chambā*, p. 212.

[3] The resemblance of some of the numerical symbols to letters is noted in part ii. Bühler (*Ind. Pal.* p. 83) states that symbols for 4 and 9 are 'ancient letter numerals'; but the 'letter numeral theory' has succumbed!

[4] *Ind. Ant.* xvii, 1888, pp. 36-37.

confirm the evidence supplied by the script (§ 126). As it is proposed to devote the second volume of this work particularly to linguistic matters I give here only a list of Dr. Hoernle's "peculiar characteristics" with references to the Chambā inscriptions [in square brackets] where the same peculiar characteristics occur.

(*a*) "Insertion of euphonic consonants." [No. 32.] (*b*) "Insertion of s." [Inscp. No. 32, p. 210.] (*c*) "Doubling of consonants." [Nos. 15 & 32.] (*d*) "Peculiar spellings." [Everywhere.] (*e*) Confusion of ṛi and ri. [Nos. 15 (p. 165), 24 (p. 184), 25 (p. 188), 26 (p. 198).] (*f*) "The *jihvāmūlīya* and the *upadhmānīya* are always used before guttarals and palatals respectively." [" It is one of the most notable characteristics of the Śāradā," writes Dr. Vogel, "that we find the *jihvāmūlīya* and *upadhmānīya* used with great regularity." (p. 58) See Nos. 13, 1·2; 15, 1·28; 32 (p. 210), etc., etc.] (*g*) "Irregular sandhi." [Nos. 14 (p. 161), 24 (p. 184), 26 (p. 198), etc., etc.] (*h*) "Confusion of the sibilants." [Nos. 14 (p. 161), 15 (p. 165), 24 (p. 184), 25 (p. 188), 26 (p. 198).] (*i*) "Confusion of n and ṇ." [Nos. 14 (p. 161), 15 (p. 165), 26 (p. 198).] (*j*) "Elision of a final consonant." [No. 14 (p. 161) where the same actual examples occur.] (*k*) "Interpolation of r." [?.] (*l*) "Etymological and syntactical peculiarities." [Passim.] (*m*) "Peculiar forms." [Passim.] (*n*) "Peculiar meanings." Dr. Hoernle's interpretation of the two examples he gives cannot be accepted.

My references to eleventh and twelfth century inscriptions here given make no pretence of being exhaustive. They are merely given to justify my non-acceptance of Dr. Hoernle's views. On this matter I shall await the verdict of those more competent to judge than I; but my tentative conclusion is that the language of the manuscript is not appreciably earlier than the script itself.

Metre.

129. In 1888 Dr. Hoernle wrote[1]:—

"It appears that the earliest mathematical works were written in the *śloka* measure; but from about the end of the 5th century A.D. it became the fashion to use the *ārya* measure. Āryabhaṭa c. 500 A.D., Varāha Mihira c. 550, Brahmagupta c. 630 all wrote in the latter measure. Not only were new works written in it, but *śloka* works were revised and recast in it. Now the Bakhshālī arithmetic is written in the *śloka* measure; and this circumstance carries its computation back to a time anterior to that change of literary fashion in the 5th century A.D."

This statement is altogether misleading. Mahāvira's *Gaṇita-sāra-saṅgraha* (9th century) is largely in *ślokas*, and the *Sūrya Siddhānta* (c. 1100 A.D.) was written in that measure; and a number of other works dealing with astronomy and mathematics written in the *śloka* measure and rather later than the *Sūrya Siddhānta* are known. Also we can point to a number of Śāradā inscriptions of the eleventh and twelfth centuries in which the *śloka* measure is employed. It is unfortunate that Dr. Hoernle's obsession regarding the age of the work led him to employ this rather disingenuous argument, for it was repeated and emphasised by M. Cantor, the historian of mathematics, in his great work.[2] The subject of the metres employed in the Bakhshālī manuscript will be more fully dealt with in the second volume of this work.

---

[1] *Ind. Ant.*, xvii, 1888, p. 36.

[2] *Vorlesungen über Geschichte der Mathematik* (3rd edition), vol. i, p. 598.

Arithmetical notation.

130. The arithmetical notation employed throughout the manuscript is the modern place-value notation. If we knew the period when this notation was first employed in the north-west of India we should have some criterion for the earlier limit of the date of the work. Note that we are not here so much concerned with the date and place of the invention of the modern notation as with its introduction and early use in the north-west of India. Obviously we should search for evidence of this in other manuscripts, on coins and inscriptions generally. Before we deal with such evidence, however, it may be as well briefly to refer to the broader aspect of the question.

According to the Hindus the modern place-value system of arithmetical notation is of divine origin. This led the early orientalists to believe that, at any rate, the system had been in use in India from time immemorial; but an examination of the real facts shows that the early notations in use were not place-value ones and that the modern place-value system was not introduced until comparatively modern times. The early systems employed may be conveniently termed (*a*) the Kharoshtī (*b*) the Brāhmī (*c*) Āryabhaṭa's alphabetic notation (*d*) the word-symbol notation.

(*a*) The "Kharoshṭhi" script, written from right to left, was in use in the north-west of India, Afghanistan and Central Asia at the beginning of the Christian era. The notation is shown in the accompanying table. The smaller elements are written on the left.

(*b*) The "Brāhmī notation" is the most important of the old notations of India. It might appropriately be termed *the* Indian notation for it occurs in early inscriptions and was in fairly common use throughout India for many centuries. The symbols employed varied according to time and place but on the whole their form was fairly consistent. They were written from left to right with the smaller elements on the right. Several false theories as to the origin of these symbols have

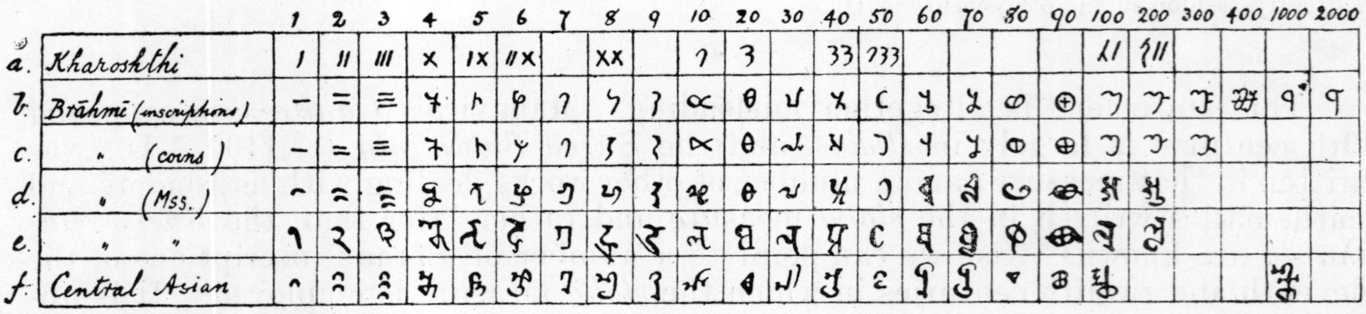

NUMERICAL NOTATIONS.

been published, some of which still continue to be recorded. The earliest orientalists gave them place-value, but this error soon disproved itself; it was then suggested that they were initial letters of numerical words; etc., etc.

(*c*) Āryabhaṭa's alphabetic notation also had no place-value. It was written and read from left to right but differed from the Brāhmī notation in having the smaller elements on the left. It may be exhibited thus :

Letters.	*k*	*kh*	*g*	*gh*	*ṅ*	*ch*	*chh*	*j*	*jh*	*ñ.*			
Values.	1	2	3	4	5	6	7	8	9	10.			
Letters.	*ṭ*	*ṭh*	*ḍ*	*ḍh*	*ṇ*	*t*	*th*	*d*	*dh*	*n.*			
Values.	11	12	13	14	15	16	17	18	19	20.			
Letters.	*p*	*ph*	*b*	*bh*	*m*	*y*	*r*	*l*	*v*	*ś*	*sh*	*s*	*h.*
Values.	21	22	23	24	25	30	40	50	60	70	80	90	100.

The vowels indicate multiplication by powers of one hundred. The first vowel *a* may be considered as equivalent to 1000, the second vowel $i = 100^1$ and so on. The values of the vowels may therefore be shown thus :

Vowels.	*a*	*i*	*u*	*ṛi*	*ḷi*	*e*	*ai*	*o*	*au.*
Values.	1	10^2	10^4	10^6	10^8	10^{10}	10^{12}	10^{14}	10^{16}.

The following examples taken from Āryabhaṭa's *Gitiká* illustrate the application of the system :

$$khyukhṛi = (2+30).10^4 + 4.10^6 = 4{,}320{,}000.$$

$$chayagiyiṅuśulchhḷi = 6 + 30 + 3.10^2 + 30.10^2 + 5.10^4 + 70.10^4 + (50+7).10^8 = 57{,}753{,}336.$$

The notation could thus be used for expressing large numbers in a sort of mnemonic form. Āryabhaṭa's table of sines was expressed in this notation, which, by the way, was used only for astronomical purposes. It did not come into ordinary use in India, but some centuries later it appears occasionally in a form modified by the place-value idea with the following values :

1	2	3	4	5	6	7	8	9	10
k	*kh*	*g*	*gh*	*ṅ*	*ch*	*chh*	*j*	*jh*	*ñ*
ṭ	*ṭh*	*ḍ*	*ḍh*	*ṇ*	*t*	*th*	*d*	*dh*	*n*
p	*ph*	*b*	*bh*	*m*					
y	*r*	*l*	*v*	*ś*	*sh*	*s*	*h*	*l*	

Initial vowels are sometimes used as ciphers also. The earliest example of this modified system is of the twelfth century A. D. Slight variations occur.

(*d*) *The word-symbol notation.*—A notation that became extraordinarily popular in India was introduced about the seventh century A. D., possibly from the East. In this notation any word that connotes the idea of a number may be used to denote that number : *e.g.* Two may be expressed by *nayana*, an eye, or *karna*, an ear, etc. ; seven by *aśva*, a horse (of the sun) ; fifteen by *tithi*, a lunar day ; twenty by *nakha*, a nail (of the hands and feet) ; twenty-seven by *nakshatra*, a lunar mansion : thirty-two by *danta*, a tooth ; etc. This notation, it is said, was used by Brahmagupta.

(*e*) *The modern place-value notation.*—The orthodox view is that the modern place-value notation that is now universal was invented in India; and until recently it was thought to have been in use in India at a very early date. Hindu tradition ascribes the invention to God! According to Masūdī a congress of sages, gathered together by order of king Brahma (who reigned 366 years), invented the nine figures! Patañjali and other early writers are supposed to make references to the place-value system. An inscription of A. D. 595 is supposed to contain a genuine example of the system;[1] according to M. NAU[2] the "Indian figures" were known in Syria in 662 A. D.; and certain other mediæval works refer to "Indian numbers"; and so on.

On the other hand it is held that there is no sound evidence of the employment in India of a place-value system earlier than about the ninth century A.D. The suggestion of "divine origin" indicates nothing but historical ignorance; Masūdī is obviously wildly erratic; the inscription of A. D. 595 is not above suspicion[1] and the next inscription with an example of the place-value system is nearly three centuries later, while there are hundreds intervening with examples of the old non-place-value system. The references to India in mediæval works do not necessarily indicate India proper but often simply refer to "the East" and the use of the term with regard to numbers has been further confused by the misreading by Woepcke and others of the Arabic term *hindasī* (geometrical, having to do with numeration, etc.), which has nothing to do with India.[3] Again, it has been assumed that the use of the abacus "has been universal in India from time immemorial" but this assumption is not based upon fact, there being actually no evidence of its use in India until quite modern times. Further, there is evidence that indicates that the notation was introduced into India, as it was into Europe, from a right-to-left script.[4]

131. As this subject is a matter of general controversy it behoves us to be circumspect about drawing definite conclusions from the occurrence of the modern place-value notation in our manuscript. As indicated above the proper procedure is to examine the evidence relating to the earliest appearance of this place-value notation in the north-west of India. Of manuscripts there are very few earlier than the twelfth century and none of these gives any example of the new notation; of coins there are plenty of examples but none earlier than eleventh century gives any example of the new notation; and the earliest inscription of this part of the world with any evidence of the new notation is not before the tenth century A.D. To suggest that the Bakhshālī manuscript was independent of all the circumstances that governed the epigraphical appearance of the new notation generally would be unscientific.

Dr. Hoernle was very dogmatic. He wrote, in 1888, "It is certain that the principle[5] was known in India as early as A. D. 500. There is no good reason why

---

[1] The figures were added at a later date. A cursory examination of the plate (*Epigraphia Indica*, vol. ii, p. 20) makes this obvious.

[2] Journal Asiatique, 1910, p. 209. M. Nau's 'authority' does not read well. He speaks of the Hindus' noble discoveries in astronomy—more ingenious than those of the Greeks and Babylonians.

[3] About nine hundred years ago Albīrūnī wrote (India, ii 211)—"They .... relate all sorts of things as being of Indian origin, of which we have not found a single trace with the Hindus themselves."

[4] For a more detailed examination of this question see my papers on (1) *Indian Arithmetical Notations, JASB* 1907, 475-508, (2) *The use of the Abacus in ancient India, JASB* 1908, 293-297; *References to Indian Mathematics in certain Mediæval works, JASB* 801-806. In 1917 in SCIENTIA (1917, pp. 273-282) Baron Carra de Vaux pursued a line of investigation completely different from that I had myself followed and came to the same conclusion, namely that the original home of the modern place-value notation could not well have been in India. For a popular exposition of the other side of the controversy see SMITH and KARPINSKY *The Hindu Arabic Numerals.*

[5] That is the place-value principle. The quotation is from the *Indian Antiquary* of 1888 (xvii), p. 38.

it should not have been discovered considerably earlier. In fact, if the antiquity of the Bakhshālī arithmetic be admitted on other grounds, it affords evidence of an earlier date of the discovery of that principle." Bühler is bolder still. "If Hoernle's very probable estimate of the antiquity of the arithmetical treatise, contained in the Bakhshālī manuscript, is correct," he writes, "its (*i.e.*, the place-value notation's) invention dates from the beginning of our era or even earlier."[1]

132. There are two other criteria suggested by Dr. Hoernle, namely (*a*) the length of the year mentioned in the text, and (*b*) the occurrence of the term *dīnāra*. Regarding the former he came to no definite conclusion but seems to have thought that there was some correspondence between the year of 360 days that occurs in the text, and the accepted estimate of the length of the year at the period of composition of the work. Had he gone into the matter further he might have come to the conclusion that the Bakhshālī work had been composed in pre-Vedic days! It seems hardly possible for any one to be misled by such "internal evidence" but it may be as well to point out that the practice of reckoning the year as consisting of 360 days for the purpose of arithmetical examples was quite common in mediæval Indian works. Mahāvīra and Śrīdhara actually give 1 year = 360 days in their tables of measures, and the latter adds the remark—"Time is calculated according to this rule in all arithmetical works."

133. Dr. Hoernle's argument regarding the use of the *dīnāra* is as follows:—In the early centuries of our era the *dīnāra* in use in India was the gold one only, and "the Bakhshālī arithmetic seems to indicate that the gold *dīnāra* and the silver *dramma* formed the ordinary currency of the day. This circumstance again points to some time within the first three centuries of the Christian era as the date of its composition."[2] The only reply that this statement calls for is to state that all the evidence of our text points to the use of a copper *dīnāra* and there is not the remotest indication of a golden *dīnāra*.[3] (See § 110.)

134. There are other indications rather than evidence of the age of the work in the material of the text. The occurrence of the square-root rule already referred to would not be an anachronism if it were found in any Indian text from the time of Āryabhaṭa onwards—but it occurs in no known Indian text until very late indeed, and its appearance in the Bakhshālī manuscript is probably due to direct western influence, possibly to Muslim influence. Such evidence is valueless to the inexpert: it can only carry weight with those who have a very fair knowledge of the mathematical field of the period. Likewise the employment of the sexagesimal notation points in the same direction, but the general reader must bear in mind that this notation had been used by Hindu astronomers from the time of Āryabhaṭa, and that the arithmetical use of it in our text is rather a matter of western fashion than the introduction of a new idea.

The employment of the *regula falsi* is evidence of a slightly different character It occurs in no Indian work until the time of Mahāvīra,[4] and it was probably, even

---

[1] *Ind. Pal.*, p. 82.

[2] *Loc. cit.*, 37. It is noteworthy that the muslim arithmetical works often devoted a section to the *Dirham* and *Dīnāre*. See E. WIEDEMANN *Beiträge z. Gesch d. Natur Wissenschaften*, xiv, pp. 29 and 31.

[3] SIR A. STEIN, in 1900, wrote: "This word, undoubtedly derived from the *denarius* of the West, is well known to Sanskrit lexicography as the designation of a gold coin usually spelt *dīnāra*. But the manifest impossibility of accepting this meaning for the passages of the Chronicle which mention sums in *Dīnnāras*, has already struck Dr. Wilson. Noticing in two passages figures are given which, if calculated in gold would be large beyond all credence, he suggested that the 'Dinars' meant might have been of copper. Curiously enough, however, none of the subsequent interpreters seems to have followed up the suggestion thrown out by Wilson, or to have otherwise paid attention to the subject." *Kalhaṇa's Rājataraṅgiṇī*, vol. II, p. 308. He then goes on to prove that the *dīnāra* used was a copper one. The whole of the dissertation should be read, pp. 308-328.

[4] See his *Ganita-sāra-saṅgraha* vii, 112. His employment of the method is rather special and limited. In Northern India the first known use of the method occurs in the twelfth century.

in the west, a fairly close predecessor to an algebraic symbolism. There is also other evidence that our work was probably produced not long before an algebraic symbolism came into use.

135. Of the evidence as to age discussed some is of doubtful value; but there remains a good deal that must be considered as giving no uncertain indication, and the period indicated is in all cases about the twelfth century. The script, the language, the contents of the work as far as they can give any chronological evidence, all point to about this period, and there is no evidence whatever incompatible with it. Bhāskara was born in A. D. 1114, Omar Khayyām was flourishing in the early part of the century, Adelard of Bath visited Cordova in 1120, Leonardo was born in 1175; and it was during this period that the Bakhshālī work was probably composed. It is possible that in the future more light will be thrown upon the script and language of the locality and period of the Bakhshālī manuscript; and that, if the manuscript be required as suggested, further internal evidence will be forthcoming. Such further discoveries may modify the conclusions now drawn regarding the age of the manuscript; but they cannot very well put the date of composition of the work back to any great extent.

THE BAKHSHĀLĪ MANUSCRIPT

PART II.

i.—*The Script.*

The Bakhshālī manuscript is written in the Śāradā character, which is a descendant of the Brāhmī, in a line distinct from but parallel with the Nāgarī line of descent. With the last mentioned Śāradā has many features in common, but, generally speaking, it is much more archaic in appearance. The Śāradā script, although not nearly so well known as the Nāgarī, has been examined and written about to some extent. Leech's *Grammar of the Cashmeeree Language* (*JASB.* xii, 1894, 399 *sqq.*) gives the alphabet. Bühler's *Indische Palaeorgraphie*, gives useful, but sometimes misleading information; Vogel's *Antiquities of Chambā State* (1911) contains the most valuable contribution—and to this I am chiefly indebted; Ojha's *Bhāritīya Prāchīna Lipimālā* gives some useful tables, but, so far as the Śāradā script is concerned, is largely based on Vogel's work. Sir George Grierson's paper in the Journal of the Royal Asiatic Soicety (1916, xvii, 677 *sqq.*) contains tables of ligatures, etc., of modern Śāradā; and in his note in the *Linguistic Survey of India* (Vol. viii, Part ii, page 254) he states that the Śāradā character is the ancient indigenous character of Kashmir, and that it is still generally used by Hindus and is taught in their schools in that country. I have not yet been able to obtain a copy of Burkhard's tables,* which, according to the editors of the *Kashmirian Atharvaveda*, contain transliterations of 340 characters and ligatures.

The principal examples of the Śāradā script that are known come from Kashmir[1] and Jammu; Gandhāra[2] (in which the village of Bakhshālī is situated); Ladakh; Kāṅgrā, Kulū and Maṇḍī; the Shāhpur district of the Punjab, and Delhi. The distribution roughly corresponds to the area between longitudes 72° and 78° east of Greenwich and north latitudes 32° and 36°.

The earlier epigraphists had rather inaccurate notions about the Śāradā script. Cunningham described it as the Gupta character, which he thought had persisted in use in certain localities;† Bühler stated that the oldest Śāradā inscriptions were the two Baijnāth praśastis from Kāṅgrā and that their date was A.D. 804,‡ whereas it is A.D. 1204; Dr. Hoernle dated the development of the script from about A.D. 500 and laid down the maxim that Śāradā characters were no guide as to age, and Kielhorn seems to have entertained a somewhat similar view.

The conservative character of the script puzzled many investigators and it seems to have retarded the formulation of any chronology based upon palaeographic considerations. Vogel, however, in spite of the weight of the opinions of Kielhorn, Bühler and Hoernle, holds that the historical development of the Śāradā script can be traced. "I believe," he writes,§ "that a close examination of the characters will also enable us to fix the approximate date of any undated Śāradā record of the pre-Muhammadan period, provided it is extensive enough to lend itself to a detailed study." Whether or not Vogel has been perfectly successful in

---

* Proceedings of the Imperial Academy of Vienna, Vol. CVII, p. 640.

[1] Kashmir is, indeed, known as *Śāradā Kshetra*, or land of the goddess Sarasvatī.

[2] *i.e.*, the Peshawar district and the surrounding hill tracts.

† A. S. R. xiv. 121.

‡ Ind. Pal. p 57.

§ *Antiquities of Chambā*, p. 49.

formulating his chronological tests remains to be seen; but, at any rate, he has provided material and detailed notes that are invaluable.

Dr. Hoernle first gave 500 A.D. as the approximate period of the birth of the Śāradā script, but later he modified this view and stated that it originated directly from the Gupta script in the course of the seventh century. The earliest known examples are on the coins of the Varman dynasty of Kashmir which start from the middle of the ninth century; while previous to this the acute-angled script was, it is said, in general use in northern India. It is, of course, possible, as Vogel points out, that the Śāradā was employed as a literary alphabet considerably earlier than the ninth century; but there is no evidence. There is a Kashmir inscription of the reign of queen Diddā that belongs to the tenth century, as also possibly does the Sarāhan inscription: and there are many dated inscriptions belonging to the eleventh and twelfth centuries. The following table gives the dates of some of the more important Śāradā inscription :—

	A. D.
Coins of the Varman dynasty of Kashmir†	855-939
Inscription of the reign of Queen Diddā	992
Devi-rī-koṭhi fountain inscription	1159
Angom inscription	1197
Baijnāth inscription (Kāṅgrā)	1204

With the Baijnāth *praśastis,* according to Vogel, the history of the Śāradā proper comes practically to an end; and the script of the later records differs so considerably from the pre-Muhammadan inscriptions that he proposes to give it a special name. "The thirteenth century," he says, "forms a blank which separates the two palaeographic periods."

The relationship of the Bakhshālī script to other Indian scripts is illustrated in its main outlines in table i, which shows the Brāhmī, Western Gupta (Bower), Acute-angled, Śāradā (Nos. 4, 5, 6, 7) and the Nāgarī alphabets. It is at once seen that the Bakhshālī alphabet is more closely connected with numbers 4 and 6 than with 7; and that numbers 4, 5, and 6 are differentiated on the one hand from number 3, and on the other from number 7. Indeed numbers 4, 5, and 6 are examples of the Śāradā group proper, which is supposed to be a direct descendant from the acute-angled script (No. 3); and number 7 is a direct descendant from the earlier Śāradā script.

The most notable differences between the Śārada and the acute-angled are with respect to the *ṇa, ta* and *śa;* while minor differences are seen in the *ga, na, pha, la* and *ha,* but in these latter cases slight variations occur in one or both of the scripts which, more or less, blot out the differences.

The differences in the cases of the *ga, ṭa, śa* are possibly due to the emphasis given in the later scripts to the overhead horizontal lines. The most important difference lies between the two example of *na*—the middle horizontal connecting line being consistently employed in the acute-angled script, and consistently omitted in all but the very early examples of the Śāradā.

Between the Śāradā proper (Nos. 4, 5 and 6) and number 7, which may be termed modern Śāradā, the differences are much more marked. The *i, e, ka, ja,*

† CUNNINGHAM *Coins of Mediæval India,* plate iv.

ḍa, ya (Table I) are particularly noticeable; while the *ṇa* in both the Baijnāth example and in the modern Śāradā shows an inturned tail on the left.

The alphabet.

Further details of the alphabet are given in Tables ii to iv. Table ii shows the consonants and their more important combinations; Table iii gives the various methods of writing the vowels; and Table iv gives most of the ligatures, the numerals, etc. It is proposed to examine the whole alphabet more or less in the order of these tables.

The Consonants. *k-, kh-, g-, gh-, ṅ-*

The normal *k-* of the manuscript differs very little from other early Śāradā examples. Indeed the variation in the manuscript is almost as much as in the inscriptions. The right-hand curve exhibits, on the whole, a tendency to close inwards towards the vertical like the modern Śāradā, particularly in the "M" section of our manuscript; but there are some half dozen examples where it is almost straight. The Sarāhaṇ and Baijnāth examples have a much shorter and more open right-hand curve, while in modern Śāradā the curve is quite closed. Vogel says that the left-hand loop is generally more rounded in the older inscriptions and in the Bakhshālī manuscript, but this is, at least, doubtful.

As the first element in a ligature *k* harks back to the Brāhmī type, *e.g.*, in *ku, kṛi, kt- kr-, ksh-, etc.* (Table ii, 1), as it does in all the Śāradā examples; but as the final consonant in a ligature it generally retains its normal shape.

In the Śāradā, *g* and *gh* have fairly constant forms, which show very little deviation from the Bower manuscript examples. The left-hand shoulder of the *g* is sometimes slightly rounded in the Bower manuscript and the acute-angled script; and in the Bakhshālī manuscript similar examples occur once or twice. In ligatures each of these letters retains its normal shape.

The letter *ṅ-* is only found in combination with other consonants, as is general in the Śāradā. The form of *ṅ-* is the same as in the Gupta period but it has developed a swelling at the right-hand end of the top horizontal stroke, and this is sometimes mistaken for the sign of *ā*. In the Bakhshālī manuscript such right-hand protuberances can, however, generally be distinguished from the *ā* sign. See *j-* and *ṭ-*

ch-, chh-, j-, jh-, ñ-

Bühler speaks of the "quadrangular *cha*" as characteristic of the Śāradā, but it can hardly be deemed to be such, for in the Sarāhan and Bakhshālī examples it is often somewhat rounded, or in the latter has the right side vertical, and in the Devī-rī-koṭhī inscription it is almost triangular. Bühler possibly had the Baijnāth and modern Śāradā examples in mind.

The akshara *chha* is also fairly constant in form, although the Bakhshālī examples are more cursive than in the inscriptions. The digraph *chchh-* frequently occurs. These letters are illustrated in Table ii, 2.

The *j-* is a conservative character in the Śāradā. In the Bakhshālī manuscript it differs very little from the Bower manuscript and acute-angled examples, except that the top horizontal stroke has developed a much more pronounced knob on the right-hand. According to Vogel this knob or wedge disappears in the Muhammadan period, but Grierson gives it for modern Śāradā. The akshara *jā* differs considerably from *ja.* The top bar disappears and a nearly vertical stroke is added to the end of the tongue; but instead of this modification of the *mātṛikā,* in late inscriptions occasionally we meet with a *jā* formed by the usual medial *ā* sign added to the *mātṛikā,* and in our manuscript there is one such example (folio 16R.) The later script seems to revert to the Bower manuscript type.

The letter *jh-* only occurs in *jjhi* and *jjhya.* It is something like the *jā* with the left-hand stroke turned backwards. It shows little alteration since the Gupta period.

The *ñ* also only occurs in combinations (*jña, jnā, ṅcha*). The superscribed *ñ* differs considerably from the subscribed form, but both forms are almost identical with examples that occur in the Bower manuscript.

ṭ-, ṭh-, ḍ-, ḍh-, ṇ-

Of the linguals *ṭh* does not occur at all independently, and it is doubtful whether it is intended as a subscript. The *ṭ-* generally, but not always, has a top bar, and always it has a well developed right-hand knob. As the first element of a ligature *ṭ* occurs twice, but as a subscript to *sh* it is common. The scribe evidently made no distinction between *sht-* and *shth-*, for the open and closed letters are used indiscriminately for the same words. The normal *ṭ* or *ṭh* as a subscript to *sh* is generally accompanied by another curve attached to the right of the *sh* near the bottom. Usually this line curves slightly inwards towards the subscript; but in several cases it is a long stroke slanting outwards. This additional stroke is absent in the Sarāhaṇ inscription (and in the late Kashmir Śāradā) but it is fairly common in other early Śāradā inscriptions.

The *ḍ-* is almost exactly like the Baijnāth and modern Śāradā examples. It differs considerably from those in the Bower manuscript, and its development is somewhat uncertain. Many examples of *ṇḍ-* occur in our manuscript.

The akshara *ḍha* is very like the *pha* but has a kink in the downward stroke. Its form has not altered essentially since the Brāhmī period.

Table I exhibits three distinct types of the Śāradā cerebral nasal *ṇ.* The Sarāhaṇ example has a horizontal connecting stroke in the middle of the letter, like the Bower and acute-angled examples; the Baijnāth example is without this horizontal stroke but has a tail turning inwards from the left, and the modern Śāradā has the same tail somewhat shortened; the Bakhshālī example has no horizontal middle line and no left hand tail, and is exactly of the same type as that exhibited in the Devī-rī-koṭhī fountain inscription and other inscriptions of the twelfth century. Vogel seems to think that we have here a reliable criterion for age, and if so the Bakhshālī manuscript places itself between the Sarāhaṇ and Baijnāth praśastis. But like all other particular tests for the Śāradā this must not be relied upon with too great confidence, for in the later Śāradā we find the Bakhshālī and Baijnāth types used promiscuously.

In the Bakhshālī manuscript the form of *ṇ* is very constant but different sections of the manuscript show slight variations. ` The tendency is to make the left-hand stroke shorter and slightly thicker than the others, particularly in the " M " section ; the third example in the table is peculiar in having a long thin tail but it only occurs once or twice in some hundreds of examples. The principal conjuncts are *ṇḍ-*, *ṇy-*, *rṇ-*.

t-, *th-*, *d-*, *dh-*, *n-*.

The dental group is quantitively very important. It is, on the whole a conservative group, the chief changes being, perhaps, in the general shapes of the *th* and *dh*, which in the early inscriptions were, more or less, crescent shaped, *e.g.*, in the Sarāhaṇ praśasti. The employment of the top bar modified the crescent shaped letters into lozenge shaped. In our manuscript the crescent shape is partially retained.

As a subscript *t* is something like initial *u* but without the thin up-stroke (See Table ii, 4) ; *th* in ligatures has different forms as a second and third element, *e.g.*, as in *tthy-*, *sth-*, *rth-*; *dh* as a subscript usually has a triangular form (*ddh-*, *bdh-*, *rdh-*, *ddhv-*), but it is sometimes difficult to distinguish *ddh-* from *dv-*.

In the Devī-rī-koṭhī fountain inscription the *rth-* is rather remarkably different from the Bakhshālī examples and so it is in modern Śāradā. The examples in fig. 1, from the Suṅgal grant, are interesting as showing the transition stage in the development of *th* as a subscript and medial.*

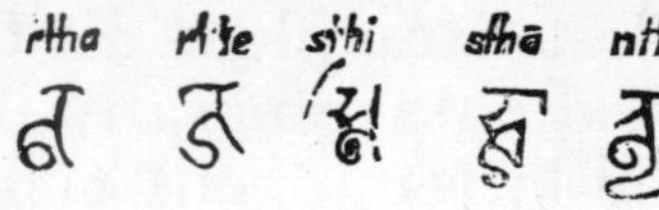

Fig. 1. Suṅgal.

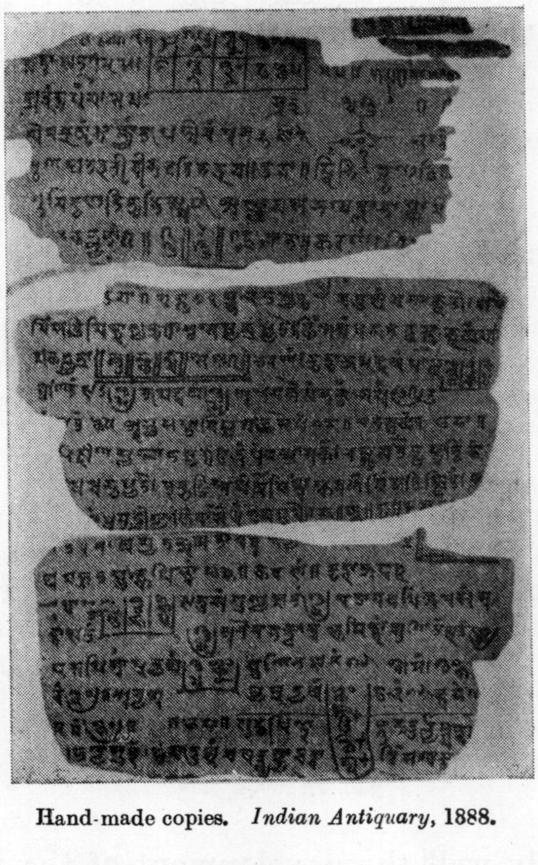

61 Verso

14 Verso

14 Recto

Hand-made copies. *Indian Antiquary*, 1888.

---

* Bühler (Table VI, column viii, 50) gives an incorrect form for *rtha*. I may as well point out here that numbers 17 (*gu*), 24 (*jñā*) and 38 (*bha*) of Buhler's table are hardly representative examples. He had only Dr. Hoernle's hand-made copies to follow, and these, as Dr. Hoernle was well aware, are not perfectly reliable. For example, the transcript of F14 recto given in the Indian Antiquary (xvii, 1888, p. 43) misrepresents *lyo* and *tu* in the penultimate line ; ¢*paṁ.*, *ddhyā*, etc., in the last line ; and that of F14 verso gives l. 1 ϕ*paṁ* ; l. 3 *tthya, tāṁ, stha* ; l. 5 *ṅśa* ; l. 6 *tpra* ; last line *śvame*, inaccurately.

p-, ph-, b-, bh-, m-.

The labial group is also very important quantitively, and this rather accentuates certain difficulties that present themselves. The principal of these are due to similarities between the *v* and *dh*, the *ph* and *ḍh* and the *m*, *d* and *s* under certain conditions. Generally in the Śāradā no distinction is made between the *b* and *v*, and in the Bakhshālī manuscript this lack of differentiation is almost, if not quite, complete. The resemblance between the *b* or *v* and *dh* in ligatures has already been noted upon; the *m* is distinguished from the *d* by having a complete vertical stroke on the right, but the ligatures *nd-*, *nm-*, *rd-*, *rm-* are sometimes rather difficult to differentiate; and badly written examples of *m* and *s* are also confusing. Final *m* (with the virāma) is very different from its mātṛikā and occasionally is only to be differentiated from the numerical sign for "six" by the length of the very long virāma stroke.

y-, r-, l-.

In table i are shown two types of Śāradā *y*, of which the Bakhshālī examples may be described as representing the transitional stage between the Bower manuscript example and the Sarāhaṇ example. Table ii,[6] however, shows that the open (Sarāhaṇ) type also occurs in the Bakhshālī manuscript, but it occurs with comparative rarity.* Most of the examples are of the type shown in table i and the first example of ii, 6; while the type illustrated as the third example in table ii, 6, occurs generally in the A_6 section of the manuscript. In most Śāradā scripts the latter (open) type prevails but in the Devī-rī-kothī fountain inscription there is exhibited a tendency towards the prolongation of the central vertical, although the left-hand loop is never actually closed, as it is in most of the Bakhshālī examples. As an element of a ligature the form of *y* is masked. Many examples are given in table iv. Occasionally the upward right-hand portion of the curve is lengthened as in *gya* (iv, 1) and *ddhya* (iv, 5): otherwise there is little to note. As the middle element of a ligature *y* seldom occurs, unless we count *ū* as one of the elements.

The letter *r-* is one of the most consistent in form in this family of scripts. The bottom serif had developed in pre-Śāradā days. This serif is generally rather larger in our manuscript than in other Śāradā texts and occasionally develops into a loop (F 24). As a final element the *r* seems to be formed by a thinning and lengthening of this serif (table iv). For the peculiar shape of *rū* see page 95 and for other modifications see below.

Two types of *l* are shown in the Sarāhaṇ and Bakhshālī examples exhibited in table i. The former has the left curve attached to the vertical by a horizontal stroke, and this seems to be a characteristic of the earlier Śāradā scripts: the latter has the left curve attached by another curve—practically always in our manuscript (table ii, 5, etc.). According to Vogel the former type was still prevalent at the beginning of the twelfth century.

ś-, sh-, s-.

Table i illustrates fairly well the development of the Śāradā *ś*, which, in its later form, is exactly the Nāgarī *s*. In our manuscript the *ś* is always square in

---

* See FF 4, 6, 38, 41, 52, 53.

shape and generally has the triangular wedge on the left emphasized (See table ii, 6). This wedge is never open but always blocked-in solid. The *sh* is consistent in form (See table i), but in the ligature *ksha* the portion of the vertical above the cross bar is suppressed* (tables ii and iv) : See also *shsha* in table iv, 4.

The *s* is like the *ś* but open at the top. In the Bakhshālī manuscript it is quite common for *ś* and *s* and *sh* to be used indiscriminately.

The letter *h* is fairly consistent but occasionally tends towards angularity, like its Acute-angled prototype.

Visarga.

The visarga is represented by the two usual dots like a semi-colon. Vogel states that in several Chambā inscriptions "composed in corrupt Sanskrit" the real meaning of the visarga is misunderstood and that it is regularly used to separate words and sentences. The same remark applies to our manuscript.

The jihvāmūlīya (χ-) and upadhmānīya (ϕ-) are often employed. In one case (fol. 33) the former seems to be used in the middle of a word, *duχkha* (?) for *duḥkha*. It (χ) is similar in shape to the *b* or *v*, while the latter (ϕ) is very much the same as in the Chambā copper-plate grant of Somavarman (l. 2). Vogel states (p. 170) that the upadmānīya dropped out of use in Chambā after about 1200 A.D. Examples are shown in tables ii, 1 and ii, 5. See also page 79.

Virāma.

The final consonants *k, t, t, n* and *m* have each a very long virāma symbol drawn through the top of the letters on the right. See table iv, 6.

The Vowels.

a and ā.

Initial *a* and *ā* are fairly consistent in the Śāradā, and the Bakhshālī examples differ very little from the acute angled and early Gupta types. The *a* has at the bottom of the right vertical stroke a small triangular wedge, while the *ā* has in place of this wedge a small curve something like the bottom part of the *t*. Both *a* and *ā* have open tops, while in modern Śāradā the tops are closed, as in the Devanāgarī.

Medial *ā*.

The symbol for medial *ā* is a wedge or serif attached to the right of the mātṛikā at the top. Examples are seen in table iii, 1. In the Bakhshālī manuscript this symbol is generally rounded and is often written with a very small flourish at the top.

* Possibly to prevent confusion with *th* ; but in ligatures *th* is considerably modified.

There are three methods of forming medial *ā* to consider : (i) The archaic method of joining the symbol by a horizontal bar attached to the *left* top of the mātṛikā. Vogel thinks this method dropped out of use about A.D. 1000, and that it may have been merely a local fashion. When there is any choice this method is not employed by our scribes. (ii) The more common method is to join the symbol to the top of the right of the letter, and this is generally done in our manuscript with each of those letters that has a dexter upright or a top bar. (iii) The third method entails a modification of the mātṛikā, but this only occurs in the case of *j*. Examples of *jā* are shown in table ii, 1 ; but there is an isolated example* where *jā* is formed by method ii, and, according to Vogel, method ii, as applied to the *j* came into fashion about A.D. 1200.

It will be noted that each of the mātṛikās of *ṅ*, *j*, and *ṭ* already possesses a mark similar to the medical *ā* symbol. In the cases of *ṅ* and *ṭ* the usual *ā* symbol is added according to method ii, but, as already stated, to form the *jā*, method iii is generally employed. In the *ṇā* the symbol is often slightly curled inwards, and this hook-shaped symbol is said to be characteristic of early Śāradā.

Note that the -*ā* symbol is joined to the left upright of the *d*-, but this is because the dexter upright is not a complete one—and thus the *dā* is differentiated from the *mā*.

The vowels *i* and *ī*.

The initial vowel *ī* does not occur. The short vowel *i* is of the form that is, with fair consistence, employed in all Śāradā scripts : it consists of two dots placed above a, more or less, semicircular loop. (See table iii, 2.)

Medial *i* and *ī* are, almost without exception, of forms that are essentially the same as in the Devanāgarī, and consist of left (-*i*) and right (-*ī*) vertical strokes, sometimes considerably longer than the mātṛikā and joined to its top by a bend inwards.[1] But our manuscript contains three or four examples in which the -*ī̃* is formed by a sickle-shaped curve above the mātṛikā and with its convex side upwards and turned slightly towards the left. (See table iii, 2.) Vogel says, "We may assume that about A.D. 1200 the superscribed medial *i* and *ī* dropped out of use."

u and *ū*

Initial *u* is the same character as in the Bower manuscript. The left upstroke, which is generally continued to the level of the top of the letter, differentiates *u* from *ta*. The long *ū*, as an initial, is the symbol for the short vowel with a streamer hanging down from near the right top of the letter. It is distinguished from *rū*, which has a similar " steamer," by the long upward curve on the left.

Medial *u* is expressed in three ways : (i) By a triangular wedge† attached to the bottom left of the right vertical, or, where this vertical is absent, to a vertical attached to the bottom of the mātṛikā ; (ii) By the addition of the initial *u* as an ordinary ligature ; and (iii) by attaching a downward steamer to the right of the

---

* F 16 recto.

[1] The *ī* stroke is often longer than that of -*i*. Compare the aksharas *stī* and *ti* in table ii, 2 : these are contiguous aksharas from F 33v.

† Often closed solid, but occasionally open.

letter near the top. Each of these methods generally applies to a definite set of letters only, as follows :—

The wedge is attached to	The curved u symbol is attached to	The streamer symbol to
d-, dh-, n-, p-, b-, y-, ś-, sh-, s-, h-	*k-, g-, t-, bh-, ś-*	*r-*

This practice is quite in accordance with that of the early Śāradā ; but in our manuscript the *śu* has either form, and this is not surprising, since the *ś-* and *s-* are occasionally employed indiscriminately. According to Vogel the wedge symbol is earlier than the curve.

Medial *-ū* is formed most generally by a, more or less, horizontal " streamer " attached to the left bottom of the vertical or to a vertical provided for the purpose (*hū, dyū*). The *rū* and the *brū** are exceptions to this method of formation : the former is very like the initial *ū* but without the left up-stroke, and the *brū* is sometimes formed by an angular attachment to the middle of the right vertical, and this is possibly a modification of the initial *ū*. In one case *brū* is formed by the addition of two hanging streamers on the right and the *r-* stroke is shortened.

ṛi

Initial *ṛi* occurs once (63 r.). Medial *ṛi* is invariably the initial *ṛi* sub-joined as a ligature. Examples are shown in table iii, 5. This symbol differs from other Śāradā examples by being less rounded. Bühler terms it angular, and Vogel states that this angularity is only found in the later inscriptions.

The vowel *e*

Initial *e* does not generally differ essentially from the Bower manuscript examples. There are, however, examples in our manuscript similar to the 12th century examples given in Bühler's table VI (11, x-xi) and these examples suggest one of the medial forms.

Medial *e* is formed in two distinct ways : (i) By a slanting stroke over the letter, touching or nearly touching the letter at the right top corner [è]. In late Śāradā this stroke tends to become horizontal. (ii) By a short thick stroke or knob attached to the left top of the consonant by a horizontal line [·e].† Vogel describes this as a wedge ; but in our manuscript it is never wedge-shaped. It is very like a reversed medial *ā*, but with two slight differences : the connecting horizontal stroke is comparatively long, and the terminating knob or stroke has no little flourish but is perfectly smooth.

These two symbols, so different in appearance, seem to be used indiscriminately, at least for certain letters ; and they seem to be very unevenly distributed. For example, F 11 exhibits the second method [·e] 11 times and the first or top-stroke method only twice ; while F 60 does not show the second.

* It is the *r* in the ligature that occasions the so-called exception.

† The symbols shown in square brackets in this and the following paragraphs are mnemonic only. They do *not* accurately represent the symbols employed in the manuscript. The tables and text should be consulted in all cases.

The table on page 97 shows that the two methods are used almost equally in the "M" section and that in the remainder the ratio is roughly 3 to 2.* The following table gives more details for sections A to L, and M :--

		ke	ge	chhe	te	ne	me	ye	re	śe	se
A to L	è Percentages	38	100	100	68	82	100	29	50	33	100
	·e ,,	62	0	0	32	18	0	71	50	67	0
M	è ,,	100	50	56	13	33	0	67	22	50	0
	·e ,,	0	50	44	87	67	100	33	78	50	100
No. of examples		*21*	*11*	*36*	*45*	*23*	*16*	*24*	*21*	*51*	*7*

The values of these ratios depend largely upon the number of cases. As a whole they seem to indicate an inversion of fashion. The M section, which is possibly the more ancient, generally shows a larger proportion of the second method [·e]; but notable exceptions are *ke* and *ye*. In one or two cases complete inversion is exhibited : for example, *me* is formed by the first method [è] only in sections A-L, while in "M" it is formed by the second method [·e] only; and the same is the case with *se*. The best test aksharas of those given in the table appear to be *ke, chhe, te, ne,* and *me;* although all may be used with fair safety. It should also be noted that *ne* is formed by the first method [è] only in all sections by reason of the shape of the mātṛikā. Similarly the first method predominates with *le*.

Vogel relates an interesting anecdote† of the 15th century which tells of a forger changing *me* into *dasa*. The original narrator writes : "In order to express *e* following a consonant the clerks used formerly to write a stroke behind the consonants. But as, in the course of time, the script became changed the writers of to-day write the stroke expressing *e* over the consonant."

Medial *ai*

Initial *ai* does not occur in Śāradā. Medial *ai* is formed in two ways : (i) By two top strokes [a̎i]; (ii) By a combination of the two *e* symbols [·ài]. In both cases the *ai* symbol is thus a symbol for *ee*.

Of these methods the latter [·ài] is the older. Possibly the earliest example of the former method [a̎i] occurs in Chambā No. 25, and Vovel assumes that the change took place about A.D. 1100. In the Baijnāth praśastis the double top stroke [a̎i] is the more common. Later it is invariably used and the top strokes are horizontal.

In the Bakhshālī manuscript both methods are employed, but the second [·ài] predominates : indeed out of 80 examples 61 or approximately three-quarters are formed by the second method [·ài], and 19 by the first or double stroke method [a̎i]; but in the M section there is not a single example of the first method [a̎i].

Medial *o*

Initial *o* does not occur. Medial *o* is formed in three ways : (i) By a large circumflex shaped symbol placed above the consonant [ǒ]; (ii) By a combination of

* *i.e.,* 3 *è* to 2 ·*e*.

† *Antiquities of Chambā* p. 63.

the *e* sign [·e] with the symbol for ā [·o·]; (iii) By the top stroke for *e* combined with the ā symbol [ò]. See table iii, 8.

Out of 252 examples the distribution is as follows :—

First method	ŏ	189	or 75 per cent.
Second method	·o·	38	or 15 ,,
Third method	ò·	25	or 10 ,,

But in the M section the percentages are remarkably different, namely 36 per cent., 32 per cent. and 32 per cent. as against 85 per cent., 14 per cent., and 1 per cent. in the remainder.

Medial *au*

Medial *au* is, in our manuscript, always expressed by the circumflex sign combined with the medial ā sign. See table iii, 9. The more ancient method, which combines the second *o* method [·o·] with the top *e* stroke [è] (and of which three examples occur in the Sarāhaṇ inscription) is not employed.

The following tables give statistics of the methods employed in the formation of the medials *e*, *ai* and *o*, etc.

Section.	Medial *e*.		Medial *ai*.		Medial *o*.			Jihvāmūliya and upadhmānīya.		Style of writing.
	·e	è	·ai	̀ai	·o·	ò·	ŏ	χ	Φ	
A	39	46	13	9	6	0	42	1	7	α_1 α_2
B	22	28	6	1	0	0	35	3	1	α_2
C	28	29	1	1	3	2	26	0	0	α_2
D	22	28	2	0	0	0	10	1	0	α_2
E	8	10	2	0	0	0	9	0	0	α_1
F	13	28	6	0	0	0	5	0	0	α_1
G	52	36	1	1	2	0	20	2	3	α_3
H	10	24	8	3	1	0	11	1	1	α_2
J	5	7	0	4	0	0	3	0	0	α_1
K	1	4	0	0	0	0	3	0	0	α_1
L	11	41	4	0	0	0	4	3	0	α_4
M	67	77	18	0	26	23	23	6	4	β
TOTAL .	278	358	61	19	38	25	191	17	16	

Percentages.

Section M . .	47%	53%	100%	0%	36%	32%	32%
Remainder . .	43%	57%	70%	30%	14%	1%	85%
Whole MS . .	44%	56%	76%	24%	15%	10%	75%

A comparison with certain dated inscriptions.

Period.	Medial *e*.		Medial *o*.			References.
	·e	è	·o·	ò·	ō	
xth century . . .	70	30	58	12	30	Sarāhaṇ Inscription.
xith century . . .	57	43	10	20	70	Chambā Inscriptions, No. 25.
?	44	56	13	10	77	Bakhshālī Manuscript.
xiiith century . . .	0	100	0	10	90	Ant. Chambā pp. 63, 64.

Ligatures

The ligatures can usually be easily analysed into their constituent elements and only occasionally give difficulty. There are about 90 different digraphs and some 30 different trigraphs.* Some of these occur very frequently, while of others (*e.g.*, *ggh-*, *ṅch-*, *ṭp-*, *ṭph-*, *shṅ-*) there are only isolated examples.

The consonants, as elements of ligatures, may be classed as (i) those in which the mātṛikā is considerably modified and (ii) those which retain their original shape. Of the former are *k*, *ñ*, *th*, *y*, and in certain cases *r*.

As the first or middle element of a ligature the letter *k* reverts to the form pertaining to the Brāhmī script (Table iv, 1 and 5), but as the last element it retains its normal shape, *e.g.*, *lka* (iv 3). The letter *ñ* occurs only in ligatures and, except as a subscript to *j*, only in isolated examples (*ñch-* iv, 1, *shñ-* iv. 4). As the final consonant its form has little resemblance to the mātṛikā.

The letter *th* as the middle or bottom element is also very different in shape from the mātṛikā. In our manuscript it always takes the form of a spiral and there is no example of the S shaped *-th* (See page 91).

In *tthya* (table iv. 5) the spiral shape is lost but see the Suṅgal example given on page 91.

As a final consonant *y* is also disguised in form. See *kya*, *khya*, *gya*, *chya*, *jya*, etc., in table iv,[1]. Generally the final *y* is of the type shown in *khya* but occasionally the bottom curve is more prolonged upwards on the right as in the *gya* shown in the table.

As a first element *r* is often shortened and sometimes loses its bottom serif, *e.g.*, *rṇa*, *rtha*, *rya* (table iv. 3) ; and in *rva* it is marked only by a small excrescence on the left curve of the *v* mātṛikā. As the final element it is changed altogether and is something like the virāma but placed at the bottom of the ligature. Nearly all the examples in our text have the stroke continued left and right (table iv) ; but as a middle element the *r* consists of a stroke slanting upwards on the left of the vertical only, *e.g.*, *tryū* and *śrya* in table iv. 5. In *brū* it is treated as a middle element and also loses the right half portion.

* *i.e.*, counting consonants only, but when the vowel signs form part of the vertical scheme they are essential elements also. See for example, *trvū* and *dbrū* in table iv, 5. Similarly the circumflex *o*, although detached, is an element of the vertical scheme. The ligatures are really examples of vertical writing while medial *i* and the left-handed ·*e* are examples of right-to-left writing.

In *shṭ-* and *shṭh* the original forms of the *ṭ* and *ṭh* are, more or less, preserved, but a "streamer" is added to the right of *sh* (iv. 4). The word *ashta* "Eight" occurs some thirty or forty times but the form of the conjunct is not consistent. There are, at least, three types : (1) that shown in table ii, 3 as *shṭ-*, (2) that shown as *shth-* in the same table, and (3) that shown in table iii, 9. The last type occurs some half dozen times in the "M" section only.

The top bar of the second element (or third) of a ligature is often omitted. See *tta, nta, pta, ptra;* sometimes it is given in a shortened form (*śla* iv. 4) ; and sometimes in its full extent (*mbha* iv. 3).

Numerals.

The numeral figures are shown in table iv. 7.* There is some resemblance between certain of these symbols and certain letters, *e.g.*, the 2 is not unlike the *-ñ* in *jña,* the 4 resembles the *k* but instead of the top bar has a loop to the left, the 5 may be said to resemble the *p*, the 6 the final *m*, and the 8 is not unlike the *ṅ-* and is also rather remotely like the *h*. But this resemblance to letters has probably only a fictitious value—the letter-numeral theory having succumbed.

There are two types of symbol for unity—the semi-circular curve and the almost horizontal line ; but generally these have separate functions. The curved symbol is ordinarily used while the straighter symbol is used in fractions. The other differences shown in table iv. 7 are a matter of style in writing. The examples in the first line are all taken from the "M" section, while those in the second line are taken from other sections. The example of 2 in the second line only occurs occasionally. The chief differences in form are shown in the cases of the 6, 5 and 9. The "six" indeed is a fairly safe criterion in differentiating between the work of the scribes. The symbol for zero besides being used as such is also occasionally employed as a sort of symbol for an unknown quantity.

The symbol for "minus" is a small cross placed after the numeral. Hoernle attempted to connect this sign with the Brāhmī *ka* and to show that its use was an indication of considerable antiquity. But he was hardly successful and the details of his arguments in this matter do not now call for any remark. See §§ 61 and 127, Part I. No other signs of operation are employed in connexion with the numerical notation.

Punctuation.

Besides the visarga the single bar | and double bar || are used, and the end of a sūtra is marked by  The virāma appears to be used to mark a pause or the end of a verse. On folio 5 recto two punctuation signs, which are possibly cancellation marks, are employed.

Cancellation.

On several occasions letters and words are cancelled by marks like accents placed above the letters (See folios 8 and 9).

---

* The figures given in the *Indian Antiquary,* XVII, 1888, p. 36, and in Bühler's table IX are hardly representative.

TABLE ii. Consonants

1. Gutturals

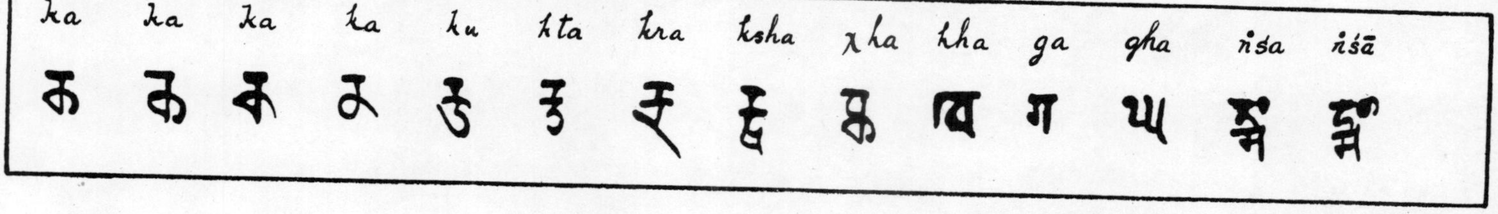

2. Palatals

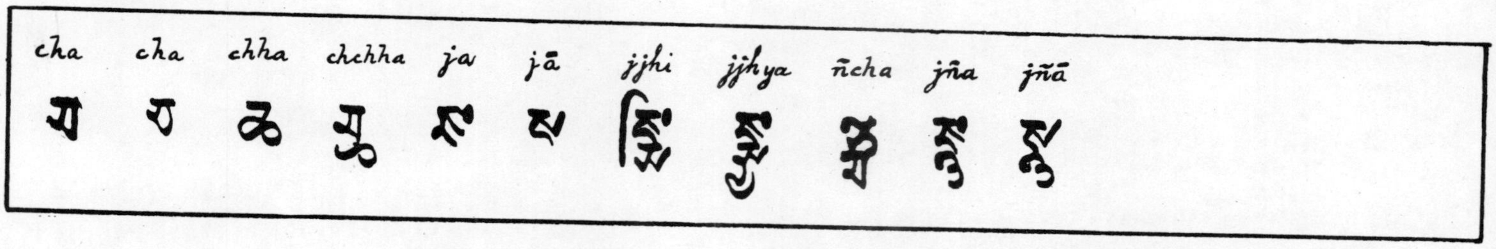

3. Cerebrals

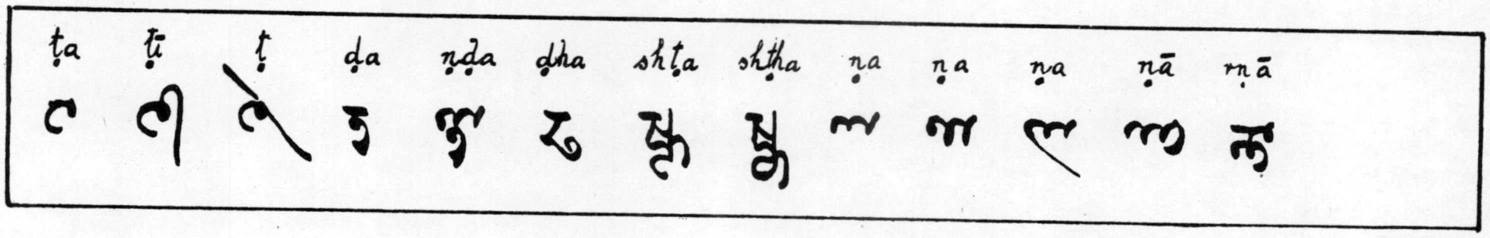

4. Dentals

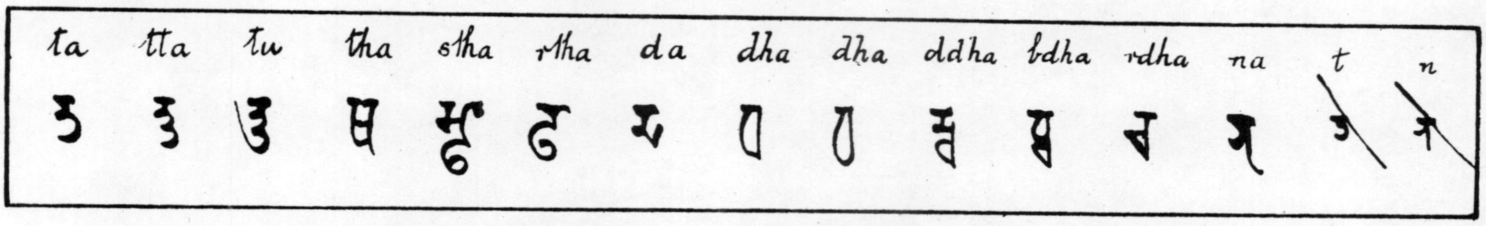

5. Labials

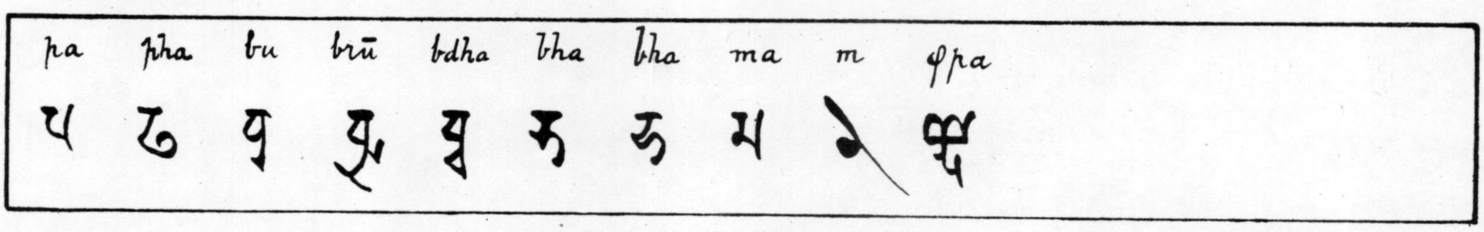

6. Semi-vowels and sibilants

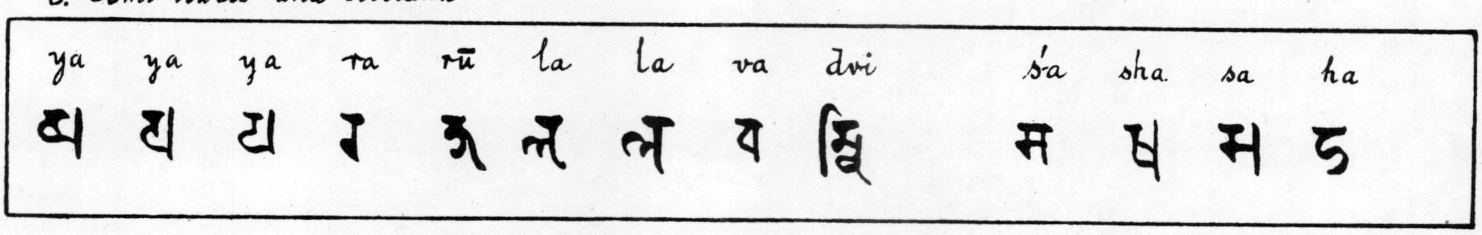

TABLE iii Vowels

1. Initial a and ā, and medial ā

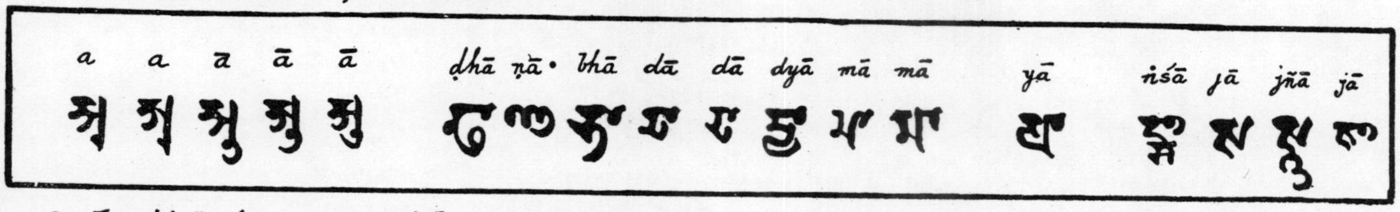

2. Initial i and medial i and ī

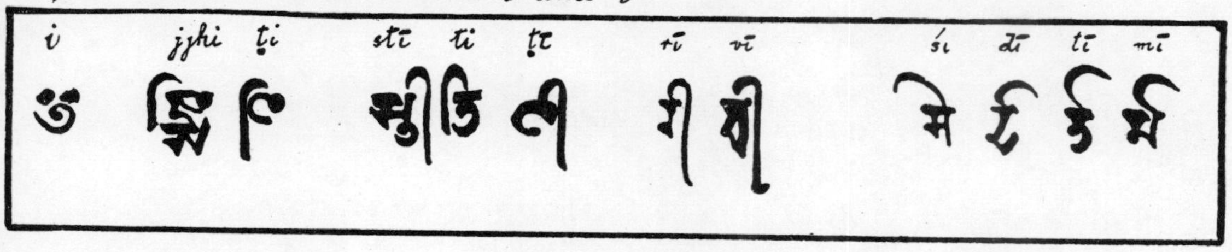

3. Initial and medial u

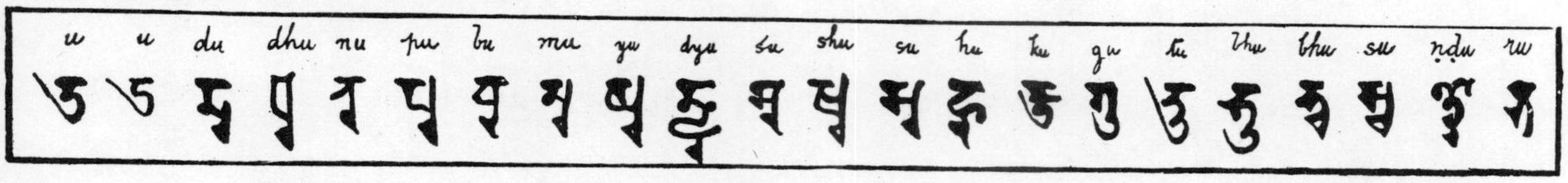

4. Initial and medial ū

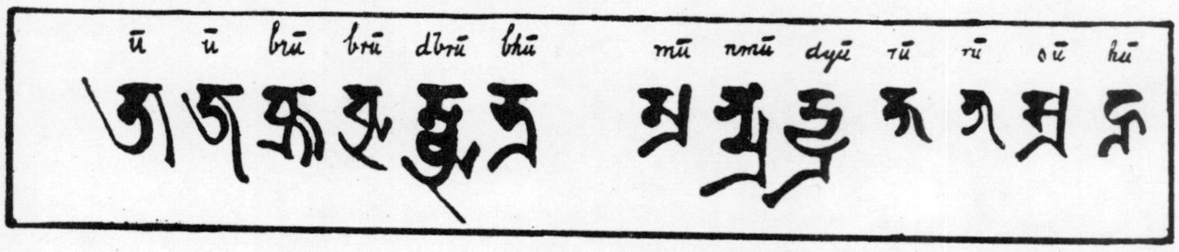

5. Initial and medial ṛi

6. Initial and medial e

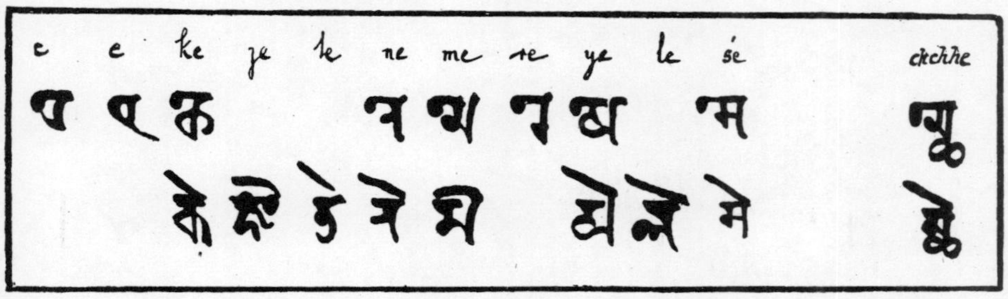

7. Medial ai

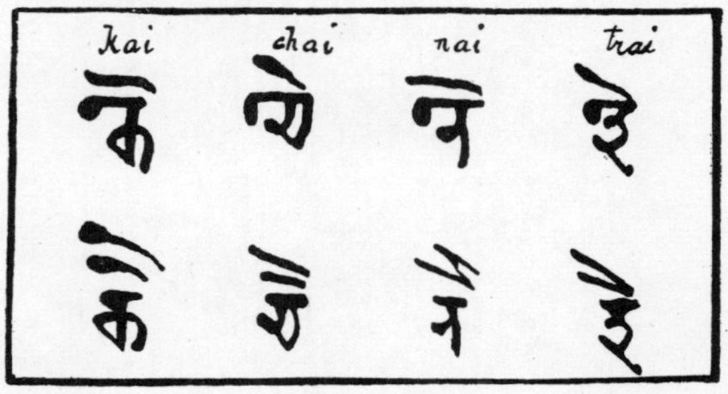

8. Medial o

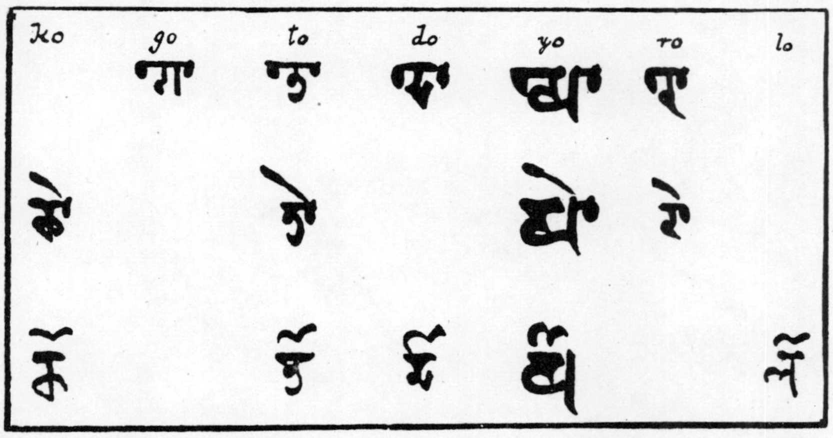

9. Medial au

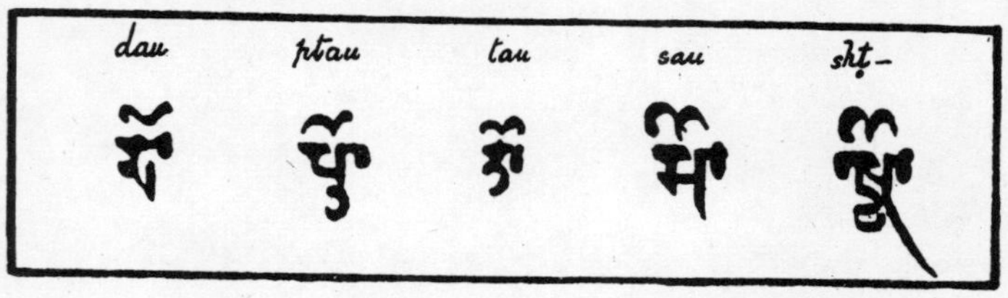

TABLE IV. Ligatures, &c.

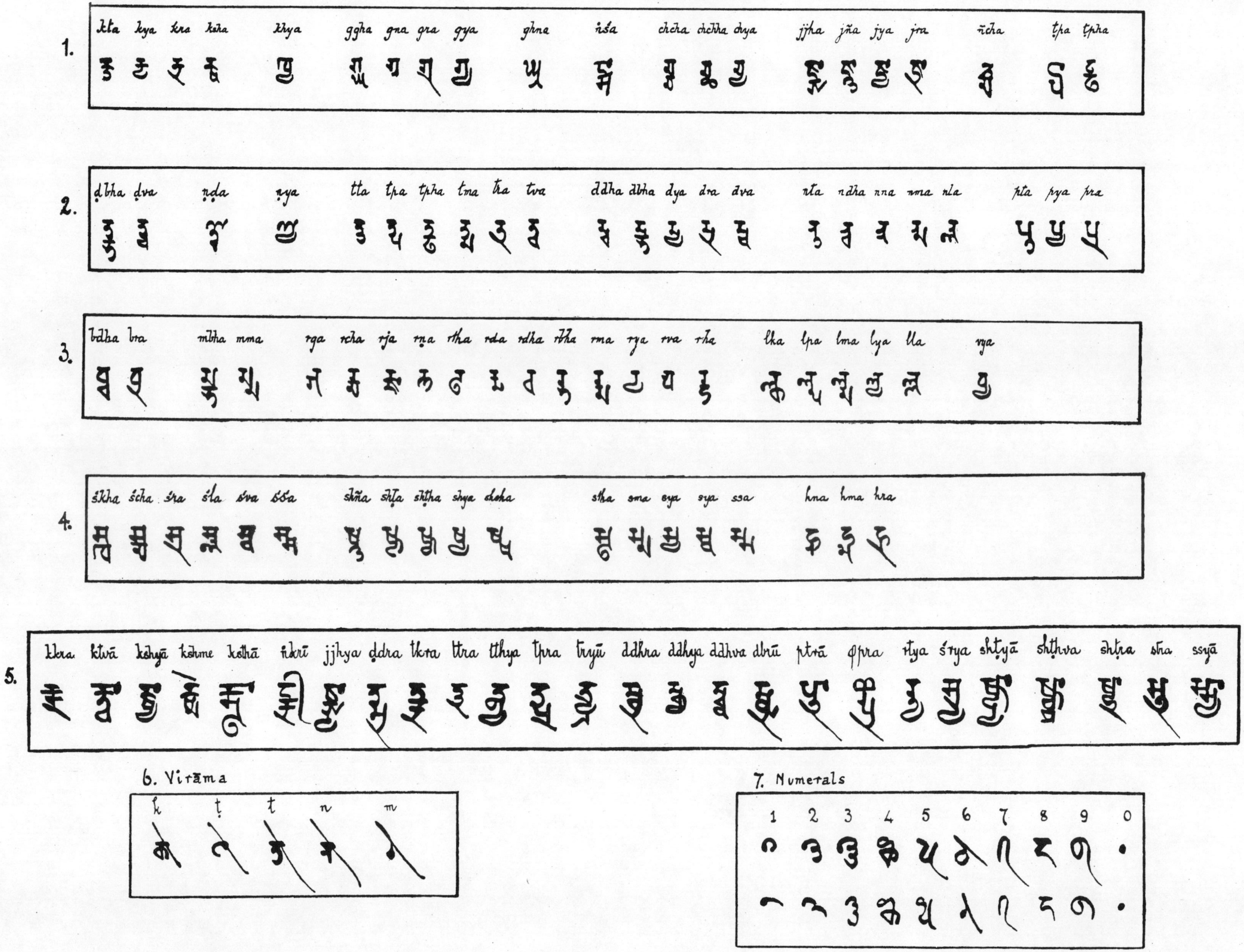

PART II.

ii.—*Transliteration of the Text.*

P

NOTE.

The transcription is printed so as to represent, as closely as it is possible with Roman type, the original manuscript. Italics denote that the reading is possibly doubtful, but the words and letters italicised have some basis in the original and are not merely conjectural. The dotted lines are similar restorations. Portions of the birch-bark that are left blank in the original are marked by asterisks. These gaps are often occupied by knots in the wood or by similar faults that make the surface unsuitable for writing upon.

The system of transliteration followed is that prescribed by the Director General of Archæology, except for the *jihvāmūlīya* which is represented by χ, and the *upadhmānīya*, represented by ϕ.

The order of the folios follows the Bodleian Library arrangement.

TEXT.

bhyāsa | 1 | rū . . . . [1r.]

turguṇaṁpaṁchaguṇaṁhastagatamdhana*ṁ* | ja . . . . . .

paṁchaguṇaṁ **25** || navamasūtraṁ **9** || || sūtraṁ || guṇaupṛi

thagrūpayutauyāchanāyuktisaṁguṇauḥ || guṇaneṇaguṇe . ai

rūpahinenabhājitau | viparītayāchanākshiptauguṇasāsyorayaṁvidhiḥ || . .

evaṁsūtraṁ || dvitīyapatrevivaritāsti || daśamasūtraṁ **10** || ||

|| sūtraṁ || aṁśāṁviśoddhyachchhedebhyakuryātatparivartanaṁ || . .

sāsyaṁtataprojjhyadhanānviśavinirdiśet || udā | paṁchānāmvaṇijāma*dhye*

*vikrī*yatekilaḥ tatroktāmaṇivikrītāmaṇimūlyaṁkiyadbhavet . . . .

. . . . dam | ardhatṛibhāgapādāṅśaṁpaṁcha*bhā*

. . . . [1v.]

. . . . sya*ṁt*atoprejjhyaḥśadṛi*ś*aṁkṛiyate |

. . . . jātā

120	**90**	**80**	**75**	**72**
60	**60**	**60**	**60**	**60**

tatraprojjhyajātaṁ **120** | . **0** |

. . | eshāṁyogakṛi tejātā **437** atosa . .

. śeshaṁ **377** eśamaṇimūlyaṁ | chaturṇāṁsaṅkasarvasvaṁ || prathamasyasaṅka ar*dhaṁ*

*8*0 | **75** | **72** chaturṇāṁyoga **317** prathamārdheṇaśashṭibhiryutaṁ **377** prath*am* . .

. *pra*thamadhanaṁ | tṛitīyachaturthapaṁchamasyadhanaṁsarvasvaṁ **347** dvitīyā tṛibhāgaṁ **3** .

. . *yu*taṁ **377** eśadvitīyasyadhanaṁbhavati || punaprathamadvitīya chaturthapaṁcha

. . svaṁ **357** tṛitīyasyapādaṁ **20** eshayutaṁ **377** eshatṛitī*yasya* . .

. . . ti || punarapiprathamadvitīyatṛitīyapaṁchas*ya* *362* . . . . .

. . . eshach . . . . ***bhavati*** || . . . .

[2r.] syadhana*m̐*bhavati || athapratha . . . .

tyaṁśashṭiśeshaṁ 377 || athadvitīya | 75 / 72 | sya | 120 / 30 / 80 / 75 / 72 | evaṁ

. . . . ṁdvitīyasyabhavati || athatṛitī ya syakṛi*yate* | *120* / *9*. / *2*. / *7*. / *7*. |

. . 377 tṛitīyasyadhanaṁbhavati || chaturthasyakṛiya te | 120 / 90 / 80 / 15 / 72 |

evaṁ 377 chaturthasyadhanābhavaṁti || paṁchamasyakṛiyate | sthā

panaṁ | 120 / 90 / 80 / 75 / *12* | evaṁpaṁchamasya 377 || eshamaṇimūlyaṁpra

|| udā || || anyonyaviditavibhavaṁ

. . . vaṇik . yaṁ | tṛi . . . . . . dalaṁ*ta*thā ||

[2v.]* . . . . ṛi . . . .

. . . . aṁśāṁ | 7+ / 12 / 12 | 3+ / 12 / 12 | 5+ / 6 / 6 | vi . .

. . . . . *ś*oddhyavis*o*dhayet riṇaṁsthitaṁ | e sha . .

. yate | 19 / 12 | 7 / 4 | 11 / 6 | kuryātatparivartanaṁ | 12 / 19 | 4 / 7 | 6 / 11 | *ch*chh*ed*e

. . ṁjātaṁ | asya | 924 / 1463 | 836 / 1463 | 798 / 1463 | projjhyajātā | 924 |

. . . . e shāṁyutiṁkṛiyate | .

. kṛit . . jātā | 2558 | chchhedaprojjhyaṁ 1095 etanmaṇimūlyaṁ

* The bottom portion of this page is blank.

[3r.]

. tvāssamadhanājātāprastamūlya*ṁ*taduchyat*āṁ* | 4/1 . | 5/1 go | 6/1 śā | e

. prastamūlyaṁ 2 | 3 | 6 | dattaissamadhanā jātā 17 | 17 | 17

śamasūtraṁ 13 || ✣ || sūtraṁ ||| ekayutānāsaṁkhyādvi

. . . *hīnā*cha || evaṁtāvatkāryaṁyāvatpurushaisamābhavati

. *ptam* . *pa*trebhilikhitasthitachaturdasamasūtraṁ 14 ||| ✣

. *traṁ* | *ga*tisyaivaviśeshaṁchavibhaktaṁpūrvagaṁtunāḥ tenaivakālaṁbhavatis*tha*

. . . . kenatu || udā || addhyardhayojanagateśatahaṁtasa . . .

*ya*syahayānnavaḥ ūshṭrādaśatṛitī*ya* . . . .

[3v.]

pradattaṁchaparasparaṁ | pṛithagdhanaṁtuvaṇijāṁmūlyaṁvāprāṇināṁpṛithakyad . . . . .

. vaktuṁtatomechchhindhisaṁśayaḥ | 7 a | 9/1 ha | ū | 10/1 | vaṇijjakā 3 d . ya .

. ṇikpiṇḍahataṁ | piṇḍa 7 | 9 | 10 | deyaṁ 3 śuddhaśeshaṁ 4 | 6 | 7 *tata*

*pa*rasparakṛitaṁguṇitajātaṁ 168 | 168 | 168 | svaśesheṇatuvi*bhak*

| 168/4 | 168/6 | 168/7 | labdhaṁ 42 | 28 | 24 | eshapratyaikamūlyaṁekaikasya

*gu*ṇitājātāni aśvaihayai ūshṭrebhyaḥ 294 | 252 | 240 | ekai*ka*

. . . . . . ṁjātā 262 | 262 | 262 | etessamadhanāj*ā*

[4r.] . t . . y . . . ch . . . .

yojanapaṁchakaṁ | saptadinānitasyaivagatasya | paratadvitīyanavayojan*aika gatake*

. . tām $\frac{1}{1}$ di $\frac{5}{1}$ yo ‖ dina $\frac{7}{1}$ gatasya | gatayojana $\frac{35}{1}$ ‖ dvi $\frac{1}{1}$ di $\frac{9}{1}$ yo ‖ **gatisyaiva viśeshaṁchaka**

. . yate | gati 5 | 9 | viśeshaṁ | 4 | vibhaktaṁ | $\frac{1}{4}$ | **pūrvagata 35 eshapādergunitaṁ** | 3

. . bhirdinaisamagatībhavantinavayojanaṁ ‖ pratyayatrairāśikena

$\frac{1}{1}$ di	$\frac{5}{1}$ yo	$\frac{35}{4}$		
$\frac{1}{1}$ di	$\frac{9}{1}$ yo	$\frac{35}{4}$	pha	31

. udā ‖ ashṭādaśayojanā ekenadineyāti | tasyāshṭadi

gatasya | **dvitīyapaṁchaviṁśeyojanādineyāti | kenakāle**

sāsyatāṁ ‖ evaṁekādaśamapattrebhilikhitapūrvepi ‖ **paṁchadaśama sūtraṁ 15**

. . . . **yorviśeshakartavyaṁ uttarasyaviśeshataḥ vibhaktamsuttare** ❖

. . . . īṇasāsyaṁg . . i . .

[4v.] . . . .

. . taraṁ | 2 | vibhaktaṁ | $\frac{1}{2}$ | ādiśesha $\frac{2}{1}$ | jātā | $\frac{1}{1}$ | dviguṇaṁ | 2 | rūpasaṁyutaṁ | 3 | esha

. . . **saṁkalitepratyayāpadaṁhīnā ubhayesthāpitavyā rūpoṇākaraṇephalaṁ** | 21. 217

. . . . . **kiṁprabhūtepilikhite** ‖ shoḍaśamasūtraṁ 17 ‖

sūtrebhrāntimasti

. . . . *ā*dyorviśeshadviguṇaṁchaya*s*śuddhirvibhājitaṁ ❖ **rūpā . i**

. . . . . *g*atisāśyaṁtadābhavet ‖ ‖ udā ‖ dvayāditṛi chayaschaivadvi . .

. dikottaraḥ dvayochabhavatepaṁthākenakālenasāsyatāṁ ‖ sthāpanaṁkṛiyate | e . .

$\frac{3}{1}$ ‖ pa $\frac{0}{1}$ ‖ dvi ‖ ā $\frac{3}{1}$ ‖ u $\frac{2}{1}$ ‖ pa $\frac{0}{1}$ ‖ karaṇaṁ | **ādyorviśesha** . .

. . . . i . . . . .

ā 5 / 1	u 6 / 1	pa 0 / 1	dha 0 / 1
10	u 3 / 1	pa 0 / 1	dha 0 / 1

karaṇaṁ | ādyorviśesha*ṁ*ādi . . [5r.]

chayaśuddhichayaṁ 6 | 3 | śuddhi | 3 | ādiśesha 5

dviguṇam 10 uttaraviśesha 3 vibhaktaṁ | 10 / 3 | sarūpaṁ | 13 / 3 | anenak*ā* . .

samadhanābhavanti // pratyayaṁ rūpoṇākaraneṇa phalaṁ | 65 / dvi 65 |

ashṭādaśamasūtraṁ 18 || dinagamana mādirahitaṁ . .

ṇaṁtachchhottareṇa ✣ saṁyutaṁ | pratinihita ātmaguṇaṁjñ*eyaṁ*

. . . . . o . i | ashṭhottaraguṇitekshepasaṁjñakodatvāmūlaṁprati . i

. . . . hata*ṁ* | 30 | [5v.]

. . . . dinagamanamādirahitaṁdinagamanayojana ɸ pañcha | 5 | ādi .

. . . . 3 | rahitaṁjātaṁ | 2 | dviguṇaṁ | 4 | tachchottareṇasaṁyutaṁ | 8 |

. . .*ātma*guṇaṁ | 64 | eśakshepasaṁjñakorāsi | ashṭottarasaṁgu . i . .

. . *lab*dharāshi | 30 | ashṭaguṇaṁ | 240 | uttareṇaguṇaṁ uttaraṁ | 4 | .

. . . . guṇitaṁjātaṁ | 960 | kshepasaṁjñakodatvā | tatrakshepasaṁjñ . .

4 | yutaṁjātaṁ | 1024 | asyamūlaṁ | 32 | pratinahi*ta* . . . .

i . m | *8* | y*u*taṁjātaṁ | 40 | u . . . . ṁ . . . .

śikepratyaya	1 / 1	5 / 1	5 / 1

phalaṁy . . . . [6r.]

. nenassaha 55 eśasamābdhānaṁ || udā || ādipaṁcha uttaraṁtr

ṇinaroyojanagamyate | dvitīyapratidinaṁssaptagatasya .

napaṁchakaṁ | kenakālenasamatāṁkatthyatāṁgaṇakotta . .

. . 5 / . . 1	u 3 / 1	pa 0 / 1	pratigati 7	

dina | 5 | paṁchadinaga . .

yojanikaṁyojana | 35 | karaṇaṁ | dinagamanamādira*hi*

tatradinagamanaṁ | 7 | ādirahitaṁ | ādi | 5 | rahitaṁ . . .

[6v.] . . . . 3 anenaguṇitaṁjātaṁ 84 . . .

. saṁjñakodatvā | tatrakshheparāshi | 49 datvājātaṁ 889 .

. . . . dānadadātisamaṁ | karaṇokriyate | sūtraṁ || akṛiteśli . .

. . . . chhedodvisaṁguṇaḥ tadvargaḥ dala | saṁślishṭha | hṛiti .

. . . yaḥ anenasūtreṇaslishṭhamūlamānayasvamatimā . . .

. *telabdhaṁ* mūlaṁ 29 pratinihitaṁ 7 anenayutaṁ 36/48 .

. . . 2136 48/58 dviguṇottarabhājitaṁ | ta . o . .

. . . . . . e . . . .

[7r.] . . . .

7 dalitā 16 cha . . 7/58 . . . .

9 sāsyeyu 60/33 li taṁ 737 padaghnā | ṭatrapadaṁ 178/29

. anenaguṇitaṁ 60/6 vi jātaṁ 58 65593/841 *slish* . . . .

. tyaśeshaṁkṛiyate | 60/śe 6/29 65569/841 bhā gehṛite 7 . . .

. . . . pratyayaṁtrai rā śikena 1/1 7/1 yo 17 . .

phalaṁyojana 42 śe 28/29 niyataṁtena

77 e . o . viṁśatima *sūtraṁ* 19

[7v.] . . . . 1 . . . .

. . matāndanaḥ ā 3/1 u 4/1 pa 0/1 nityadatta 7/1 ādiṁviśoddhya . .

3 niyataṁ 7 viśeddhya 4 uttarārdhenabhājitaṁ | uttaraṁ 4 a . .

. . . . . jitaṁ 4/2 jātaṁ 2 labdhaṁsarūpa | esharūpādhikaṁ 3 eśakāla

. . . . 3/1 u 4/1 pa 3/1 rūponākaraṇenaphalaṁrū 21 || dvitīyasyatrairā .

1/1 di 7/1 3/1 di pha rū 21 || eshassamadhanājātā || *udā* ||

. . . . . ka uttaradvayaṁdvitīyapaṁchapratyahaṁ | kenakālenasama*tām* .

. . . ṇakottama ā 1/1 u 2/1 pa 0/1 niyatanityaṁ 5/1 ādiṁviś*oddh*

|| sūtraṁ dviguṇaṁprabhavaṁśuddhādviguṇā daya . [8r.]

ttareṇabhajechchheshaṁlabdhaṁrūpaṁvinirdiset || udā || va .

bhṛitakaχkaśchitatraikodashamāśakaṁ | pratyahaṁkurutetatrakarmaṁbha . i

mānavaḥ dvitīyaṁkṛiyatekarmaṁdvyāditṛitayaruttaraṁ | padaṁtatra .

. kenakālenasāsyatāṁ | ā 2/1 | u 3/1 | pa 0/1 | prati 10 | . .

dviguṇamprabhavaṁśuddhā | prabhavaṁ | 2 | dviguṇaṁ | 4 | niyatapunadvi .

. . . . o . . taṁ | 16 | (uttarārdheṇabhājayet)* uttaraṁ . .

*The portion within the brackets is deleted in the MS.

gantata | ekā —śchagamanajñeya yutāssaṁguṇya . . . . . . [8v.]

|| niyorathośvairdaśabhiryujyatehayapaṁchakaṁ | gaṁtavyaṁyojana . . . .

. . ośvakimudbhavet | ha 10/1 | hayalagnarathasya 5/1 | gantavyoyojana . . . . .

. . . yorvibhajyagantavyaṁ | tatrahayā | 10 | gantavyaṁyo | 100 | atobhāga . . .

. . bdha | 10 | tatrayuktāśva | 5 | etaissaṁguṇyapariyogajātaṁ . . .

. . . yojanānyaikośvarūḍha | pratyayaḥ paṁchabhiśśatasaṁguṇyajātaṁ . .

. . . kṛiyate || yadida . . shayojanā ϕ paṁchas . . . .

. . . . r . shṭh . . ch . . . . udā || [9r.]

. . . . . māptaṁdvijanmabhi | tatpunastesamaṁbhaktvāda .

. ssamāptavān saṁkhyāyaχ katim*māchakshukativiprāχkati . r . . ṁ

. 1/1 | u 1/1 | pa 0/1 | labdhaṁ 10/1 | karaṇaṁ || labdhaṁdviguṇitaṁkṛitvātatra

bdhaṁ | 10 | dviguṇaṁ | 20 | tathādyūnaṁ | 18/1 |

. . . . vibhājitaṁ | atrottaraṁ | 1 | anenabhaktvājātaṁtade

. . kaṁ | 19/1 | ayamprashṇāprāhūṇā ekonaviṁśati | sthāpa

. . *pratyayaṁ* | ā 1/1 | u 1/1 | pa 19/1 | rūpoṇākaraṇenaphalaṁ | 1

*The first m is deleted in the MS.

[9v.] *yojana* . . . . . . *serathagama* . . . .

. yo 6/1 ‖ śayo 1/1 ‖ yo 70 gantavyaṁ ‖ adhvasaṁ*yoga* | 7/1 | *bhaktaṁ* | *1*/7 |

. enaguṇitājātānlabdha | 10 | dviguṇaṁ | 20 | eshālpasyaḥ ‖ atha .

. nā ayaṁkālojñeyaḥ anenakālenashshaṭ yojanānigantavyaṁ | ā

. . yamekayojanikasyasamāgamobhavati ‖ tadyathātrairāśi . e . . .

. . . yadyekasyashaṭ yojanātadāviṁśānāṁkiṁ | 1/1 | 6/1 yo | 2/1 . . . .

. . . saptatiśoddhyaśesha atrassaptati | 70 | āgatapaṁchāśa | 50 | *adhva*

. . . . . . i . . ṁ ‖ 1/1 di | 1/1 yo ‖ 20/*1* di ‖ pha yo . .

[10r.] sūtraṁ 24 . . . .

. . *dhānta* ⁜ saṁguṇa . . . . pravṛit*tirgu* . . . .

vinirdiset ‖ udā ‖ tṛibhāgamalada*gdh*asyatridhāntasyaiva . .

*ashṭ*ottaraśatānidattaṁkiṁśeshaṁvadapaṇḍita ‖ 108/1 1/1/3+ 1/1/3+ 1/1/3+ ‖

kṛitvārūpakshyaṁpārthajātā | 32 | śesha ‖ pra

thamabdhāntekshayaṁ | 36 | śeshaṁ | 72 | dvitīyabdhāntekshayaṁ | 24 | śe*sham*

tṛitīyabdhāntekshayaṁ | 16 | śeshaṁ | 32 | pratyayaṁkṛiyate | sthāpanaṁ

0/1 1/1/3+ 1/1/3+ 1/1/3+ | bhā śeshaṁ 32/1 | phalaṁmūlā 108 ‖ atha .

majātikṛiyā*karaṇ*aṁ . i . .

. . . . *pratyayaṁ* [10v.]

. . . . ṭṛibhitryashṭabhāgasaṁyutaṁ .

. . . . tadā | 3+ 1 3+ | shṭottaraśatā χ kiṁ | 27 8 | 1 1 | 108 1 | pha śe .

. . . . yadyekasyatrayastraya ashṭha

bhāgātadādvātṛiṅśānāṁkimiti | 1 1 | 3 3 8 | 32 | phalaṁ 108 1 | udā ||

. . kṛiddhāntasyalohasyadaśāṁshākshīya | testrayaṁ | saptatedviguṇa

. . śchakiṁśeshaṁvadapaṇḍitaḥ | 3 10 | 140 1 | kṛitvārūpakshayaṁpāsthamiti

. . rūpaṁ | 1 | 3 10 | kshayaṁkṛitvā jātaṁśesha | 7 10 | mūlaṁ

40 | anenaguṇitaṁjātaṁ | 98 | kshayaṁ | 42 | evaṁ | 140 | *pra* . .

. . . . . na | 7 . . | 1 1 | 98 . | phalaṁ 140 | . . . .

. . pratyayaḥ | 0 1 1 3+ 10 | . . . . [11r.]

palākrīte pa latṛibhāgaṁkshayavrajati | ashṭādaśa . . . .

. . th*a*tāṁbrū hi || | 1 3 3 2 | 18 1 bhā | karaṇaṁ | addhyardhapalaṁśchhe .

idaṁ | 2 9 | kṛitvārūpaksha yaṁrūpaṁ | 1 | kshayaṁkṛitvājātam | 7 9 | 18 1 |

guṇitaṁjātaṁ | 14 | kshayaṁ | 4 | pratyayatrairāśikena || addhyardhapala

krītetṛibhāgaṁkshayagachchhati | ashṭādaśapalakrītākiṁkshayaṁvada

ṇḍita | 1 1 2 | 1 3 | 18 1 | phalaṁ 4 1 | punatṛibhāgādivardhaṁtadāchatu .

kimiti | 1 3 | 1 1 2 | 4 1 | phalaṁ 18 1 | || udā || chaturbhāgamala . . . .

rṇaśatapaṁ . . . . . .

[11v.] . . . . athapratyay . .

. . . 1 | 4 1 4+ | phalaṁmūlā 500 || punarevaprastārakra

5 maṁ | 500 1 | 1 1 4+ | 1 1 4+ | 1 1 4+ | 1 1 4+ | phalaṁ

. . . 1 tediti

4+ 64 shaṁsthāpya

1

4+ śesha 158 to 1 śe | 1 64 | anyaṁchaturthapratyayaṁkṛiya

1

4+ te | 0 1 | 1 1 4+ | 1 1 4+ | 1 1 4+ | 1 1 4+ | bhāśesha 158 1 5 1 64 | phalaṁ 500 ||

*ādyaṁkshaya*ṁ 125 śeshaṁ 375 || dvitīye kshayaṁ 93 to

. . . āśa 9 śeshaṁ | 281 1 4 | kshayaṁ | 70 5 16 | śeshaṁ | 210 15 16 | kshayaṁ | 52 47

. . *shaṁ* | 158 | eśasarva trakartavya ||

...

[12r.] . . . . . *tuprasthama* . . . .

. dhunāstathāḥ ambh*asa* . . . . | . . . .

kṛitvārūpakshayaṁpāstamiti . tatrakshayaṁ . pāstamiti . tatrakshayaḥ rū .

nyaśeshaṁ | 3 3 3 3 4 / 4 4 4 4 1 | gadyūtigadyūtigatvātprasthaṁ . . t

gadyūti yojanaṁ | chatuprasthai āḍhakaṁ | tadādhāntaśorguna

. . tataḥ | 81 256 | āvṛittipravṛittirguṇanaṁ*ta** *taḥ || 4 ||anena . .

ṇ . taṁjātaṁ | 81 64 | eśamaddhubhāgābhāgehṛitelabdhaṁ | madhuprastu 1 .

śe | 1 16 | aṁbha bhāgāprasta 2 kuḍava 2 śe | 15 16 | . . . . .

. ktiprakshepake āḍhakāśo*d*ashaku*d*avābhavanti | 16 | atoma . . . . .

. . . . śeshaṁ 12 . . . .

. . . . **16** e . [12v.]

. . . . prasthakuḍavā | 4 | 3 śeshachatvāra .

. . kuḍavaḥ | 2 2 / 1 1 / 4 4 | śeshāchakuḍavāpītā | ma | 7 / 1 / 4+ | 9 / 1 / 4 | punacha*tvār* .

*kuḍavābhu*ktaṁśe sham | 81 / 16 | 175 / 16 | jalabhā gaṁ | madhukuḍava 5

jalakuḍava | 10 / 1 | śe | 15 / 16 | evaṁkuḍava **16** || || udā || datvāśulk*aṁ*

chaturbhāgaṁashṭau āṇītakuṁkumā | chatuśulkaśālaistukiṁśeshaṁvad*a*

. ṇḍ . ta | 8 / 1 | 1 / 1 / 4+ | karaṇaṁ | kṛitvārūpakshayaṁpāstaṁ | pāstaṁ | 8 3 / 1 4 | gu . i

. . . . taṁ | 6 | śulke | 2 | śeshaṁ | 6 / 1 | 1 / 1 / 4+ | ane nagu

. . . . ṁ . kshayaṁ | 1 / 1 / 2 | śeshena | 4 / 1 / *2* | 1 / 1 / 4+ | datvāguṇitajātā | **27** . . .

. . . . . . *ta* [13r.]

. | 1 / 1+ / 4 | 1 / 1+ / 4 | 1 / 1+. / 4 | 1 / 1+ / 4 | guṇitaṁjā*taṁ* | 81 / 32 | || . u . y . | 8 | 4 | 4

phalaṁ | 81 / 32 | phalaṁ || punānyaṁ | 8 / 1 / 1+ / 4 / 1+ / 4 / 1+ / 4 / 1+ / 4 | phalaṁ | 81 / 32 | punapratyayaṁ | 0 / 1 / 1+ / 4 / 1+ / 4 / 1+ / 4 / 1+ / 4 | bhā 32

phalaṁkuṁkuma 8 || udā || tribhāgashaḍbhāga pamchā .

. guḍapiṇḍāshṭabhārakaṁ | kiṁśeshaṁdattabhirbhavet . .

| *8* / *1* | 2 / 3 | 5 / 6 | 4 / 5 | guṇita*ṁ*jātaṁ | 32 / 9 | etatphalaṁ || || u .

chatu ϕ paṁchakalā bhenadaśadroṇātprayojit*a*

. dvaitṛibhistukiṁlābhaṁkatthyatāṁguṇakottama | 10 | 5 . | 5 . | .

. ṁ . . . **125** |

[13v.] . . . . 0 1 1 4 1 4 1 4 phalaṁdro 19 || udā || kasyā

. . . . 4 1 4 1 4 . . . . jjakasya . u . vṛiddhyātṛibhā gena ā 2 4 dro pra 0 4 ā pra ku 2 ku 4 prasthi 10 1 shashṭhisvadale svapādena nakshayaṁgata tatojjhitaṁ | .

ddhyātupaṁchabhāgenastathā vṛiddhidvayogataṁ | kāvṛiddhi .

. . . . . . syākiṁvāśeshaṁtaduchyatāṁ 60 1 | 1 1 2+ | 1 1 3 | 1 1 4+ | 1 1 5 rūpalā .

. . . tā 36 || pratyayaṁpu nāsyaiva

1 1+ 2 | 1 1 3 | 1 1 4+ | 1 1 5 bhā 36 1 phalaṁ 60 || punānyaṁpratyayaṁ 60 1 1+ 2 phalaṁ . .

. . . . syevamūlaṁnajñāyate || 0 1 1 2+ 1 3 1 4+ 1 5 pha lam

[14r.] *lat* . dhan . | yasyatanmayatāchakshu . . y . . . . . d . . . . .

. pahṛitaśulkapiṇḍaṁ 24 || karaṇaṁ || kṛitvārūpāksha . . 3 . . .

. pāsta 2 3 | 3 4 | 4 5 jātusaṁguṇyajātam 2 5 etāvadapirūpasaṁśu

. ājātaṁ 3 5 anenabhaktvāśulkapiṇḍaṁguṇitaṁjātaṁ 40

eśapiṇḍaṁpratyayaṁ 2 40 5 1 guṇitajātaṁ 16 śeshaṁ 24

. ṁ 40 | anyaṁa syapratyayaṁ 40 1 1 3+ 1+ 4 1 5+ phalaṁ 16 kshayaṁ 2 . .

. vaṁ 40 || || udā || guḍapiṇḍa jñātatulyośchatvari . .

. avyeguḍaṁ | tṛichatu ϕ paṁchashaḍvṛiddhyā chatvāriṁśavekshaya .

. . . . e . . . . . e . . .

. . . . | 3 | u . . . . [14v.]

. . . udā || ajñā*tara*mbhal*o*hasyatṛi chatuϕ paṁchakākshaye | sapt .

viṁśatipiṇḍasyatṛidhāntaśeshyadṛishyate | kiṁsarvaṁvadatatvajñakshayaṁcha .

makatthyatāṁ | 1/3 | 1/4 | 1/5 | śe 27/1 | karaṇaṁ | kṛitvārūpakshayaṁpāstha | 2/3 | 3/4 | 4/5 |

guṇitaṁjātaṁ | 2/5 | rūpakshayaṁ | 3/5 | anenaśeshaṁbhaktaṁśeshaṁ | 27 | bha

. ṁ jātaṁ 45 asyasaptāviṅśa | patyaśeshaṁ 18 || etakshayaṁ || udā ||

rikshīnasyalohasyatṛidhāntaṁpaṁchamāśakaṁ | najñāyatetpravṛittikāṁ*na*

. . shapradṛiśyate | pravṛittiśeshaṁyopiṇḍaṁkevalaṁviṁśatisthitaṁ | a .

. . . ṁpravṛittīsyākiṁvāśeshaṁvadaśvame | 1/*3* | 1/4 | 1/5 | kṛit*vā* . . .

. . . . *kshayasyacha* . r . . . . . . *śaḥ* . . . .

[15r.]†

yaṁkaśchiyadiśakyastaduchyatāṁ || etanmesaṁśayaṁprājñaddhdhāntaksh*ayaṁ*

vichāraṇāḥ | 2/3 | 3/4 | 4/5 | kshaśe 32/1 | karaṇaṁ || bdhāntasaṁguṇyagu*ṇi*

. jātaṁ | 3/5 | rūpaṁdadyā | 8/5 | bhāgehṛitelabdhaṁbhaktve*ti*

. jātā | 5/8 | 32/1 | phalaṁ 20 eśasāpravṛitti | śeshaṁ 12 . .

. . 32 || paṁchaviṁśatima sūtraṁ || 25

† The bottom portion of 15 recto is blank.

. . . . y . . pravṛittibhav*etsakhe* ||

[15v.]

. *3* | . *3* | 1/3 | 1/3 | śe 16/1 | karaṇaṁ || dhāntaśoghātitaṁtena | rūpa

kshayaṁkṛitvājātaṁ | 2/3 | 2/3 | 2/3 | 2/3 | guṇitaṁ | 16/81 | bha.

| 81/16 | śesheṇaguṇaye | śeshaṁ | 16/1 | guṇitajātā | *8* .

. . pravṛittirityarthaḥ || athānyavidhikalāsavarṇe | chaturdhānta

. . lohasya ekāśītiśchadattavān kiṁśeshaṁvadadharmajñaya .

. . ṇitekṛitaṁśramaṁ | 81/1 | 1/1/3+ | 1/1/3+ | 1/1/3+ | 1/1+/3 | phalaṁśe 16 || pu .

. . . yaṁkṛiyate || mūlaṁnajñāyate

. . 1/1+/*3* | 1/1+/3 | 1/1+/3 | 1/1+/3 | bhāśe 16/1 | phalaṁlohapala 81 ||

[16r.] *vi*bhakt*aṁ*jātaṁ | 2/9 śe 10/1 | dh . . . . | . . . . . .

| 9/7 | anena guṇitaṁjātaṁ | 90/7 | | 2 bhāgehṛitelabdhaṁ 12 ||

. asyapratyayatrairāśike | na | 6 | 1/1/2 | 10/1 | pha 12/6 ||

|| udā || mākshikagghaṭakasyaivadvitṛibhāgapravardhi | 7 | . .

. īy*e*dvipaṁchamobhāgotṛitīyedvisaptako*dbhavaṁchaturthe* . .

. vaṁbhāgaṁevaṁjātapalatrayaṁ | babhūvāsaulkikaihṛitvā . .

. rvaṁvadapaṇḍita | 2/3 | 2/5 | 2/7 | 2/9 | śe 3/1 | dhāntaśo*ghā* .

. ātimiti | kṛitvā . . . . .

[16v.] . . . . *sūtraṁ* || idānisuvarṇakshayaṁvakshyā

. yasyeda*ṁ*sūtraṁ || kshayaṁ —⁘— saṁguṇyakanakāstadyutirbhāj*a*

yettataḥ saṁyutairevakanakairekaikasyakshayohisaḥ || udā || *e* .

. . . ssaṁkhyāsuvarṇāmāshakairiṇai | ekadvitṛichatussaṁkh*yā*

* *rahitāsamabhāgatāṁ || sṭāpanaṁkriyate | eshāṁ

. .+ | 2+/2 | 3+/3 | 4+/4 | karaṇaṁ || kshayaṁsaṁguṇyakanakādi . i

. . kshayenasaṁguṇyajātaṁ | 1 | 4 | 9 | 16

. . . ti | eshayuti | 30 | kanakāyuti 10 anenabhaktvā la . .

[17r.] || udā || ekadvitṛi | 1/10/1 | 1/30/1 | ./4/1 | phamāśe 12/1 | . . . . . .

*va*rṇaprojjhitā ime . . . . māśakadvi

tṛitāṁchaivachatu*ϕ*paṁchakarāṁśakaṁkiṁkshayaṁ | 1/1/2 | 2/1/3 | 3/1/4 | 4/1/5 | k .

. . kshayaṁsaṁguṇyakanakā eśasthāpayate |

| 1/2 | 2/3 | 3/4 | 4/5 | stadyutirbhājayetatataḥ harasāsyekṛiteyut*aṁ*

| ·63/·0 | saṁyutai*χ*kanakairbhaktvātadākanaka | 10 | anenabha*ktaṁ*

. . . jātaṁ | 163/600 | eśa ekaikasuvarṇasyakshayaṁ || . . .

. . . śikenakar . . . ||

.	.	.	.	200	. . *srūṇuśvame* | kr*amena*dv*ayam* [17v.]
·0	163	4	pha	163	shādi uttare ekahīnatāṁ | suvarṇaṃ*me* .
1	60	1		150	sammiśryakatthyatāṁgaṇakottama ‖ sthāpan*aṁ*

4+	5+	6+	7+	8+	9+	1+	2+	3+	kshayaṁsaṁguṇya
.	6	7	8	9	10	2	3	4	jātaṁ |

. . . 30 | 42 | 56 | 72 | 90 | 2 | 6 | eśāṁyuti | 330 ‖ kana

. . . ṁyuti 45 | anenabhaktvā labdhaṁ | 330 45 | paṁchadaśabhāgechchhe*da*

| phalaṁ | 7 śe 1 3 | eka ekaikamāśa kakshayaṁ | pratyaya . .

· · · 330 | 1 1 | phalaṁ 22 | evaṁsarveshāṁpra*t* .

. kaḥ kā*ṁ*chanaiyadbhavelabdh*a*sakshayajñātamāśakā ‖ udā ‖ . . . . [18r.]

. shakoprāptodvau . prāptaṁchapaṁchabhi | tryaśchakatibhi ϕprāptāshaḍeva

. nikevalaṁ | chaturbhimāshakairhīṇaṁkaṭidṛishṭḥvāmayāsakhe | trayaścha

katibhi ϕprāptāsuvarṇāṁmaśakovadaḥ

1	2	3	6	karaṇam
2	5	0	4+	

aprāptasaṁguṅākaṭiditi | 6 1 |

. i . tvāra | 4 | saṁguṇyajātaṁ | 24 | kāṁchanānitatojjhitaṁdvābhyāṁ

. . kapaṁchabhidvayaṁsaṁguṇyajātam 2 | 10 1 | tadyuti 12 | hi . ā *2* . .

. . hitvājātaṁśeshaṁ 12 | aprāptagaṇḍikai . . i . . . .

[18v.] . . . . shṭaviṁśatimasūtraṁ

|| sūtraṁ ūnaissaṁguṇyakanakātatpiṇḍaṁchaviśodhayet su

varṇakanakābhyastārāśisheshaṁvibhājayet aprāptagaṇḍi

. śaśuddhenakanakenatu | yallabdhaṁtatpramāṇaṁtugaṇḍikāyā vi . i

. . t || udā || ekadvitṛichatussaṁkhyā aprāptamāśakāni tu

. kadvitṛichatuśsaṁkhyā ekatrāvartitākilaḥ gaṇḍikājñā . .

. . kā ūnaikādaśamāshakai | aprāptajñātakanakaipra . .

. . . . yaḥ	1	2	3	4	0	karaṇaṁ	ū	ai	
	1	.	.	.	.				

[19.] *Folio 19 is blank.*

. . . . [20r.]

. . . . . †. | asyaivaϕprashṇasyā . . rovi

. . 20 rakti 1 | dhā 1 a 1 0 ya 4 | su 1/1 chhe 80/1 raktisurakti 1/1 | pha 4 y . dhā

. 4

. *pā* 1 || punatṛi | tīyāsyaiva | 2 20 / 1 1 / 4 | 1/3 | 1/1

. chhedaṁ 6 dhā | 1/4 | draphadhā 4 ya 1 | 1/2 1/2 chhe 1/2

pā 2 mū . || suvarṇasyamāṇaṁsama | 1/4 chhe 1/4

. dā || sapaṁchanavabhāgāṇidinānitrayodasaḥ

. . . . . . ṇāṁkiṁ || ||

. . . y . | ·4/1 | 5/1 | 1 chhe 7 *chhe* . . . . [2ov.]

pa 12000 | . . . .

. . *araṇaṁ* | sarposhṭādaśahastopraviśatyārdhāṁgulaṁśanava

. . . i ekaviṁśatibhāgaṁmapaharaṁti | pratidinenaḥ kiṁkāle

. . vilaṁsaṁprāpyate | 1/2 1/21+ | 1/1 1/360 | 18/1 chhe 24/1 aṁha

. . . 2 mā 4 di 10 | 1/9

. . . || kīṭa | χ kilārdhāṁgulaṁdivasedivase

. . . . *paṁchāśā* . . . ke . .

. . . . *tribhāga* [21r.]

. . . *dine*tatha | *trirūpapaṁchabhidinai* | . shāṁda

. 1/1 | rū 1/1 | rū 3/1 | dṛishya 100/1 | karaṇaṁ ||

. . | *di* | 5/1 di | kṛitvā | 3/1 | 2/1 | 3 | dṛi 100/1 |

.

. . . . *bhaktaṁ* | 5 | *dṛishyena*

†This leaf consists of six scraps besides the portion marked (*a*).

[21v.](a)

. . . .

. . . vārdhaṁtṛitīyasya jīvalokāt eshāṁdīnār .

. . . . kasyakiṁbhavati | 2 dī 1 2 | 3 dī 1 2 | 4 dī . . |

. . rivartanaṁkṛiyate || | 1 1 di 2 | 1 1 di . | 1 1 *di* 4 |

| 36 10 | dṛi | 500 1 | prakshe

[22r.] . . . . | 47 . 5 947 |

*r*titajātāphalaṁdī 500 || asyapratyayatrai

. śikena

2 dī 1 2	1 di 1 2	100,000 947	phalaṁdi 60,000 947
3 dī 1 2	1 di 1 3	157,500 947 dī	phalaṁdi 60,000 947
4 dī 1 2	1 di 1 4	216,000 947 dī	phalaṁdi 60,0*00* 947

[22v.] . . . . dviguṇaṁdvitīyasyaprathama*ā*

. . . . . | prathamāchaturguṇaṁchaivachaturthechaivadattavān cha

. . . . *aṁ*śatamekaṁdvayānugaṁ || vadasvaprathamedattaṁkiṁpramāṇaṁ

. sya . . || 0 1 || 2 1 || 3 1 || 4 1 || dṛishya 200 1 || śūnyamekayutaṁkṛitvā 1 | 2 | *3*

. kshepayuktyāphalaṁ | 20 | 40 | 60 | 80 | evaṁ 200 || eshāṁ

. . . | ā 20 1 | u 20 1 | pa 4 1 | rūpoṇākaraṇenaphalaṁ 20 .

. . | || sūtraṁ || yadṛich*chhāp*inyaseśū*ṇye* *tadā* . . ṁ

. dāchatṛiguṇa*ṁ*d . . . . [23r.]

*pratha*masyatukiṁbhav*et* | 0 / 1 | tadā | 2 / 1 | *tadā* . . | . | .

yadṛi*ch*chhāvinyaseśūṇye

. . chchhā || 1 || tadāvargaṁtukārayet | . / 1 | 2 / 1 | 2 / 1 | 3 / 1 | 6 / 1 | . .

. kshipeguṇitaṁ || 1 | 2 | 6 | 24 |

*pra*kshiptaṁ 33 || dṛishyaṁvibhajet | 132 / 33 | *v*artyaṁjātaṁ | 4 / 1 | . . . . .

. ṇadattaṁ || atonyāsaḥ || 4 | 8 | 24 | 96 | . . . .

eshavargakramagaṇitaṁ || || athayutivargaṁ*kṛi* . .

s . taṁ || . ā . i . aṁś*ūṇ*yevinyastaṁtadāchaivakrameguṇaṁ

. . . . kṛitvāchaturtha [23v.]

. . . . prathamasyatukiṁbhavet . .

. . | 3 / 1 | 3 / 1 | 12 / 1 | 4 / 1 | dṛi | 300 | kāmikaṁśūnyepiṇyastaṁ

mikaṁ 1 || eshaṇyastaṁ*pra*

rāś . tadāchaivakrameṇaguṇitaṁ | 1 | 2 | 9 | 48 | eshāṁyu .

. . | 60 | anenadṛishyaṁbhājitaṁ | 1 / 60 | 300 / 1 | jātā | 5 | e .

. . syadhanaṁ ... anenakshepaṁguṇaye | 5 | 10 | 45 | 240 | e .

. . yutivargagaṇitaṁ || || udā || prathamasya .

. . . thaṁdattaṁchavaidhanaṁ | sachadvārdhay*uto*dattaṁ . . .

śata*ṁ* chatuśchatv*aliṁśādhi* . . . . [24r.]

dattaṁchaivachaturguṇaṁ | *kiṁ*prathamasya . . . . .

| 0 / 1 | 1 / 1 / 2 | 2 / 1 | 2 / 1 / 2 | 3 / 1 | 3 / 1 / 2 | 4 / 1 | 4 / 1 / 2 | dṛi | 144 / 1 / 2 | . . . . .

śūṇyeśu 1 / 1 .

yutaṁchaivaguṇaṁtataḥ yutaṁchaivaguṇaṁkṛitvākārayegaṇa . . . .

. . natu | 5 / 2 | guṇaṁ | upare uparaṁadhe adhaṁguṇaye | 10 / 2 | sārdhadv . . yutaṁ

. . tīyarā śyāguṇaṇaṁ | sārdhaissaptabhitrīṇi | 45 / 2 | sārdhatrayayut*aṁ*

. chaturtharāśiguṇayeshshaḍviṁśatibhi | jātā | 208 / 2 | sārdhachatvāriyu .

. . . . i 289 *evaṁ* dṛi śyaṁ | sarvaṁtadevajātaṁ

* This phrase is deleted in the MS.

[24v.] . . . . tṛisārdhayu .

. . . . . . *ur*guṇ*aṁcha*turthenanavārdhayutaṁdattaṁ |

. . . tādvāviṁśādhikākimatraprathamasyadattāsi*t*

· *3*/*2* | 2 5/1 2 | 3 7/1 2 | 4 9/1 2 | ekatraṁdattaṁ 222 | śūny*a*

. *erū* . datvā || 1 || yutaguṇitayutakrameṇajātaṁ || sthāpa .

. . | ·/2 | 67/2 | 357/2 | dṛishya 222 | prakshepenajātāṁ 222 || . .

. . ntāḥ dṛiśyāḥ 222 || jā

|| udā || prathamaṁnajānāmi | divardhayutaṁ . .

ṇaṁpaṁ*chārdhayu*taṁpratham*ā* . . . .

[25r.] · · | *3*/1 7+/2 | ./. ./. | dṛi 78/1

. . . . .

. . *rūpaṁdīyate* | yutaṁjātaṁ | 5/2 | dvitīyaguṇaṁ | 10/2 | pārya

. *tṛitīya* ekatreguṇitaṁ | yutena | | yutaṁ || 10/2 | 23/2 |

. yutaṁ | 3*3*/2 | guṇitaṁ | 132/2 | riṇaṁjātaṁ | pār*ya* .

eśanyāsa | 5/2 | 5/2 | 23/2 | 123/2 | dṛishya 78/1 | e . .

. | . 6/2 | vibhaktavyaṁ | 2/1 . . | 78 . . . . . .

raṇ . . śūṇyasthāne . . . . [25v.]

rūpaṁdatvā | 1 | yutājāta | 5 / 2 | . . . .

. | 15 / 2 | prathamātṛitīyasyatṛi guṇaṁyutaṁjāt*aṁ* . . . . . .

. rguṇaṁnavārdhayutaṁjātaṁ | 29 / 2 | ekatranyāsā | 5 / 2 | 15 / 2 | .

. 9 | dṛi 71 / 2 | prakshiptaṁ | 71 / 2 | bha ktaṁdṛishyaṁjātāṁ 1 | .

. anena sarvaṁguṇitaṁtadeva | 5 / 2 | 1 . / 2 | 22

*e*katraṁ || uda ||

. shāṁaparovidhiḥ || prathamadhanaṁdattaṁnajātaṁkiṁtudivardhayutaṁ | tadādvi

*igu*ṇaṁdattaṁpaṁchārdhahīnaṁ | tadātṛitīyenatṛiguṇaṁdattaṁsa*ptārd*ha

. . thāchaturtheṇachaturguṇaṁnavārdhahīnaṁdattaṁekatraṁta . . .

. . . . im . . ṁ . . . . . || 2 5 / 1 2+ || 3 gu 7 / 1 . || 4 *gu* / 1

. haraṇ*aṁ* || śū*nye* . . . . [26r.]

. . rūpaṁdatvāḥ yutaṁjātaṁ | 5 / 2 | p . . . .

. ṁ | 5 / 2 | prathamātṛitīyaṁ tṛigu*ṇaṁ* . . . . . .

mā chaturthaṁchaturgaṇaṁnavārdharahitaṁ | śeshaṁ | 11 / 2 | *e* . . .

| 5 / 2 | 5 / 2 | 8 / 2 | 11 / 2 | dṛi 29 / 2 | prakshepayuktiḥ | 29 / 2 |

bhaktaṁ | 2 / 29 | 29 / 2 | jātaṁ | 1 |

. . guṇitaṁtadeva | evaṁriṇarāśībhavanti | tṛiprakāraṁ

. samāp*taṁ* || śūṇyasthānerūpaṁdatvā | tadanuyuktaṁ | guṇita .

[26v.] . . . . phala . .

. . . athadvau **4** | **30** asyadalam*pha* .

. . . *athā*shṭha **8** | **32** | dalaṁ *pha* .

. . . bhu **36** **16** | **28** dalaṁ *pha* .

16 **4** **24** | **16** dalaṁ *pha* .

. **24** **4** athatrīṇi usārā da .

28 **4** **36** | **20** | **4** | asya tṛi .

. **32** **4** **32** | **20** | **8** atho chchhed .

36 **4** **28** | **20** | **12** puna .

bhu **36** **24** | **20** | **16** .

[27r.] . . . . chh . y . *tas*mātki . .

. . . karaṇaṁ | pṛithakrūpaṁvinikshipya | pṛithak rūpaṁ kshiptaṁjātaṁ .

. *63* bhyāso | tatraguṇa | **3** | **4** | abhyāsaṁ | **12** | rūpahīnaṁ | **1** . .

*abhāgā*chatu*ϕ*paṁcha *45* | atrakshiptaṁjātaṁ | **15** | **16** | eśatṛiguṇ .

. . . tāmūla . . nichatu*ϕ*paṁcha || **5** | **4** | esha

[27v.] . . . .

. masyadhanaṁ | eshaṁanukkrameṇapūrv*okt* . sth . . ṛi . . .

9 pra | **7** dvi | **10** tṛi | **8** cha | **11** paṁ | yutaṁjātaṁpratyai*k* . .

. ṁ | **10** tri | **8** cha | **11** paṁ | **9** pra | **16** | **17** | **18** | **19**

. . . . aṁ sarvatra kārayet ||

17

[28r.] **(a)*** 25 dā

taṁdviguṇaṁdviguṇaṁbhāraṁas .

labdhaṁ **14** || punakṛiya

*This is not homogeneous. There are portions of at least two leaves. There are other small scraps.

(a) . . . . vet || . . . . . [28v.]

. . . guṇaye | $\frac{1}{16}$ | $\frac{1}{8}$ | guṇi . jātā $\frac{1}{9}$

. . . . āhutva aḍho . . shaguṇa

sṇ . abhāgasyadivardhāχkiṁ $\frac{1}{96}$ $\frac{11}{2}$ phalaṁ

phalaṁ 5 . . o an

(a)* . . || sūtraṁ || gu*ṇau* . . . . . [29r.]

. . . kadhanaṁ || guṇanyāsorūpahīnaṁlabdhaṁrūpaṁ

(b) prathamaṇyasya | tatrechchhāpaṁchaḥ | 5 | tatprathamacha .

4 | 15 | tadādī*śś*odhayetkramāt *ādina* . . . *paṁchaśadv*

(c) | $\frac{1}{1}$ 4 | 1 . | . . . . . kr . . . t . i . ch . ś .

*eta*chaturdaśabhiśoddhyaśeshaṁ | 6 | e . *npaṁcha*

(d) . . diset || udā || dhanā . . . .

. syadvitīyayonmiśraṁdhanaṁtatrattrayodashaḥ dvitīyatṛitīyayonmi .

. . . *śa* | ādyatṛitīyayonmiśraṁdhanaṁpaṁchadaśasmṛitaḥ ekaikasyadh*a*

. . . . . . chchhichekatthyatāṁmamaḥ | $\frac{13}{1}$ | $\frac{14}{1}$ | $\frac{15}{1}$ |

(b) 1 | $\frac{19}{1}$ | $\frac{2}{1}$ | karaṇaṁ || ichchhā [29v.]

. . dānīśodhayetkramāt tatrādī 16 śud . s . e .

(c) . tīyāyaṁśoddhya 7 chaturthāyaṁśoddhya 12 paṁ .

(d) y . . *dvitīyay* . . *miśraṁdhanaṁ*

yatṛitīyay . nmiśraṁdhanaṁsaptādashasmṛitaḥ tritīyaśchaturthay .

. . . chatuϕpaṁchakamiśraṁtudhanaṁekonaviṁśati | *prathama*

. . titatrachā ekaikasyadhanaṁkiṁssyādvechchhi . .

. . 17 || 18 || 19 || 20 ||

(a) . . . . tr . bh .

*Besides the pieces (a), (b), (c) and (d) are two other small scraps in the envelope. It is doubtful whether (a) is in the proper envelope.

[30r.] . . | 1 . . . . . .

. . pra . . . .

sūtraṁ | ekayutanarasa . . .

sarva*shshaḍbhipala . . . anenalabdh . .

hītāpra*thaṁ . | 36 | 42 | 48 | 54 | 6 .

. . . . | | *sh*. | 78 | 7. | .

[30v.] . . . kā . . | . .

. . *śdā*tsadṛiśakṛi | . .

. . bhāgahāraṁkriyate | 234 70

*tatu*lāḍhe 3 | mudgāḍhe 1

. *kri*yate | 24 70 | 47 . .

[31r.]* . . | *8* 10 | . . . .

trairāśikena | 3 1 | dine | . *2* 1 2 | *168* 11 *dinā* . . . .

īyasya kṛiya te | . . .

. . *draṁ* | 168 11 | dinā | phalaṁdraṁ | 140 11 | prathamenadattaṁ .

. . ssaptaḥ | dattaissamadhanā *jātā* | h .

. shadṛiśaṁ | 77 11 | 294 11 | pātyaśeshaṁ | 217 11 | dvitīyasyadatta . .

. . 77 eśassamadhanājātā || || punānyaṁ*sarvabhā* . .

. . bhā 4 1 4+ dine | draṁ 15 jīvyā || dvitīyasya | bhā 3 1 3 dine | draṁ

*F. 31 consists of two leaves stuck together. The transliteration given is that of the outside surfaces of the two leaves only; but it is not always easy to differentiate.

. . . . 2 [31v.]

. . . . bhāga

. . . kāranaṁ 2 | 1 ||

. . . chchhedasaṁguṇe draṁ 1 4 / 1 1 / 2 taṁ | draṁ 16 / 1 / 2 mudgā

. . *rdha*yutīhṛitaphalaṁ || asyaguṇākāradvay | 2 / 9 / 2 | 1 / 13 / *2* |

. . uparaṁgunaye | *ā*dautava *5* | dva

. . guṇaya | 2 | . . . . . dvinavabhāgesu . . .

chanmanibhāṇa . . . . [32r.]

. tugaganaṁnita* *rāvaṇe | rā . . . . . .

. yaṁtyaktasutayaśetayā | sākaikenaparāvartaṁdhanurbhāg*aśapa*

. . vaṇepatamāṇāsaudaśabhāgaṁnidhāryate | evaṁtatparimāṇa

. hiyamānaṁtunityaśaḥ kiyatastuparāvartaibhūmiṁprāpyayateja .

dha 1 / 1 / 5 | 1 / 10+ | parā 8 | yoja 30 / 1 chhe 8000 / 1 yoja | phalaṁparā varta

. 18181 śe | 9 / 11 | udā | nāgaśvachchhandagāmi*dratachadaśa*

. . i . . . naṁ . . . . . .

. . . . *cha* . . *ḍ* . . . *ḍ* . . [32v.]

. . . . yā*svaga** **n*dh*i*te etadviko . .

. . . | 100 / 6 / 8 / 3 / 4000 / 5 / 24 | ūrdhachchhe 768000 ayo | adhachchhedaṁ 1 .

phalaṁva 429867 mā 1 di 4 || udā || vr*a* . .

*cha*rīśvā ktāpatitaṁbhūmitalepaṭaṁ || triśatāṁ . .

nāṁtusa ptayojana*h*īyate | chaturdaśaśtuko*ṭṭī* . .

yatapaṁ chaśashṭicha | kaidinaibhūtaleprāpyava

. ko *tt*amā || nyāsasthāpanaṁkṛiyate | . .

[33r.] . . . . *bhi*sārdhaivā*

. . . . 1 chhe 9 guva | gha . . . .
. . . 2 | 1 | udā || dīnārakonāmaviś*ā* . .

. trīdu χ khārjanīyaṁsukhabhojanecha | tasyārdhamardhaṁchayadardhamardhaṁ*tane*

. guru prasādaṁ | 1/1 | 1/2 1/2 1/2 1/2 1/2 1/2 | 108/1 | phadī 1 dhā 8 . .

. *paṇa*dhanabhu ktaṁ || ardhaṁ . .

. . stāraṁnavaromaśatānicha | dvādaśastīticharmāṇikatiromā . . . .

. . . . | 900 | 12/1 24/1 | 12/1 24/1 | 12/1 24/1 | pharoma

[33v.] . . . . . * nyā . sth . . .

. . . | *1*/1 36*0* | *gha 12/1 chhe 33/1 . .

. 112 udā || sumerupṛithivīśaṁkusurāṇāṁparimāśrayaṁ || āru .

. . . χ kaśchitarasāsuramadiraṁ || satataṁsaptasārdhānāṁsap*āga*dhyase .

. . || satribhāgatṛipaṁchāṁśanityamevaṁchagachchhati | yojanānāṁsa*taguṇa*

. . chaturāśītiruchchhṛitām | kenakālenasaugachchhevadametaśuniśchi .

. . . 7 | *di* 1/1 | yo 84000/1 | adhachchhedaṁ 360 di .
. . . .

[34r.] . . . . sasya . jat

. . . . rthinaḥ khage . k . d . *bh*uktaprasṛitiṁchaivamevacha | .

. . shṭauvadasakhekiṁ*kha*gaṁvadasundari | pra 1/1 | kha 11/1 | khā

phalaṁkhaga 63360 eshabāhupramāṇaṁ ||

kaśchitpumāṁsuvarṇastukalāpādayutaṁyavaṁ || pratyahaṁsūlinesū .

kiladattavāṁ | paṁchā*bd*aimāśamevaṁtu diṅaipaṁchadaśastathāḥ datvā

syasarvāyajñātumichchhāmitatvata | di 1/1 | 1/1 1/4 | 6/1/4 bhā | 5 . .

. . chchhedaṁ 192 yavatola |

. . . yu*kta* . . . . [34v.]

chattṛitāmgai | tānī*yatāṁsaraparayā*rju*ne*nagṛiddhra . r . .

tayāspṛiśaṁnti | 1 8 1 9 1 5 | śa 1/1 | śakā | yo 777 8 | ja 222 1 1 7 chhe | phala .

940 || mā

rdhayutovya*kt*

. stapaṁchapaṁ chāśasatereṇavajramaṇa ilabdham

. trakathayaśva mūlyaṁśāṇachaturbhāgasyasiddhārthapaṁcha

. . gasya | ku 1 1 2 | chhe 128/1 | māku 1/2 | māchhe 40 | simā | sa 55/1 |

. . . . . i . . || ū . . ṁ || . . . .

(a) . . . . jātā | purushaḥ | 1/4 | 3/2 | 3/1 | eshā*ṁ*sadṛiśe . [35r.]

. | | dhanaṁ 19/1 | anenaguṇitaṁjātaṁ | 4 | eshaprathamasya*dhana*ṁ

. * dviguṇaṁ 12 dviyutaṁ | 14 | *etadvi*tīya*sya*

. guṇaṁ | 21 | dviguṇaṁ | 42 | tryūṇaṁ | 39 | eshaḥ nyāsaḥ pratye*ka* . .

. daśamagravṛindānāṁ chaturdaśa ekonachatvāriṁśa | ta*t*pādārdhatṛibhāgān . ṁ

. . | 4 | 1/4 | pha 4/1 | *evaṁ*dī 21 . | eshaprashṇa etainmī . . . . . .

(b) . . . . . t . . . .

. smātkāraṇā | tayoryogaviyogaṁsvāviy*o*gas .

. . *ru*śabhājitāpurusha 15 anenabhaktvā dhanaṁ | 9/15 |

. . . . . tad*nū*paddhayasahitaṁ || . . . .

. . . .

[35v.] (a) . *u*shaḥ | 4 | 6 | 5 | sa*dṛi*shayutikṛi . *ājātā* | 60 |

. . . || eshagavāśvamahishīpratyaikaśāleshu . *bhāga*

. . .	. . śā	180 gā	1 4	phalaṁ 4*5* . .	
. . .	1 1 śā	180 1 aśvā	1 6	phalaṁ 30 4+	2*6*
. . ṁ	1 1 śālā	180 mahi	1 5	phalaṁ 36 5+	*9*

(b) . . . mūleṇa | *1* 1 2 | etadviguṇaṁ | 3 | dviyuta .

yasyadhanaṁ | tadevasārdhaṁ | 3 | asyārdhaṁ | 1 1 2 | yutaṁnyāsa ||

. . . . d*vi*tīyas . . . . .

[36r.] . yāchatu*r*da . . . . . .

hyāpaṁchatṛiguṇi * *tāsakhe | 1· 1 |

r*o* 21 eshadeśapramāṇaṁ samāptaṁ || || sa . i

lav*a*nasy*a*rāshekoshṭhatāṁvākṛitāṁrharai | eshāṁchaikāṁrāśipunasu .

*pta*dhānītā | saptāṇāṁmapichaikārāśistulitāṇi | paṁchasaptatyā .

hasraṁbhavet saptāshṭaguṇaṁkiṁ | rā 1 1 | 1075 1 | 56 1 | adhachchhedaṁ

2000 pabhā | phabhā 30 pa 2 00 || esha

. . śilāvaṇapramāṇaṁ || kākinīdaśabhāgasyadadyādashṭādaśī .

. . . ā | tasyaṁviṁśatibhāgasvaśatabhāgaṁprayachchati | narovaksha .

. . *l* . . [36v.]

. . yāchayacha * * 1 . 1*5002* .

. . yojapaṁ*chha*kaṁ | nyāśa*sthā*

.*2* / 1 | 1 1 / 1 360 | yo 5/1 chhe 4608000 yayo | phava 2 . .

mā 4 . .

yojanasyatṛibhāgārdhaṁsatṛibhāgapadonakaṁ | yānaudinattṛibhāge .

genagachchhati | śāpuna ϕpaṁchabhāgārdhaṁyojanasyatathāshṭamam

tinivartaṁtevāyuvegavalāhatā | yojanānāṁshṭootaraśataṁ

. . lenagach*chh*ati || di 1/3 bhā 1/3 1/2 gu 1/5 1/2+ 1/8 3 . .

. . . . k . . . . [37r.]

. . . . . sūyamāṇasyadivākarasyaghaṭikai χkiṁprayātasyava*da*

. suniśchitaṁ | 30/1 muchhe 2/1 ghamu | 500000000/1 | gha 1/1 | *pha*

yo 8333 *3*

. 73 | rbhāṇorathaṁsuramahoragasiddhasaṁhaividyādharaiϕparivṛita*ṁ*

. rātrau | koṭīśatārdhaṁsarathaṁprāyāsyāt tadbrūhiśāstrakuśalo

. . vartaṁ || muhūrtamekenakiṁgachchhebrūhimeganakottamā ||

. . . . || 500000000 | *gha 2* || . *yo* 16666 . .

. . . . | | | stubhāga . . . . . [37v.]

bhāgebhavedrāśi | ūrdhachchhedaṁ 108000 viliptāṇāṁ . .

liptā 5 paṁchārdhasaṁvatsarebhukterāśaikāyadibhānujaḥ brūhi

ka tatvajñasamaśvavāsareṇakiṁ | 2/1/2 | rā 1/1 | 1/1 aṁ 1/360 |

ūrdhachchhedaṁ 108000 viliptāṇāṁ

rāśi | adhachchhedaṁ 1/60 viliptāliptà || phalaṁviliptā 2 || e

shagrahagatiṁ || udā || rājāyudhishṭhironāma ϕ pāṇḍuvaṁ*saśa*

. . . . ṇripat . ṁ . m . . . . . .

[38r.] (a) | 14

dvitīyapaṅkyāyoga | 75 / 15 | vartyaṁ 5 / 1 pānīyaṁ || tṛitīyapaṅktyā | 24 / 11 | *20*

kṛiyate yogaṁ | 90 / 15 | vartyaṁ jātaṁ 6 payasaṁ | 1 . . .

. . ṁ . ī . ṁ . . . .

(b)* kshetraṁ 100 / 1

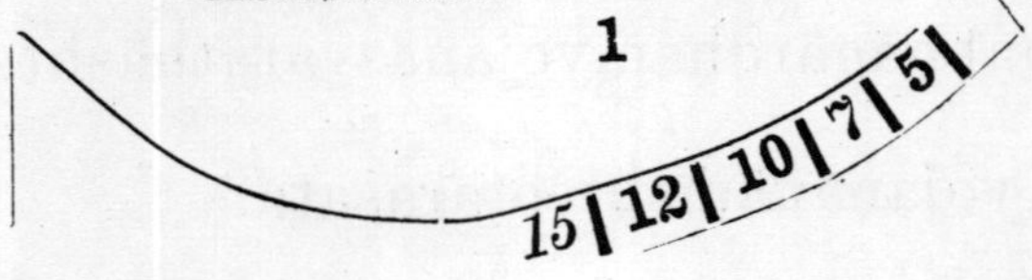

| 300 *bh* . v . d

vaipulyādyogaṁ

(c) bhāgā vimsas chadasaguṇā | saptam . . ksh . . . jñā . . .

. . . . dṛishṭhaṁchaśatāni . . ai

. . . . o

[38v.] (a) | 7 / 1 | 9 / 1 | . . / . | dṛi 60 | prakshepayukti 30 *vibhaktaṁ* 1 / 30 dhanaṁ

ṇitājātā 14 | 18 | . . . . e . . *60*

(b) k . mūlyaṁ : . . tall

l . ā || 1 / . *jāru* . / . mūlye || 6210 maṇale ā iṁ || pha

(c) | pha 1*40* / pha 1*60* |

*(b) Should go with 99 (a); and (a) possibly with 39 (e).

(a)*. śadvādaśanṛiśakaśtathā **|** sapta*p* . ch . . ch [39 r.]

. dhāsaptapaṁchānāṁtṛidvimeka*ϕ*prakalpitaṁ **|** tasyavāhasya kiṁ k . m .

. *tatr*āmama **|** kshetrasyasthāpanaṁkriyate **|** **|** karaṇaṁ *kshettra*

(b) dṛi*shṭhṭha*ksh . . .

. . . daśa **|** chaturdaśatṛitīyasyachaturthasya *n* .

. . . . *sṛi*tushṭhā . . s . shṭh .

(c) *tt* . sth . *n* (d) . *shatkasāne*

5	9 ksh	
15	15	

(e)

ṁsthāpya	4 · 15	5 . 15	. 15	. | 4 | 5 | 6 |
. . syakṛi	4 15	5 15	6 15	yate **||** chaturbhipaṁchabhishshaḍbhi *gu*
. *thamarā*	4 . .	5 15	6 15	śi yoga | 60 15 | vartyaṁ 4 madhūgha

15	*20*	. 000
1.	*15*	
15	*15*	*12 a*

(a) . . . . chh . . . u . [39v.]

. . ll . . . . . *sabhi* maṇḍalaka*ai*ṭallikā **|** eśachchhedaṁbhava*ti*

yathechchh . . kāryā **|** śuthaṭalakṛiyā udāharaṇam ṭalasya*y* .

. . mekaṁ ta*dv*ā śa*shṭiśatānāṁdaśā* . i . *ā* . *ṁ* . *iṁ* .

(b) . . . . . . *dhanaṁ* **1** . . . . **10** **||** esha ekaikabhāgāguṇitajā

. . . . *1***60** **|** **180** **|** **200** **|** **300** **|** evaṁ

. . . . dhanaṁ **1200**

(d)

pha	**180**
.	**200**

(c) . . . . iṁ

(e) *yet śeshekshepa **16** anena trabhāga | **32** / **16** | *labdha* | **2** |

. **40** **|**

pha .

| **12** / **1** | pha **120** | ev . labdhe*r*bhāga*ṁ* | **28** / *2* | jātā **14** lab*dhakshe*pa*ṁ vi* . .

*The portion to the left of *yet* possibly belongs to another leaf.

[40r.] (a)* . ch . . yatra . . . . ś

. āgaṁchaivakārayet kshetravaipuly . . . t . . . y

. || pṛishṭhaśatadvayaṁchaiva ucharesatamekataḥ *vaipulyādvi*

ṁ . . e . . ṁ . i

(b)* āgā*viṁśaśchad* . s . . t (c)* sy*a*tudv .

yaṁśataṁ | sarvemiśrāpi sh . ṁbhāgāstasyaivapaṁcha .

(d)* ṛi . . . .

bdhāṁ bupayasoghaṭaḥ ekamiśrīkṛit . ri . . r . . . k . . . . .

karaṇaṁ | havyatulyaṁdhinikshipyaḥ | $\frac{4}{4}$ | $\frac{5}{5}$ | $\frac{6}{6}$ | kuruprakshepakaṁ tata | praksh

*This should possibly go with 39r (a).

[40v.] (a) . . . . . | . h . ntyachaivatatphal .

guṇit*ā*jātā | 6210 | eshavāhasy*a*kāṇḍapramāṇaṁ

ś . kemūlyaṁ kartavyaṁ | *a*dhach*chhe*daṁchat . . . i . . . i

(c)* *dhanaṁ* . . 00 pratyayatrairā (b)* [Blank]

(d)* *k*ṛitva ṇetu | eko .

kritaṁ | 1 śatatrayarpichabhi*ɸ*purushai*rl*abdhaṁkimādyaṁprathamādhanaṁ

| 120 || d . *225* || . . ṇaṁ . . . ṁ

* These consist of portions of two, or more, leaves stuck together. The knot on b verse does not appear on b recto.

[41r.]* *drammā* . shṭh . dvāchatvaliṁśabhirdinai | tatsaptat .

. sya $\frac{42}{1}$ dine | draṁ $\frac{8}{1}$ jīvyā | $\frac{70}{1}$ purushā 4*2*

. drammā 560 || yadipaṁchaśataśashṭyādhi*ka*

. ālimśabhi | tadrammai *ashṭa*bhi katidināṁ .

. . . . . . ṁ . . .

*Besides the portion transliterated the envelope contains six small scraps.

. . . . . . uparād . . . . | 2 / 1 / 4 | . . . [41v.]

. . ya 2 adhedāpayedattaḥ | 17 / 8 / 2 / 3 | adheno | parisaṁ .

. . parimarāśīdvayagu | ṇaye | 51 / 6 / 2 / 3 | .

. *yet* upariyuktakṛiyate ekapaṁchāśānāṁ

. . . sthāpanaṁ | *1* / *1* | 53 || *2* . ṁ | phalaṁ ā 17 u 2 || 5 / 6

| 1 / 3 | | *1* / 1 / 2 / 1 / 2 | pha 54 . . kṛiyate | | i | i | . . [42r.]†

saṅsvāsārdhayuktetrayodaśasārdhaṁbhavati | 40 / 1 bhā | 160 / 1 | 13 / 1 / 2 | .

. . pi eshāṁchchhedāṁkṛitājātā ekeṇa . .

. *ri* sārdhatrayodaśabhikimiti | 1 / 1 | 4 / 1 | 27 / 2 | pha 54 eshāṁapa .

. . . ekenalabdhachatvārishshaḍbhi sampadyatekathaṁ | 1 / 1

. . || *ḥ2* . . ekola*bhati* chatvāriśaṅsardhasyatukiṁbhavet

. . . . ṇi . m . .

| . | . | 2 [42v.]

. sajātā 54 | śaḍtri 24 | . . 12 || ardhā 18 || ekattraṁ 54 || esha*mū*

tr*ai*rāśikakaraṇapratyekamūlyavidhi || aparaṁvakshyāmi | viṁśano*m*di

. kiṁprathamekhandhakeśvayobhilikhita | apasyaprashṇāvidhi | 20 / 1 | 1 / 1 / 2 | 1 / 3 |

. śiguṇaye | guṇitājātā | 20 / 1 | 3 / 2 | 1 / 3 | chhedaṁ | 20 / 1 | 1 / 2 | 1 / 1 | bhā*ge* .

jātaṁphalaṁrū 10 || eshaviṁ śānāṁ *di* .

bhavati | atra uparimāśkhandhakasya eshaguṇākāraṁbhavati | *asya*

. *khiśyā* . . . . . *laṁ* . .

†Besides the portion transliterated there are five very small scraps.

[43r.] . . . va . . . . .g . . .

. . . ārayet sārdhadva . ś . tubhoja*n*amadyamuttamet .

satṛibhāgatrayastṛiṁśaidinaidvāṇijyakasyatu | bhāṇḍāredvāda

śaśatava*jā*rāṇāṁsthitāsyavai | eshāvyayasamutpattauka*χ*kālaṁ brū

hipaṇḍitakaraṇavidhānenadvādaśaśatasyabhāṇḍāresthitatā

10	2	bhā	13	3	bhā	13	8	bhā	1	3	bhā	1	1	bhā	1	5	bhā	2	
1	1		1	1		1	1		1	1		2	1		1	1		1	1
2	3		3	8		4+	2		3	5			2			3		4	4

2	33	bhā	1	1	bhāṇḍā	1200	guṇita . i	
	1		1	360		1		
	3							

[43v.]

2	10	800 .	
1	1		
3	2	727	

a . . chchhedaṁ 360 di*va* || bh

re ṇasaha || āyupiṇḍaṁ 2982 / 727

adhunāvyaya piṇḍaṁ di 1 / 1 | 1807 / 240 | 800 / 727 | 486 / 727

ūrdhachchhedaṁ 360

diva daṁ pha | 2982 / 486 / 727 | puna laṁ | 800 / 727 | 2982 / 486 / 727 | 1 / 1 | a . . . . 360 di*va* . pratidina .

. . . 1807 evaṁsarvatrairāśikena | udā || . . .

[44r.] | 2559 / 1 / 61 | di 1 / 1 | . . . . vyayapra | 223 / 144 | evaṁ . . . . mānaṁ || udā || eka .

*ś*ārdhamu*tp*atisatṛibhāgadinadvayet pūjārthaṁ satṛibhāgaṁcha . . .

. tāśchayet sāshṭabhāgadinātrīṇivā*s*udevasyachārchane .

pādonatrayoda* *śāṇāṁcha ashṭasārdhadināni *kh*et ||

brāhmaṇābhojanedadyā paralokahitārthinaḥ satṛibhāgaṁ

* *. jjāraṁsapaṁchabhāgadinattrayet pa .

. . . *lachendaṁ*chakaXk . lakhanavalikā | *sardhaṁsārdhaṁdine*

. . . . [44v.]

. . . ptaśatānāṁkaχkālaṁ ārjanabhakshaṇe || nyāsasthāpanaṁkṛiya

bhā 8/1 di 5/1/3 bhā 1/1 pa 32/1 bhā śū 2/1/2 1/4 36/1 bhā 1/1 1/360

. rāśi 223/144 āyarāśi 280/61 etatkālena ārjanabha . .

. di 1/1 223/144 280/61 ūrdhachchhedaṁ 36 .

phalaṁvaij

. . 2559 śe 1/61 eshavyaye ||* *

. va 4 mā . *di* 2 śe 28/61 atha āya di 1/1 . 1/1 . 28 .

. *225*/*32*800 dalitā diyutaḥ 10*225*/65600 . . . . . . , 108625/65600 padaghnāpa . . [45r.]

. *tā* 6455040625/3227520000 a topaṁchavi*ṅśā* .

. . ti uparāḥ 6455040/3227520 . .

. . te labdhaṁ 2 eshadhanaṁ || ||

. ṁ ā 1/1/2 u 1/1/2 padu 0/1 dhanu 7000 ashṭo .

. i . . o . i . .

. . . . . chaturṇāṁ . . . . .

. . 7 . . y . *384* asyavarga 147456 a*sya* [45v.]

. *4*3 271936 eshasarvaguṇitākaraṇipā*ti*

. kṛitvābhājitajātaḥ 1158 + 671250 aṁśairaṁśāguṇaye*t* .

, . . . āśivarjyajātaḥ 579 768 1158 29* *4912 .

. . 579/515520000 294912 +/777307500 777307500 .

. . 777307500 śeshaṁ 579/51522508*8*

śaṁ

. . . . . 67588 || dvayenamūle 0

[46r.] . . . 4 6 6 3 8 4 5 . . .

4 6 6 3 8 4 5 0 0 0

. . 2 1 7 9 0 2 5 0 0 4 4 dalitā e
. . 5 5 4 6 1 5 0 0 0

1 1 0 8 9 5 1 2 5 0 2 2 ādisaṁyuta . .
. . 3 2 2 7 0 4 7 5 2 2 1 5 5 4 6 1 5 0 0 0 padaghnāk . .
. . *1* 5 5 4 6 1 5 0 0 0 5 0 7 5 3 3 8 3 7 6 2 7 4 6 7 4 3 2 7
7 2 5 0 4 8 3 3 9 4 6 7 5 0 0 0 0
. . rākaraṇipāta

. . *2* 1 7 4 3 2 7 1 9 3 6 pātitajātā uparānyāsasthāpa

. . . . . 2 5 0 0 0 0 0 0 0 0 0 0 || bhā

[46v.] . . . .

| *maḍe* 8 | maḍe 6 | maḍe |

. . kā | 20 |

. aparaprashṭaḥ pārā ā i e vīhujaṇavīhānhai . .

. ṇa | gorejāmacha uppaṇesāmale a . . *dha*pa

. ḍhale āpotḍīṇe āgaṇevīhujaṇa ehuvī

karaṇaṁ | traigo*d*evārehahipaṇehi sā . .

. . . . . a *īśūḍhed* . *hivo*

[47r.] . . . . ṛi . . r . . . . .

vichakshaṇaḥ chamūs*tu*pṛitanāstisrastisrasścha*mū* . . . .

nīkīnidaśaguṇāmāhuraksh*o*hanībudhaḥ || akshohi . . .

ra	1	e	3	3	3	3	3	3	3	10 gu			guṇitajā
ga	1	sha	1	1	1	1	1	1	1	1			
na	5	pa			ratha	21870		esha akshohi					
tu	3	ti		ṇī	gaja	21870		pramāṇaṁ ||					||

dā || kaśchidrāja nara 109350 kumāraśatruda*ma* |

haya 65610 . i . . o . .

. . . . . 1 | 15 . | 10 śi | phal*aṁ* . 15*0* || ud*ā* [47v.]
. . . tāvyū | 1 | 1 | 1 | hapārthaṁhehayakī

. *na*ghnatasāyakaiśchaiva ⌀ pattisvapādadalaśoḍasai | a .

*an*vāchatasrāvaihatātenamahāṅsavāṁ || śarānāṁchaparīmāṇaṁ . .

*tra*viśārada | śi 1/1 | 16/1/1/4/1/2 | 4/1 achhe 21870/1 | phalaṁśarā

2624400 || anyā

īpramāṇaṁ || sūtraṁ || ekorathogaja .

ki | di . | ra . 3 bhā ksha *8* chhe . [48r.]
ya 1 1 ya 1 1
5 3 1
daṁ 480 ka 1 6 bhā mā 3 raktipa
1 1 ṇitaṁjātaṁ
. . gu 4 12
pā 1 *di* 1 41994*2*
. . ṁpa 4 36 pala 30 115200
śe 1
. . . 3 to 8 pale to 3 *to*lenā *sti*dh*ā* 1*2* .

. . . ā 7 dhāṇenāsti aṁ 4 dhā || aṁ 2

. . . . phalaṁbhā 2 . . [48v.]

. . . . *sti*pala 2000 bhā | pa 270 || pa

. . . 8
dhā 2 | to 8 tolapal to 6 tolenāstidhāṇa 12 *dh* .

. . 8 chhe 12 yadidinamekena eshadataṁtadvādaśavarsheṇa
aṁ 3
. yāsa chhe 5 || di 1/1 | 216 bhā | varshe 12/1 3/1 .
ya 2/1 | 3/1/5 bhā | 270 pa
phalaṁ | bhāra | 2000 chhe | 9*3* . .
6 to | ṇaṁ
varṇapramā 8 chhe . .
. . . . śiraseśi 8 dhā
. . . . *12*

[49r.] . ya 3 yavan*āstika* $\begin{smallmatrix}6\\1\\4\end{smallmatrix}$ . . . . da 4 kalānāstipā . .

pādanāstimūdri 4 pāmu 8 mū 2 || udāharaṇaṁ || . .

śūkhyairyajaṁtidevīpratimahnikechit dadāmidevyā*rva*

kaṁ*cha*ḥ krītvādīnāraśatānichatvāritadhānakā aṁ*ḍ*ikā*rakti*

kāyavākalāpādamūdṛikācha | etatramūlyaṁvadametatra

syakiṁ $\frac{1}{1}$ to $\frac{12}{1}$ mū $\frac{400}{1}$ *dhā* 1 phalaṁdī 50 dī*nāra*

aṁ $\frac{1}{4}$ stidhāṇe 12 dhā

ra 1 $\begin{smallmatrix}1\\1\\4\end{smallmatrix}$ bhā

. 10 dhāṇe anāsti aṁ 4 *ya* 1 . . . . aṁ 1 ||

[49v.] . . . . *hāsobh* . . . .

. . . . . raktikshaya . . . th . . . . *yasyati* paṁchaguṇ*aṁ*

. . . . sa*rva*paṁchchassamaṁ | divasāviṁśatikaṁkiṁ*śuṁdya*timahāṁ

vadaniśchayaṁ $\frac{1}{1}$ to 3 ksha 4+ va 25 chhe $\frac{360}{1}$ ma*dh*ya

rāsiguṇitaṁ mā $\frac{2}{12}$ ya 60 mā $\frac{5}{12}$ jā

$\frac{62321}{19200}$ ksha aṁ $\frac{3}{4}$ si $\frac{4}{8}$ di $\frac{20}{30}$ yaṁśodhya

ya $\frac{3}{4}$

$\frac{60881}{.9200}$ a sa ka 1 pā 1 6 bhā . dhachchhe*daṁ* | 2

rvaguṇitaṁ || 7 tolepalaṁ : 558 . 7*8*

[50r.] . . . . . $\begin{smallmatrix}1\\1\\4\end{smallmatrix}$ . . 4 . . . .

ntikāvati . . . . *vaśishṭaputraha*

sikasyārtheputra pautra upayogyaṁbhavatuḥ likhi

taṁchchhajakaputragaṇakarājabrāhmaṇena | sarveshāṁmevaśāstrāṇaṁ

*ga*ṇitaṁ*mūrdhni*tishṭati | ādyāvasānesaṁsāre utpaṁnna gaṇi

. . hit paśchāśṛishṭitadākartuṁśiveṇaparamāt*man* .

*y*ādyaṁchasutpaṁnnaṁgaṇitaṁsakhyakāraṇaṁ | yacha

. . . . hi*naṁ* . . *yadivet* *k* . *āṁ* . i . . . . .

guṇayebhājaye *ekar* . . | udā || dramme . . pusaśataṁ*n̄labdham* [50v.]

ardh*eṇa*labhyateχkati | eka*rāśi*stukalanāgaṇitaprakṛiyā

kuruḥ | 1 dramme | phalaṁ 50 aparaṁ | udā || sārdhad*v*aye

*tra*ya | 100 dra pusā | sārdhadivardhelabhyateχkati | $\begin{matrix}2\\1\\2\\1\\2\\1\\2\end{matrix}$

*sū*traṁ | $\begin{matrix}1\\2\end{matrix}$ || ardhenoparisaṁguṇya

. . . . v*arṭh*akrameṇacha | ardheṇa ūrdhaṁguṇayema

. *pa*ṁchasaṁguṇe | bhājayelabdha*panya*ṁ . . e

(a) . . . . [51r.]†

rā | 1 | 2 | 3 | $\begin{matrix}1\\1\end{matrix}$ | $\begin{matrix}1\\1\end{matrix}$ | $\begin{matrix}1\\1\end{matrix}$ | $\begin{matrix}1\\1\end{matrix}$ eshāṁyu*kta* $\begin{matrix}6\\1\end{matrix}$.

sha | 48 | śeshāϕpurushasa 4 || anenabhajitārlabdhā | *1* .

. . . sya bhavati | 12 | 13 | 14 | 15 | ekatraṁ 54 || . a

udā || kaśchidrājādadedānaṁsaptapaṁchāśakaṁbudha | paṁchā .

*cha*pravakshy*āmy*a*nupūrvasḥah* dviguṇadviguṇaṁchaivarūparūpottare

. prathameprāptaṁkiṁprāptamaparejane $\begin{matrix}0\\1\end{matrix}$ $\begin{matrix}1\\1\end{matrix}$ . . . .

(b)* . bhāja*to*hitvā | tatrottarā 1 | 1 | yutaṁ | 2 | . $\begin{matrix}3\\1\end{matrix}$ $\begin{matrix}3\\1\end{matrix}$

. . *9* | . eshāϕ*p* . . sh . . . . . .

(a) . 27 | 81 | *dṛi 329* / 1 | karaṇaṁ | uttar*ā bha*jya . . [51v.]

tatrottararāśīnāṁyoga 87

eshadhanādṛishyāśodhanīyājātā 242

. . jatā | purusha || *1* | 9 | 27 |81 | yoga 121 ane .

. . jātā | 2 | eshadvauprathamasyadhanaṁ || 2 | 6 | 18 | 54 | 162

uttararāśīsaṁyutaṁjātaṁ $\begin{matrix}2\\1\end{matrix}$ | $\begin{matrix}15\\2\end{matrix}$ | $\begin{matrix}48\\2\end{matrix}$ | $\begin{matrix}147\\2\end{matrix}$ | $\begin{matrix}444\\2\end{matrix}$ eshāṁ

. . . . || udā || . .

(b)

. . 6 || . . || . . . sy . . yoga 1 1 1 śeshā ϕpurushabhāji . . .

. . . . *jātā* | *37* | bhājitā |*60*

† ?Two leaves.

* The left portion or (b) is not homogeneous. There are also three other small scraps that are possibly out of place.

[52r.] . . a . . . . .

. nā $\frac{57}{1}$ *ts*edaṁjātaṁ 3 anenachāl*iṁ*śaguṇayaṁjātā 12 .

. . *sva*vaṁśurāṇāṁ ‖ pratyayatrairāśikena

‖ udā ‖ dhanāsvamardhosaṁśoddhya

chaturīyakaṁ | tatseshāpaṁchamobhāgote

$\frac{1}{1}$	vaṁ	$\frac{120}{1}$	$\frac{1}{10}$ 1		
$\frac{1}{1}$	vaṁ	$\frac{120}{1}$	$\frac{1}{8}$	$\frac{1}{1}$	.
$\frac{1}{1}$	vaṁ	$\frac{120}{1}$	$\frac{1}{4}$	30	a

śatadvayaṁ | aśītyādhikaṁdhanaṁchaivakiṁādyaṁprathamaṁdhanaṁ |

[52v.] . . ū . . . *200* ‖ asyadvayānāṁśatānāṁpā*da*

. . *dhik*aṁśataṁbhavati 150 atrāpipaṁchabhāga 30 ‖ evaṁ .

. . $\frac{1}{2}$ | $\frac{1}{4}$ $\frac{1}{2}+$ | $\frac{1}{5}$ $\frac{1}{4}+$ $\frac{1}{2}+$ | phapiṇḍu 280 ‖ paṁchamījātikaraṇaṁkṛi .

pi 280 ‖ aṁśayuti $\frac{28}{40}$ bhaktaṁ $\frac{40}{28}$ dhanu 280

400 eshaphalaṁbhavati ‖

[53r.] *y* . *tā* *ṇaṁ* ‖ | 1 1 2 | yo *1* 1 | di | *6* . . . . . 1 . . . . .

syai . . viśeshaṁtu | tatragati | $\frac{3}{2}$ $\frac{2}{1}$ | viśeshaṁ | $\frac{1}{2}$ | vibhaktaṁ | $\frac{2}{1}$

sarvagatī | yojana | 9 | a nenaguṇaye | 18 | anena . .

. *gata* | *bha*vishyati ‖ pratyayatrairāśikena | $\frac{1}{1}$ di |

. yo 27 *pūgata*dina 6 a . oyojana 9

. *ya* $\frac{1}{1}$ ‖ $\frac{1}{1}$ di | $\frac{2}{\mathit{1}}$ yo ‖ . . di . phalaṁyojana

. . . . 1 | $\frac{pha}{12}$ ‖ 6

. . . ṛí . . y . 2 . [53v.]

	18 1 yojana	20 dina 20 ghaṭike 35 ghadina	phalaṁyo 25 2
3	25 yo	20 . . 20 ghaṭike 35 ghaṭikedina	phayo 36 7

. . . . . kaṁsya . . . .

.

(a)*. . . . ksh . . y . . χk . . y . . ksh . ri . . [54r.]

yasya ‖ . . . yadhanaṁ | *dh*anateyātā | tayoryogaviyo

1 1	2 1	4 1	. 1

. . . kṛitāṁrāshayaḥ

2 1	1 1	2 1+	9 1+
1 1	2 1	4 1	8 1

dṛi 82

. . . . bhājyahi tveti | tatra uttararā . i

. . . . uttaraṁriṇaṁjātaṁ

(b) sūtraṁ jātaṁ 76 eshaprathamasya

(a) . . . dviguṇaṁchatṛi*ū*nachatṛitī*ya*syadhanaṁbhavet tryūna [54v.]

. | . saṁyutaṁ | ekaviṁshati*bhi* χ*krī*todīnā*raittara* . .

āyatudaṁsāpṛithag*va*daḥ ‖ karaṇaṁ ‖ yasyapadaṁnajñāyate

. . . . . yā 1
1 ‖ eta*dpra*thamasyadhanaṁ |

(b) . . . . . | 4
1
2 | et . . . . . chadattavān hasta

. . . rveshāṁ ‖ 0 ‖ 2 ‖ 2+ ‖ . . . . . .

*Besides (a) and (b) there are eight small scraps; and (b) appears to consist of two unrelated pieces stuck together.

ū . . ṛik .

[55r] . . va $\frac{5}{1}{3}$ to 1 dhā $\frac{1}{12}$ aṁ $\frac{1}{4}$ pha va 6 . . . . śe $\frac{9}{10}$ asya . .

guṇitaṁ 7227 / 1200 athaśaḍdrammako kaidramam śa

jjāradvidhāna

. . viṁśatipā ra 1 1 / 1 4 bhā lāhataidhānakā | asyaivaska

to 1 / 1 ya 1 3 / 1 5 bhā va 5 / 1 / 3 to 1 / 1 dhā 1 1 / 1 12 aṁ 1 .

ra 1 1 / 1 60 ya si 1 2 / 1 2 bhā 1 1 / 1 192 si 1 1 / 1 480 ka

. . . ka 1 2 / 1 2 bhā

. . pā 1

[55v.] . . . . kasyaya . . .

paṁ | 10 | chatṛiṁśataṁ | *divardhatolaṁkasyadivardhamāṁśa*kasyadi

rdha*mā*ṇḍikādivardhayavasyakiṁmūlyaṁ | nyāsa ||

to 1 / 1 | 35 / 1 | 1 to / 1 / 2 phadraṁ 58 śe 31 / 128 punā*nyaṁ*

1 1 / 1 6 / 2 mā 1 1 1 1 1 1 1 / 1 1 12 1 48 1 192 / 2 2 2 2

to 1 / 1 35 / 1 1 1 / 1 2 / 2 . ṁ

. ṁ . 58 1 1 / 1 2 / 2 ya śe 31 / 128 tathā . . ā . ṁ

. . . .

. . y . |___| . . | 20 / 21 | . . . . aḥ tastāt . . . [56r.]

. kṛityūnāṅśeshachchhedodvi saṁguṇaṁ | tadvarga . *dala*

. shṭhaḥ hṛitiśuddhikṛitikshayaḥ || śeshachchhedodvisaṁguṇakṛi

. | 21 / 20 / 21 | 400 / 441 | dala | 1 / 2 | saṁslishṭhaḥ | 21 / 20 / 21+ | bhā | śeshaṁpānya .

. tvābhāji . .

. . dhaṁupare uparaṁguṇitavyaṁvargaṁyāvarjaye . .

. . ṁ | 4 2 5 0 4 2 | 400 . . śeshaṁ 4246

. . . . .

. 880 / 84 | 964 / 168 | guṇitajātaṁ pṛithaksthānāṁva | 848320 / 14112 | chatvāriṅśa rgaṁ | 160 . . . [56v.]

sha uparāpātyaśeshaṁ | 846720 / 14112 | vaṙtyajātaṁ | 60

||

* The lower half of this page is blank.

| 5 / 8 | kṛitikshya | 11833 / 1848 | hṛi | . . 184 / 1848 | kṛitaṁ : eśamūlaṁ || tan*mūlaṁ* [57r.]

. . nmūlaṁekaṁ 1 eshasadṛiśepātitajāta | 9985 / 1848 |

. . saṁbhaktaṁ uttaraṁdviguṇaṁ 2 anenabha

. *9*985 | eshapaṁchakasyapadaṁ || || asyapratya

. *36*96 | sūtraṁ || ekonarāśidvidhāsthāpya ūpase . .

[57v.] . . . . . . r . . . . . .

shṭ*o*ttaraghneguṇite 40 dvighnamādicha yā

. . nikshipya | 41 | mūlaṁ | 6 5 6 | śeshachchhedodvi saṁguṇi

śuddhaḥ tas*māt* akṛiteślishṭha | kṛityūnāśeshachchhedo dvi

. . . dalasaṁślishṭhahṛitiśuddhikṛitikshayaḥ akṛiteślishṭha

. dādvisaṁguṇakṛita | 6 5 12 | ta*dvarga* . 12 | 6 5 12 | 25 144 dala

[58r.] udā || śaḍviṁśaśchatṛipaṁchāśa ek*o*natrimśevacha | dvāśa

śhaḍviṁśachatuśchatvāliṁśasaptati | chatushshashṭina*va* .

. . . . ṁśanaṁtaraṁ | tṛirāśīti ekaviṁśa ashṭa . .

. . . . pakaṁ |*

. . . . *2653*296226447064994 . . . *83218*

. . . . eśa . . .

[85v.] . . . . sthāpanaṁkṛiya*te*

. shya	1 1	yuvī	1 1 1	sūḍha	1 1	dṛishya	20 ja
. . . ḍo		maṁ	1 2	maṁḍa	1 2	maṁḍe	2 .

. taḍattajātaṁmaḍa 2 | ya 5 | sū*ḍhe* . .

. . . || udā || korāsipaṁchayutā . ū . . sā rāśissapta . [59r.]*

*m*ūladakosorāśiritiprashṇaḥ | 0/1 5/1 yumū 0/1 || sā 0/1 7+/1 mū 0 |

ṇaṁ | yutahīnaṁchamekatvaṁ | 12 | ta

laṁ | 6 | dvihrīṇaṁ | 4 | dalaṁ | 2 | vargaṁ | 4 | hīneyutiṁchakartavyā

. . | 7+ | anenayuti | 11 | eśasārāsi || asyapratyānayane .

. 11/1 yu 5/1 mū 4/1 | 11/1 7+/1 mū 2/1 | paṁchāśamasūtraṁ 50

. || sūtraṁ gavāṁviśesha kartavyaṁ dhanaṁchaivapuna . .

*The left-hand page of this leaf is blank.

. . . . . *rū* . . . . [60 r.]

. . ekonaviṁśatima | gāvī 10/1 | rūpa 8/1 | . v .

. . vivaritāsti ||

. . chāśamasūtraṁ 51 || sūtraṁ || āyavyayaviśeshatuvibha

. . *dṛ*shyasaṁgu —⁘— ṇaṁ | yallabdhaṁsābhavetkālaṁayaṁprashṇe

. ya . vidhi || udā || dvidine ārjayepaṁchatṛidinenavabhakshaye

. āṇḍāgāraṁtasyatriṅśakiṁkālaṁārjabhakshaṇaṁ | dī 5/di 2 || dīnāra 9/dina 3 || *dṛi* 20

. . naṁ | āyavyayaviśeshantu | tatrāyaṁ | 5/2 |

. . . . e . ṁ . . . .

[60v.]
dvipaṁchāsamasūtraṁ **52** || | . . . **60** di | phalaṁ **180**

|| sūtraṁ ahadravya ✣ harāśautatadviśeshaṁvi .

yallabdha | dviguṇaṁkā laṁdattāsamadhanāprati || *udā*

. . dine ārjayepaṁchabhṛitakomekapaṇḍitaḥ dvitīyaṁpaṁcha .

. *va*seva*s*amārjayatebudhaḥ prathamenadvitīyasyasaptadattā . i

. . taḥ datvāsamadhanājātākenakālenakatthyatāṁ **5** rū || **6** . .

. . . . . . ṁ . . ṁ . *au* . . . .

[61r.] . . *anenakālenasama*dhanābhavanti || pra*tyayaṁtrairāśik . kṛı* . .

5	**30**	pha **50**	
1	**1**		
6	**30**	**36**	*43*

prathamedvitīyasyassaptadattā | **7** | śeshaṁ **43** | **43** | etesamadhanājātā || udā ||

. japutrodvayokechinṛipatissevyasan tivaiḥ mekāsyāhnedvayashsh . .

. gā | prathamenadvitīyasyadaśadināra | dvitīyasyadivardhakaṁ

. ttayān kenakālenasamatāṁgaṇayitvāvadāśūme ||

. . | **3** | dattaṁ **10**/**1** | karaṇaṁ || ahadravyaviśeshaṁcha | ta*trā*

. . . *2* | *3* | *pha* . .

[61v.]

1	**13**	**30**	
1	**6**	. **1**	pha *65*
1	**3**	**30**	
1	**2**	**1**	pha **45**

. . . . . . *prathamenadvitīyas* .

rdattajātā | **55** | **55** || samadhanājā .

sūtraṁtṛipaṁchāsamaḥ sūtraṁ **53** ||

yenakrayaṁ bhājyaṁrūpahīnaṁpunarbhajet ✣ la . e .

guṇayetatranīvībhavatitatracha || udā || dvibhi χkṛiṇāti . .

ptavikṛiṇātitṛibhishshaṭ ashṭādaśabhavedlābhā χkā . . .

. . katthyatāṁ || **7**/**2** | **6**/**3** | **18**/**1** lābhā | karaṇaṁ | vi .

. . gu*ṇitaṁ* jā*taṁ* . . . . . . nīvījātā | syapra*tya* . . . [62r.]

. na || yadidvibhissaptalabhyate | tadāchaturviṁśatibhiχkiṁ | 2/1 | 7/1 | ./1

laṁ*rū* 84 || asyavikrayaṁkṛiyate | yadishshaḍbhitraya

. *ṇā*labhyatetadāchaturāśītibhiχkiṁ | 6/1 | 3/1 | 84/1 | phalaṁ 42 ||

*mū*laṁ 24 | pātyaśeshaṁ 18 eśalābhāḥ chaupaṁchāsamasūtraṁ 54

*vi*krayaṁbhājayechaivaguṇayetkrayapiṇḍatāṁ | rū

. mūlaguṇayelabdhalābhaṁchaprāpyate || udā

. *bhi*krīṇātiyassaptavikṛiṇātitṛibhishshaṭ mūlācha . .

. . . . . . || 7 6 . . . .

. . | 2/1 | 7/1 | 24/1 | pha 84/1 | *athavikrayaṁ* | 6/1 | 3/1 | 84/1 | pha 4 . [62v.]

. 24 | pātyaśeshaṁ 18 | eshalābhaṁ || paṁchapaṁchāsamasūtraṁ

vikrayaṁbhājayechaivaguṇayetkrayapiṇḍavat vi*bhaktam*

sachakartavyaṁguṇayemiśrakaṁbudhaḥ yallabdhaṁsābhavenmū

. śeshaṁlābhapiṇḍatāṁ || udā || tṛibhiśchalabhaterashṭaucha

*vikra*yaṁshshaṭ samūlalābhamutpaṇṇāśataṁśashthivimiśritaṁ | *kiṁ*

*laṁ*kaśchalābhaṁchakathayedgaṇakottamaḥ | 8/3 | 6/4 | miśra 160/1 | .

. . . krayaṁbhājayechaivaguṇayet

. . . . | 1 | | |

. . *pratya*yaḥ | 6/1 | 4/1 | 240/1 | phalaṁ 160/1 | mūlaṁ | 90 | pātyaśesh*aṁ* | 7· [63r.]

. . chāsamasūtraṁ 56 || vikrayaṁchavibhaktavyaṁguṇitaṁkra

. śivat kṛitvārūpakshayaṁchaivavibhaktaṁmūlam*a*

. . yāt || udā || paṁchabhiśchatuvargaṁtugṛihītaṁkenamānav*e*

. kenashshaṭ vikrītaṁshshaṭptaṁchāśariṇaṁkṛitaṁ | krayavikrayasaṁ

. . uvanīvistasyaivakatthyatāṁ | 16/5 | 6/1 | riṇaṁ 56+/1 |

. . . bhājayechaiva | 1/6 |

] 63v.] . . . *84* ⁞ *punāsyavi*kraya | 6/1 | 1/1 | 384 | *phalaṁ* | *6*

mūl*aṁ* | 120 | chatushshashṭi . pātyaśeshaṁ | 56 | eśariṇaṁkṛi .

. . ptapaṁchāśamasūtraṁ 57 || vastraśulkaṁyadbhavatita .

. . *hṛi*tavastratāṁ | trai | . rāśikavidhānenaśulka .

. . yatatvataḥ || udā || paṭasyaśulkavimśāṅśaṁka . i .

. . . *triśśa*taṁ | paṭakānāṁpaṇakṛitedvaupaṭauhṛitaśau . i

. . . . paṇadaśastathāḥ kiṁmūlyaṁ

[64r.]

. . 5 2 8 0 . 8 7 2 4	4 4 4 0 0 4 3 8 7 2 4	tayardhe . tus . . . . . . ardhamkartavyaṁ . ta . .
. . 5 2 8 0 *3* 8 7 2 4	4 4 4 0 0 4 7 7 4 4 8	saṁguṇyajātaṃ | a . . hrarāhareshugu*n* . . .
1 7 9 9 4 5 9 4 1 1 2 0 2 9 9 9 0 9 6 3 5 2		asya ūrdhāṁka . . *shashardhaṁ* . . .

1 6 0 0 0 0 + |

[64v.] . . . shachchhedodvisa*ṁ*guṇaṁ | 6 | śéshapaṁchakaṁpṛithak . .

. . aṅśasvaṁśaṁ | 77/12 | tanmūla | 5/12 | varjitaṁ | tanmūlaṁ 1 .

5/*1* | dviguṇottarasaṁbhaktaṁ | 65/24 | eshapadaṁ ||

. . . nayanaṁ | ā 1/1 | u 1/1 | pa 65/24 | rūpoṇa 41/24 . . . .

. . . . . ḥ dalitā | 41 | ādisaṁy . . . . . .

. . | . . . . etatk*āla* [65r.]

*bhir*manushya a . . . . . . lagyantiḥ

|| aparaprashṇaḥ yadyekapurushasyadrammāshshaṭ

bhirdinaijīvalokā | tatkāryaṁprastu*tā* . śsaptatīnāṁ .

. . *ṇaṁ*pākarākshakānāṁdrammaishshaḍbhikatidinājīvalokaṁ*bhav* . .

. . karaṇaṁ | ādautāvayadyekapurushasyadrammāshshaṭ triṇśa . i . .

. . jīvyā | tatsaptatīnāṁkiṁ | 1 pudraṁ 6 / 1 1 | 30 di | 70 pu / 1 | pha .

. . . didrammāṇaṁt*rī*ṇiśatāsār*dh*āsa . . . . ṁ . . ṁ . .

. . . . *u*

. chyate | ā 1 / 1 | *u* . . . . . 6 . | *karaṇaṁ* [65v.]

. shṭhottaraghneganṇite | ashṭhaghanaṁ | 480 | uttaragha*na*

. . va | dvighnamādi | ādidviguṇa | 2 | chayojjhitaṁ | cha

. uttaraṁ | ato uttaraṁpātayitvā ekaṁbhavati | 1 | var*dhi*

. *tra*nikshipyadhanasya | 481 | mūlaṁślishṭhakaraṇyā | 21 / 40 / 42 |

. . vaṁśaṁ | 882 / 40 / 42 | śeshaṁchatv*ā*r*i*ṁśapṛithaksthāpya | 40 |

. . yojyaṁ | 922 / 42 | tanmūlavarjitaṁ || tanmūlaṁ

. . . . ṇottara . aṁ . . . 8 8 0 . .

. . . . . yukt*abhāgaha* . . . . [66r.]

*nye*kārghaṁtupaṇyānāṁekadvitṛichatushshaṭ *sa*

*mpā*ṇāṁsa . . ṇyānimānayaḥ sthāpanaṁkṛiyate

1	1		draṁ	1	2		draṁ	1	3		draṁ	1	4		draṁ	1	6
1	1			1	1			1	1			1	1			1	1

. . . . . ṇetu*sa* . * * saṁ

. . . . . 1 6	4 2	||	śe *draṁ*
1 *3*.	.			

[66v.] . . śikena | 1 draṁ 1 | 1 rū 1 | 12 draṁ 1 | phalaṁ rūpa 1 2 .

. draṁ | 2 rūpa 1 | 6 draṁ 1 | phalaṁrūpa 12 | . . .

. draṁ | 3 rūpa 1 | 4 draṁ | phalaṁrūpa 12 | . . .

. . . *rū* . | . *2 draṁ* . . rūpa 12 . . . .

[67r.] . . . . 8 8 | jātā | 61 8 | . .

. . | sadṛiśaṁ | . . 1 2 11 8 | . u . . . | 16 8 61 8 | jātā | 77 8 | .

. . *sadṛ*iśaṁekasya | 16 16 | yutaṁ | 77 16 | jātaṁ | 93 16 | eshaphalaṁbhavati | . r . . y .

. . . pratyayaḥ | 93 16 | 1+ 1 2 1 2+ 1 | 2+ 1 2 1 3+ 1 | 3+ 1 2 1 4+ 1 | 4+ 1 | .

. || huṇḍikāsamānayanasūtraṁ || dinabhaktaviśeshaṁchavi .

. . . *kṛi*yatechaivakālameshāṁvinirdiśet trairāśikavidhānena . .

. | . . ādattaṁchapātavyaṁsūkshmedattaṁchatatsamaṁ || udāharaṇaṁ ||dvi

. . . . ṁ . e . . . . o . . . *dhanaṁ*

[67v.] . . *re*

. *naṁ* . . . . . y ksh *viśe*

. achchheshaṁtadviguṇaj . *tā* | nirgachchh*a* . . .

. *n*pr . viśamānechatvāridattaḥ punadviguṇa*ṁjātaṁ* . .

. . āsūṇyahaśtaṁgataṁtasyakimatumūla pana*sya* . .

. . bhā 2 1 | 2 1 2 1 bhā 3 1 || 3 1 2 1 bhā 4 1 || 4 1 2 1 bhā *5 1* || *mū*

. . . . . ś . *yutaṁ* y . . ṁ . . guṇaṁ*chā* . ri . . .

(a)

168
4

. (b) . . deśadv. . . . [68r.]*

. pāṭyajātāśeshaṁ 21

ekatraṁ 29

. . hārika . . draṁ 2 .

. *diy*

*The reverse of this is blank.

. yojana . . . . [69r]

2	2	2
5	*6*	*7*

dṛishya . . . . ‖ a*to*sadṛiśa hakaṁ | uparimāṁ śaṁtaṁ*ttu*lābhavanti

tvālimśa | dūnāchaurāśī . . . *tataṁktulā* | dvichatvāriṁ

. ṁ . . ṁ . . vanti etevrīhakāsarvattraḥ sthāp*anaṁ*asya

. . . yatrairāśikena . . yasyakṛiyate

5 *ā*	2 . ṁ	210	pha . m 84
1	1	1	

. . . . yate ‖ 7 . | 2 . | *210 ā* | *pha* . ṁ

. . . . k . taṁ 105 [69v.]

. . ‖ *udā* ‖ . . . . tṛibhirdatt*ai*tṛiguṇatṛiguṇenatu | vi

. . mi*veṅsi*taduchyatām

1	3	9	dṛishya 130
1	1	1	1

prakshepa

10	30	90

e katraṁ 1

. . . . . *vān* ‖ taṁ *śa*taṁ*tṛi*bhirdatyai*pa*ravaptrāpavaptṛi . . . .

phalaṁ . . chh . . . y .

4	6	9	dṛi 190
1	1		

	60	90

ekatraṁ

[70r.] (a) tola *5* . . *akbhito* (b)

| 35 / 2 | etebhāgā *5* |
| . 17 | 117 / 70 |

(c) . . . . e . . . .

. . . 2 to 0 rītā *7* pala 2 tola 1

. . trampala 6 ‖ ‖ udā ‖ samā . i

. . . . ṇipesīkṛitā*nicha* | *dva*chatisraś . ṁ

. . tisrasamā*dā*yatulitānitray*o*daśe |

. *e*kaikasyaś*ārdhaaḥ* ‖ $\frac{1}{2}$ ‖ $\frac{1}{3}$ ‖ $\frac{1}{4}$ ‖ d .

[70v.] (b) . . . . . ri . . . . . . ri (a) prakshepayuktyāphala*ṁ*

. . guṇyaphalarāśī

(c) . . . katra*ṁ*pala 8 ‖ ‖ udā ‖ ardhatṛi . .

. . . . . dāṅśaṁpaṁchaśashṭinṛipodadau | sevakānāṁtud .

. . . . . $\frac{1}{2}$ ‖ $\frac{1}{3}$ ‖ $\frac{1}{4}$ ‖ dṛishya $\frac{65}{1}$ ‖ sadṛi

. . . . *30* *20* .

. . . . ‖ ‖ . . .

PART II

iii.—*Facsimiles of the manuscript.*

THE BAKHSHĀLĪ MANUSCRIPT AS PRESERVED IN THE BODLEIAN LIBRARY
Folios
35 VERSO | *37 RECTO*
36 VERSO | *38 RECTO*

Plate II

1 RECTO

1 VERSO

2 RECTO

2 VERSO

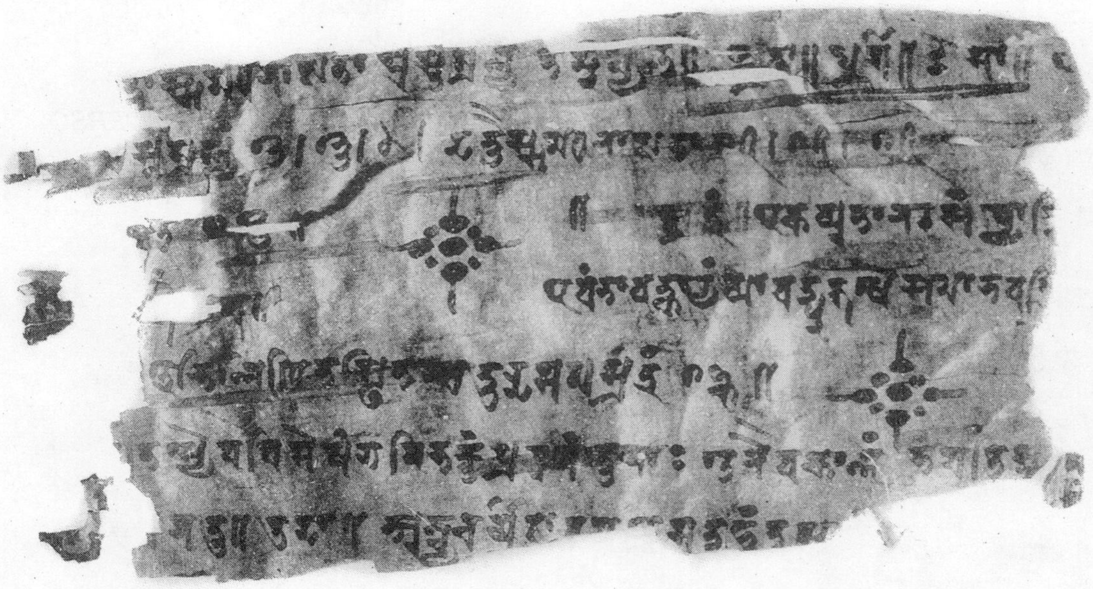

3 RECTO

3 VERSO

Plate IV

4 RECTO

4 VERSO

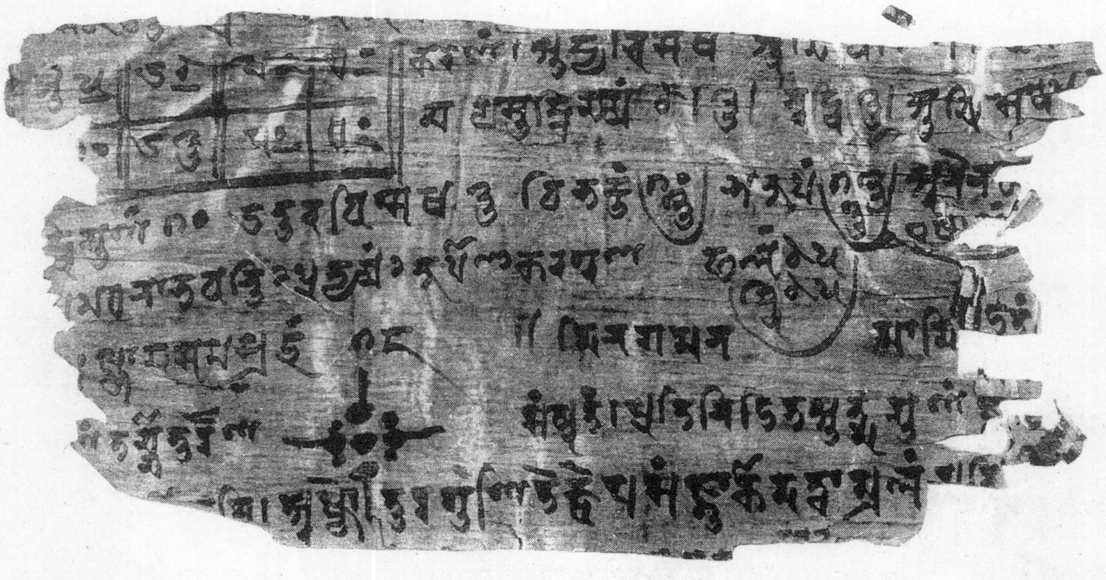

5 RECTO

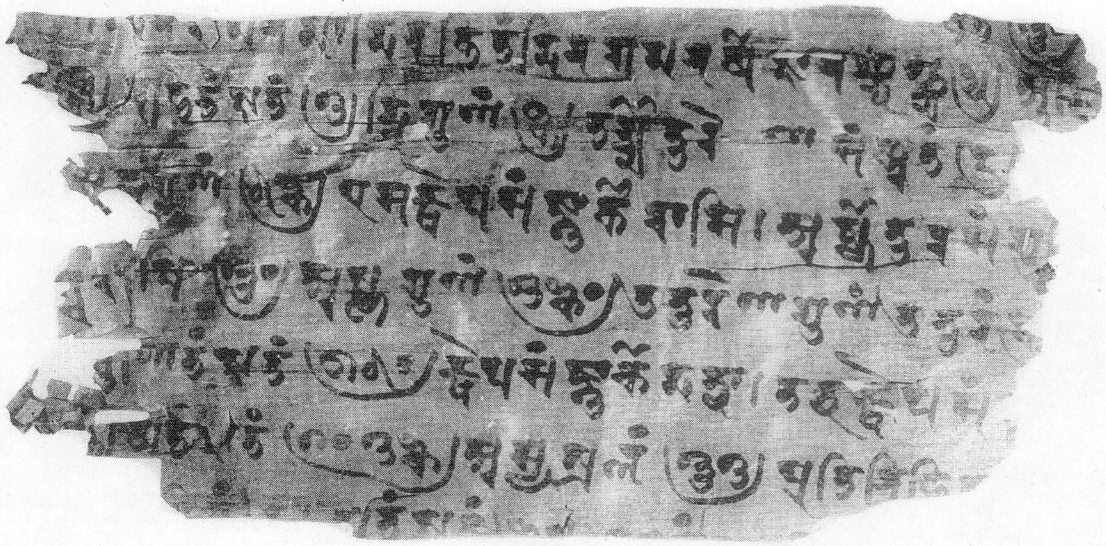

5 VERSO

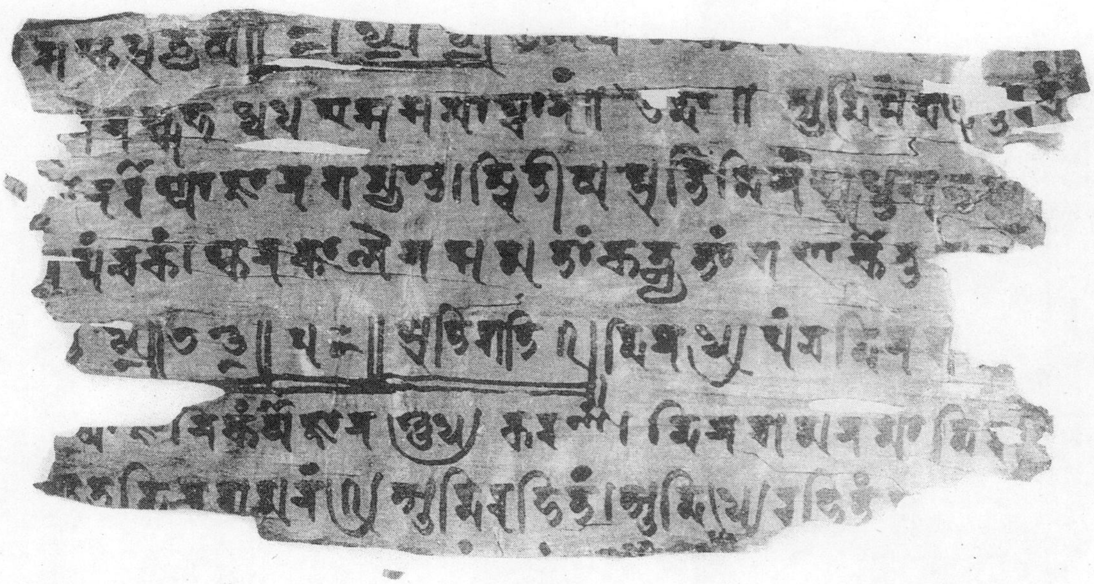

6 RECTO

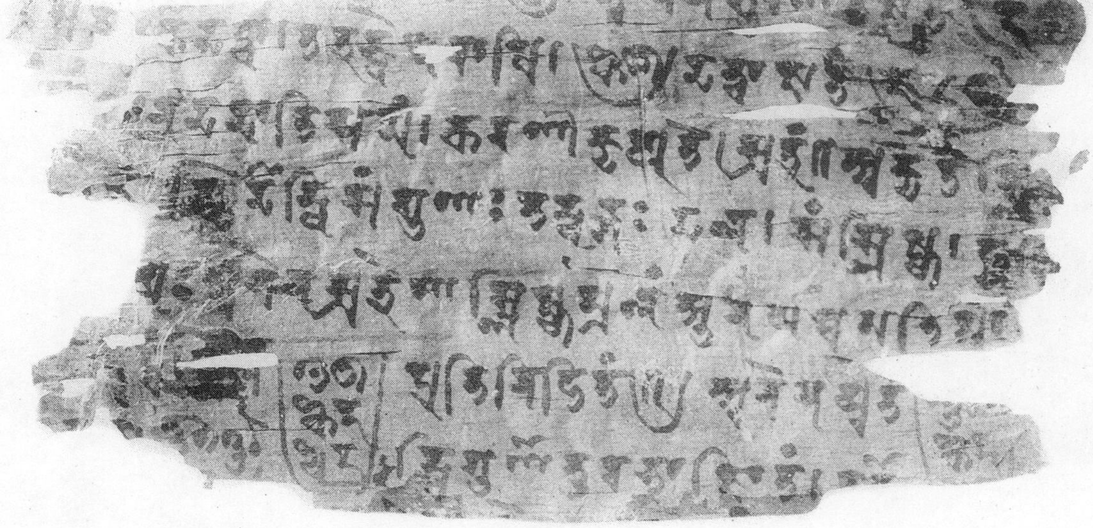

6 VERSO

Plate VI

7 RECTO

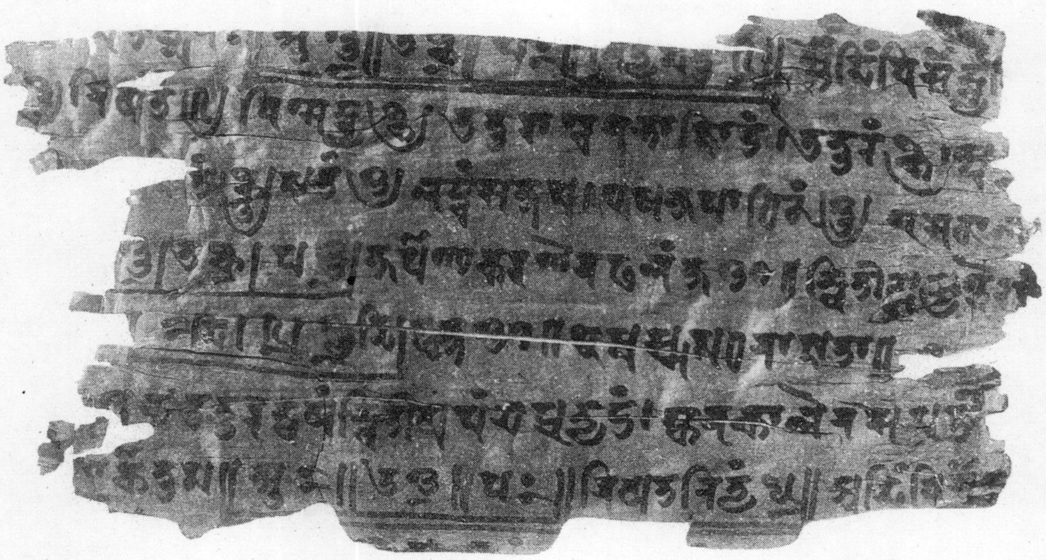

7 VERSO

8 RECTO

8 VERSO

9 RECTO

9 VERSO

Plate VIII

10 RECTO

10 VERSO

11 RECTO

Plate IX

11 VERSO

12 RECTO

12 VERSO

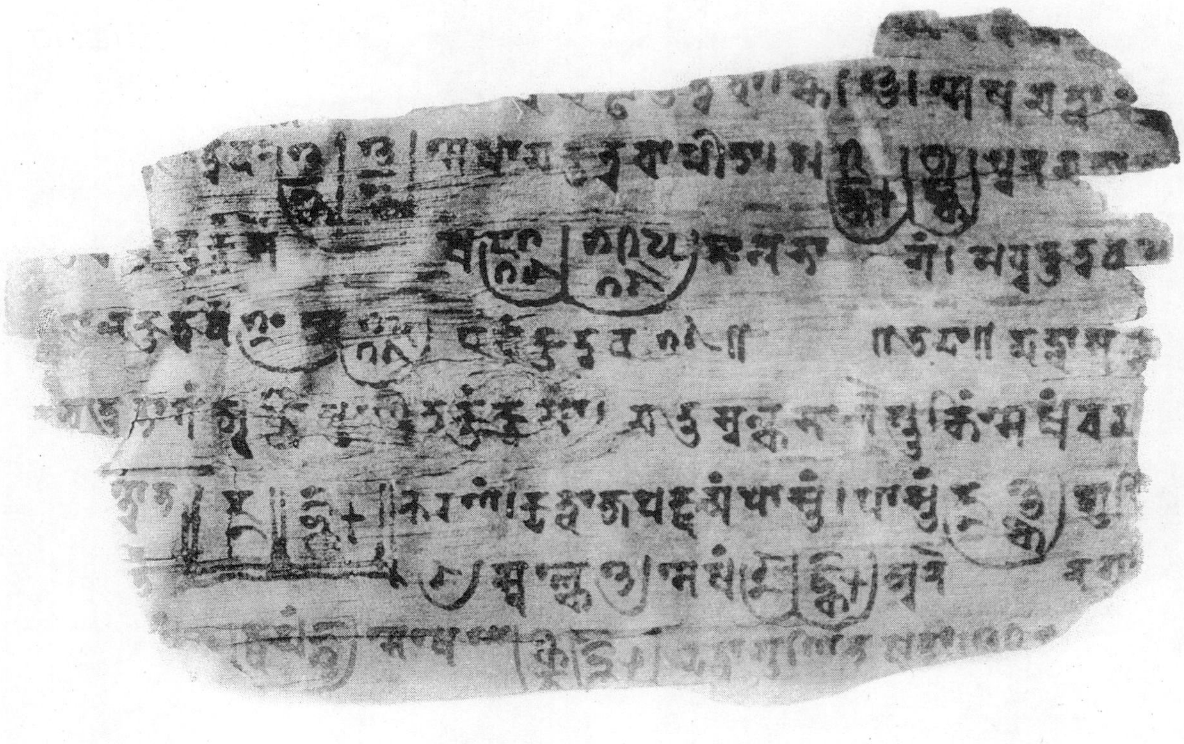

Plate X

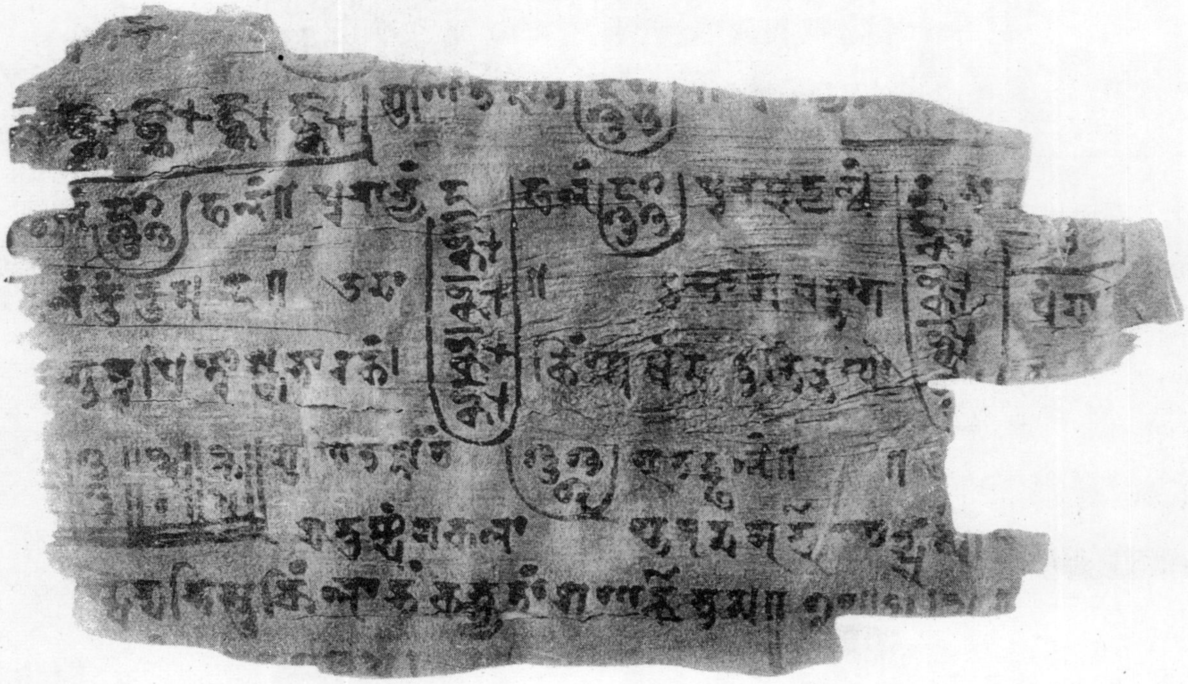

13 RECTO

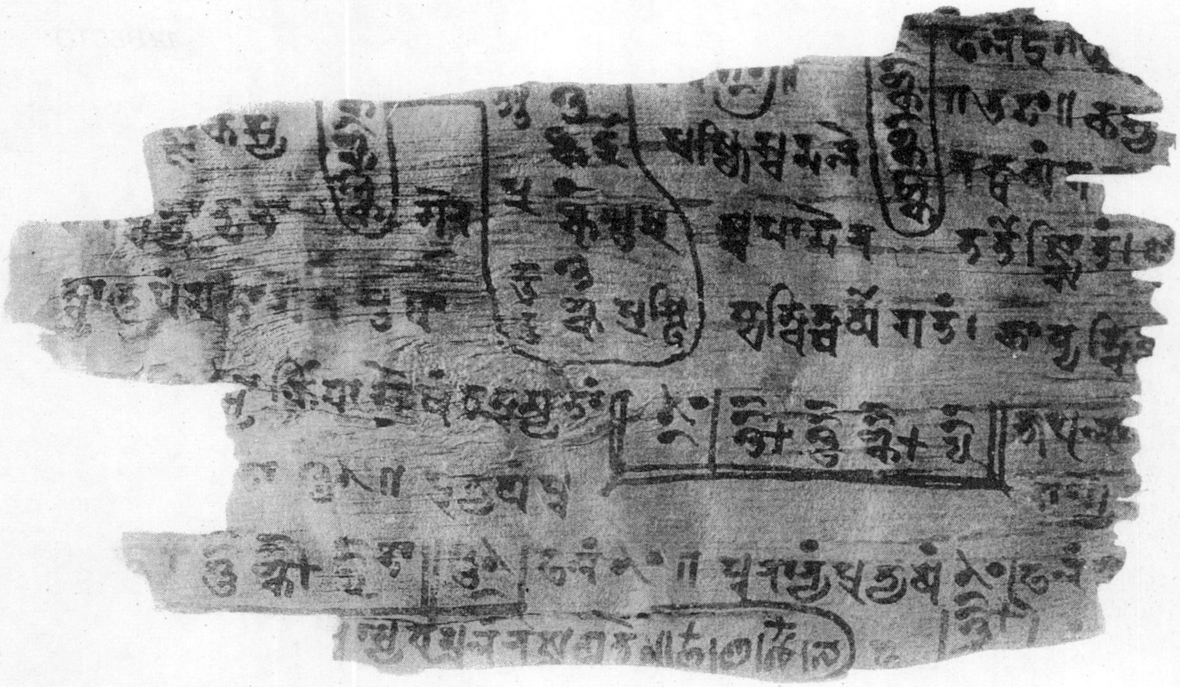

13 VERSO

14 RECTO

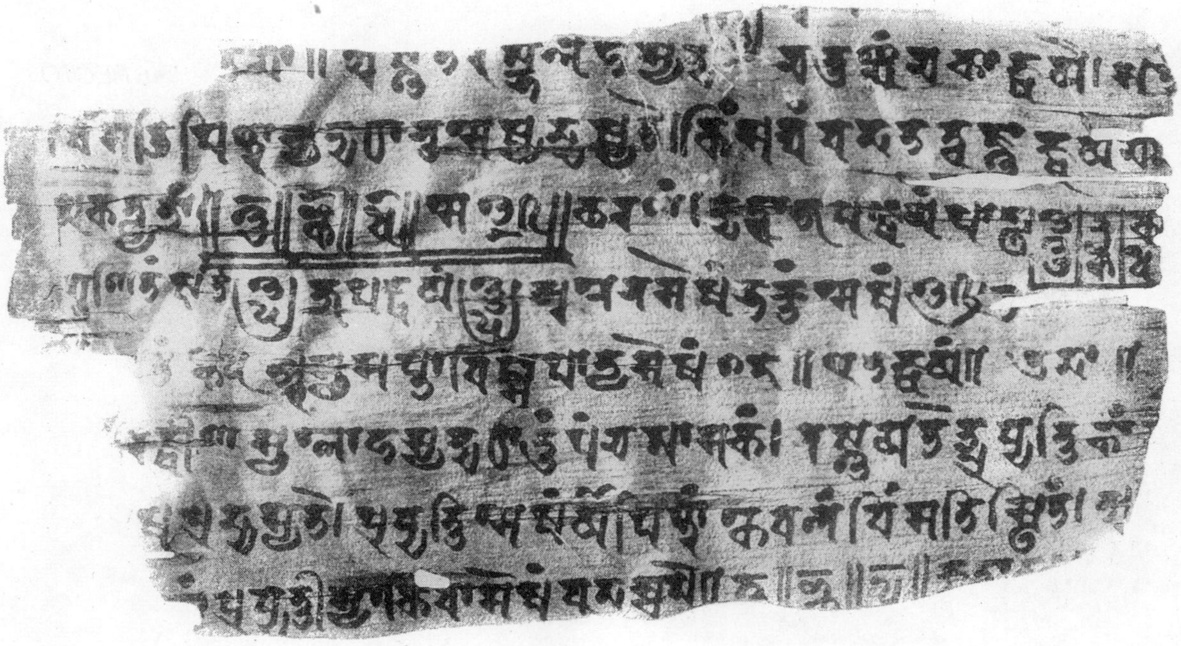

14 VERSO

15 RECTO

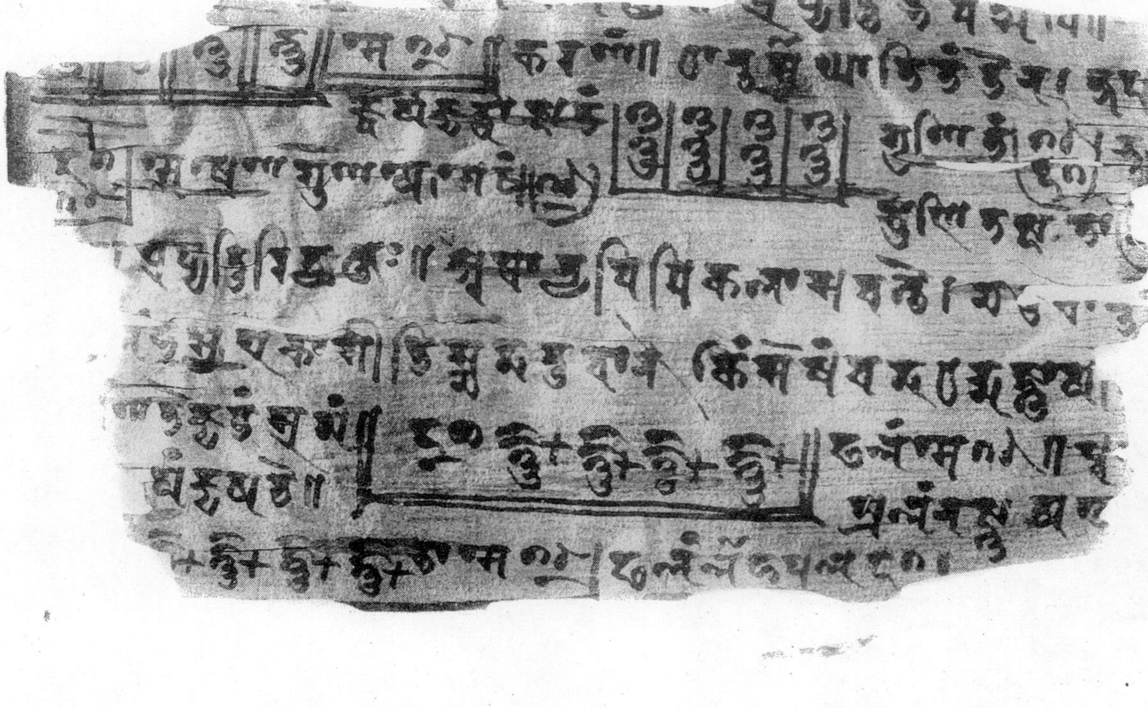

15 VERSO

16 RECTO

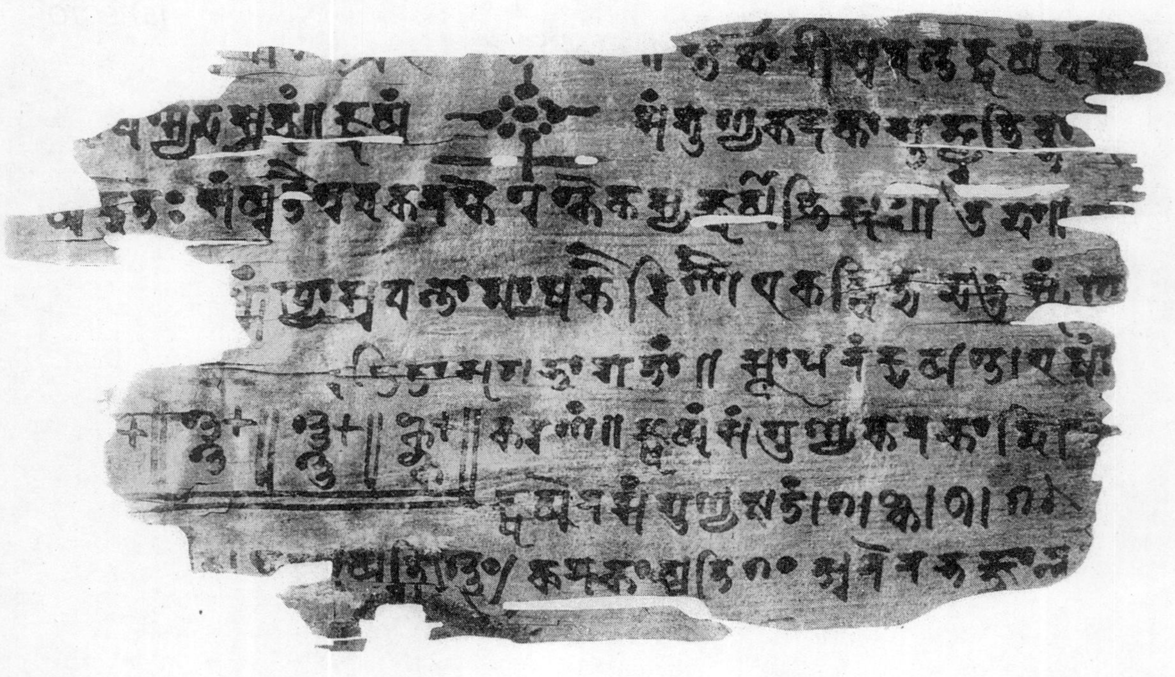

16 VERSO

17 RECTO

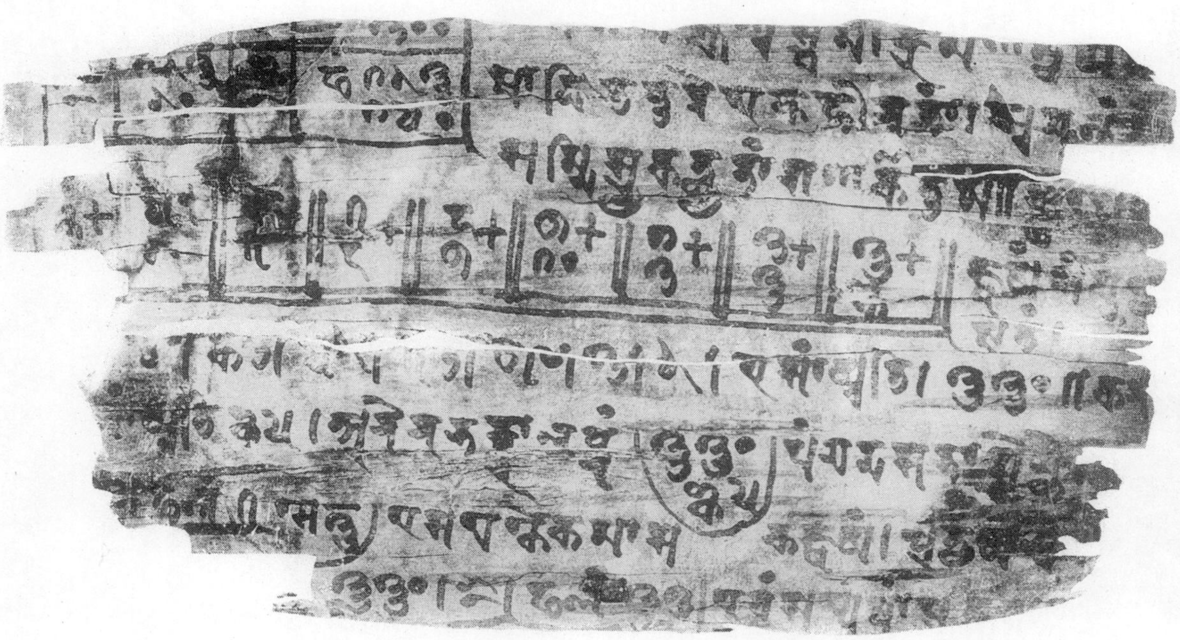

17 VERSO

18 RECTO

18 VERSO

Folio 19 is blank

Plate XIV

20 RECTO

20 VERSO

21 RECTO

21 VERSO

22 RECTO

22 VERSO

Plate XVI

23 RECTO

23 VERSO

24 RECTO

24 VERSO

25 RECTO

25 VERSO

Plate XVIII

26 RECTO

26 VERSO

27 RECTO

Plate XIX

27 VERSO

28 RECTO

28 VERSO

Plate XX

29 RECTO

29 VERSO

30 RECTO

Plate XXI

30 VERSO

31 RECTO

31 VERSO

Plate XXII

32 RECTO

32 VERSO

33 RECTO

33 VERSO

34 RECTO

34 VERSO

Plate XXIV

35 RECTO

35 VERSO

36 RECTO

36 VERSO

37 RECTO

37 VERSO

Plate XXVI

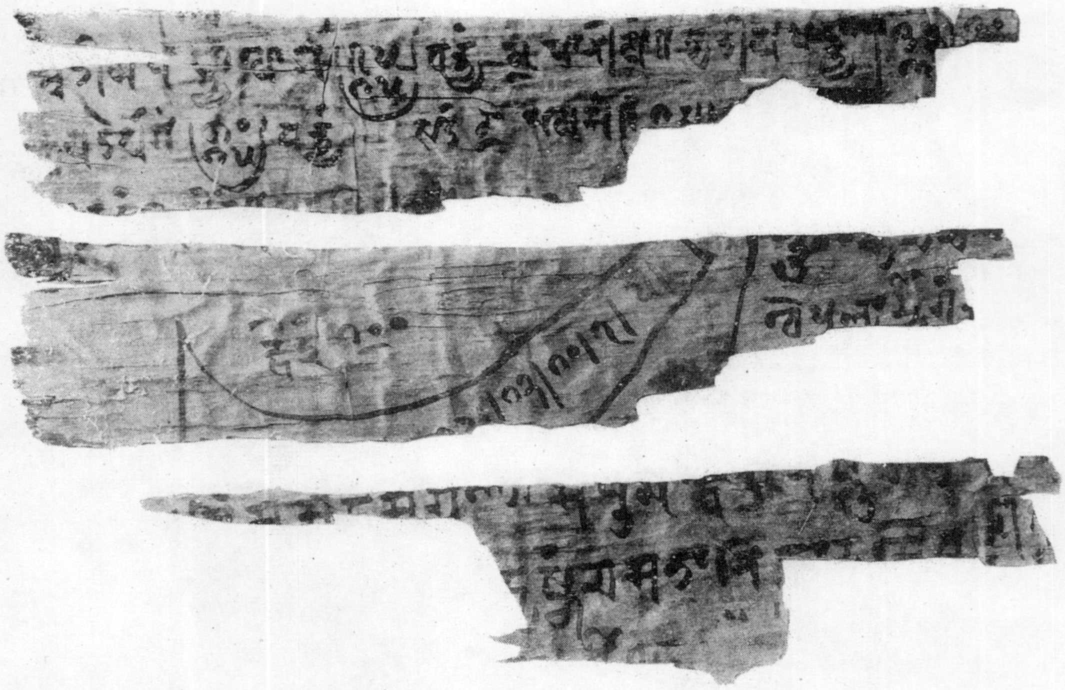

38 RECTO

38 VERSO

39 RECTO

39 VERSO

40 RECTO

40 VERSO

Plate XXVIII

41 RECTO

41 VERSO

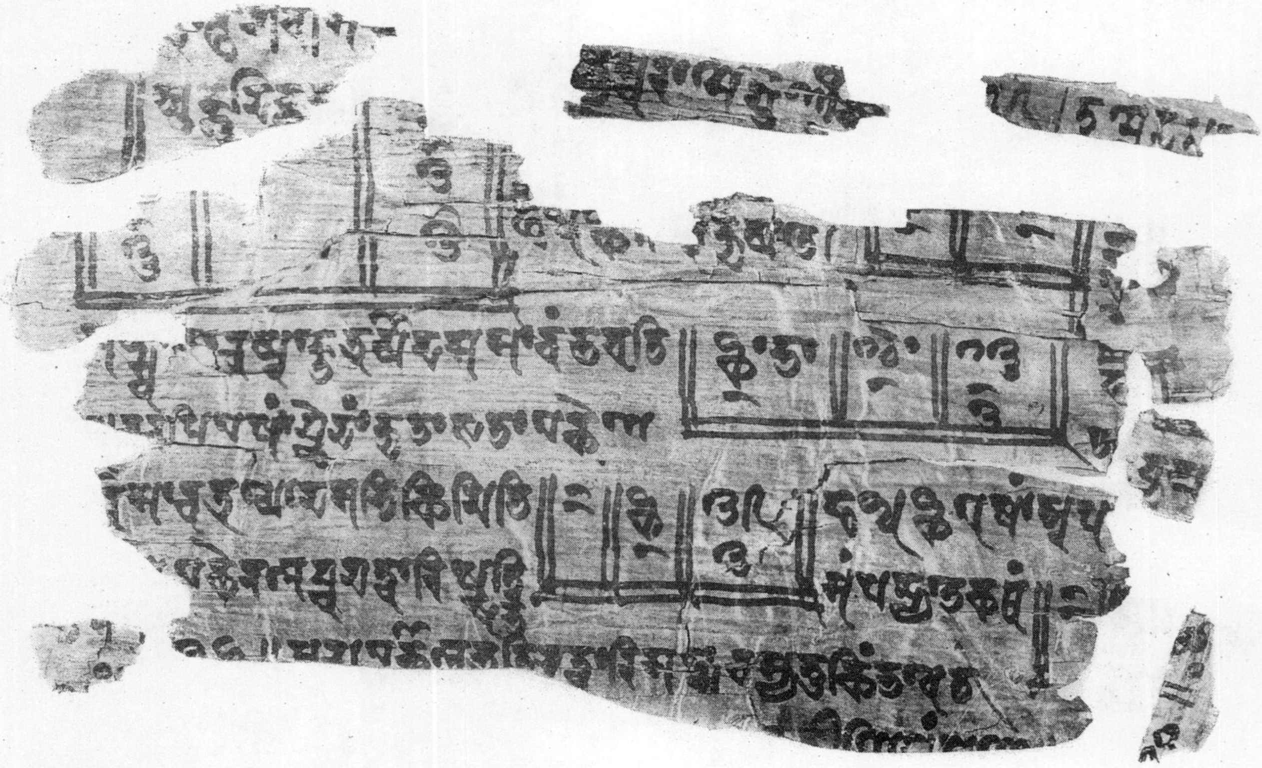

42 RECTO

42 VERSO

43 RECTO

44 RECTO

44 VERSO

45 RECTO

45 VERSO

46 RECTO

46 VERSO

Plate XXXII

47 RECTO

47 VERSO

48 RECTO

48 VERSO

49 RECTO

49 VERSO

50 RECTO

50 VERSO

51 RECTO

51 VERSO

52 RECTO

52 VERSO

Plate XXXVI

53 RECTO

Facsimile not available in original (see page 3)

53 VERSO

54 RECTO

54 VERSO

55 RECTO

55 VERSO

56 RECTO

56 VERSO

57 RECTO

57 VERSO

58 RECTO

58 VERSO

Plate XL

59 RECTO

9 Verso is blank

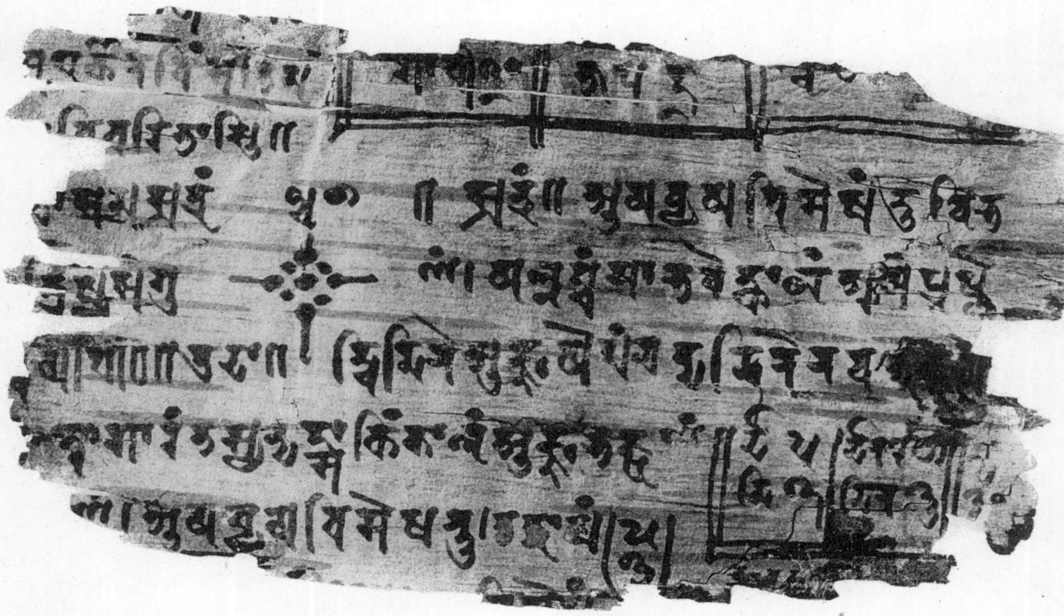

60 RECTO

60 VERSO

61 RECTO

61 VERSO

62 RECTO

62 VERSO

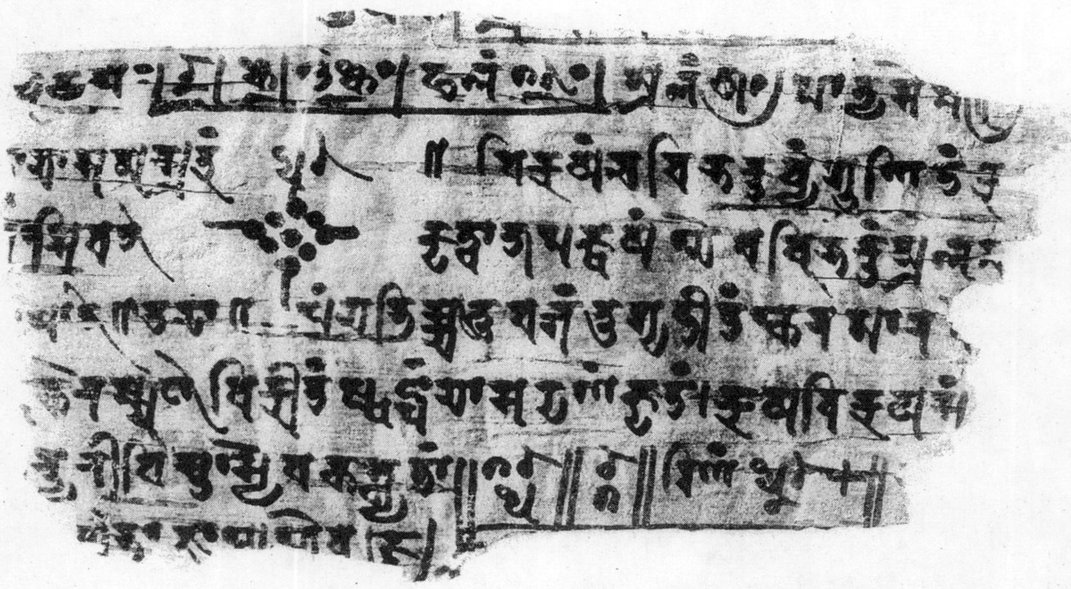

63 RECTO

63 VERSO

64 RECTO

64 VERSO

65 RECTO

65 VERSO

66 RECTO

66 VERSO

Plate XLV

67 RECTO

67 VERSO

68 RECTO

68 Verso is blank

69 RECTO

69 VERSO

70 RECTO

70 VERSO

INDEX

TO THE

BAKHSHĀLĪ MANUSCRIPT

INDEX

(The numbers marked 'Fol.' refer to the folios of the text. The other numbers refer to pages.)

A

B

C

C—*contd.*

D

E

F

G

H

I

J

K

L

M

M—*contd.*

N

O

P

Q

R

S

S—*contd.*

T

V

W

Y

MGIPC—L—X-3-5—5-10-27—550.

ARCHÆOLOGICAL SURVEY OF INDIA

New Imperial Series. Vol. XLIII

Part III

The Text re-arranged

THE
BAKHSHĀLĪ MANUSCRIPT

A Study in Mediæval Mathematics

BY

G. R. KAYE

DELHI: MANAGER OF PUBLICATIONS
1933

PART III.

1.—The Text Re-arranged.

Quotations in the text are distinguished by daggers† . . . . . †, and abbreviations by ° superscribed. Asterisks attached to numbers denote change-ratios (See § 103). In the foot-notes angular crotchets < > indicate that the portion enclosed formed part of the argument or was implied in the original text, but is now missing.

On pp. 13 and 14 of Part I are tables equating the Bodleian Library order with the revised arrangements.

The notes attached to the revised arrangement are very crude and are presented with considerable diffidence ; but they are the result of much labour and will possibly save the student of the MS a good deal of spade work.

G. R. K.

Owing to Mr. Kaye's unfortunate death, the last proofs of this part have been prepared for the press by Mr. K. N. Dikshit, Deputy Director General of Archaeology for Exploration, who has also made a few emendations.

A 1.

. . . yatra y . g . . . . bhāgaṁ chaiva kārayet kshetra vaipulya . . . 40ª recto.

. . pṛishṭhā śata-dvayaṁ chaiva uchare śatam ekataḥ vaipulyād vi . . . .

. . śa dvādaśa nṛi śakas tathā | sapta paṁcha bhavet chānaṁ bhakti sthāne 39ª recto.

. . r . . . . dhā sapta paṁchānāṁ tṛi-dvi meka ϕ prakalpitaṁ | tasya

vāhasya kiṁ ka . . . . . tatrā mama | kshetrasya.

sthāpanaṁ kṛiyate |

kshetraṁ 100

. . . . .	15	12	10	7	5

karanaṁ | kshetra . .

300 . . *da*ṁ cha . . . . 38ᵇ recto.

vaipulyād yogaṁ . . .

esha shaṭ . . . . 39ᵇ recto.

. . . . chaiva tat phal . . . . guṇitā jātā 40ª verso.

6210 | esha vāhasya kāṇḍa pramāṇaṁ . . . śake mūlyaṁ *karta*vyaṁ |

adha chchhedam chatus sashṭhi . . . . la | *sutha dvi* triṁśa*bhi* maṇḍalakai 39ª verso.

tallika esa chchhedaṁ bhavati . . . yathe chchh . . kāryā | *su*tha ṭala

kṛiyā udāharanaṁ . . ṭalasya . . . mekaṁ ta dvā sashṭi śatānāṁ

daśādhīkānāṁ kiṁ mulyaṁ : ṭalla . . . tale a° 38ᵇ verso.

| 1 . . rū 1 mūlye | 6210 maṇale abhim | pha . . .
| 1

A 1. [38-40.] Plates xxvi and xxvii exhibit some sixteen fragments all placed out of order. Some of these have now been pieced together. (See the illustration facing page 4.)

This grouping is not final because some of the fragments consist of portions of two or more leaves stuck together, and until these are separated no exact order can be achieved.

We should naturally expect the first leaves of the manuscript to be comparatively more damaged than those in the middle of the book, and the 'find order' and the writing indicate that these fragments are probably portions of early leaves; but neither of these criteria is rigorous and it is quite possible that we have placed the fragments in their wrong places.

[38-40 recto.] These fragments appear to relate to a geometrical problem concerning an area whose width (*vaipulya*) is increased.

[38-40 *verso*.] A fragment of a problem connected with the area of a circle or the surface (*ṭala*) of a sphere. The phrase *esha vāhasya kāṇḍa pramāṇam* ought to be illuminating but is not. The change-ratio 64 is possibly connected with a "square measure." See Part I §108(*b*). The number $6210 = 3^3 \cdot 230$ is said to be the product of certain quantities.

A 2.

. . . . **ksh** . . . . daśa | chatur-daśa tṛitī- 39b recto.

yasya chaturthasya . . . . bhāgās tasyaiva paṁchama . . . 40c recto.

. . . *bh*āgā viṁśas cha dasagunā | saptama ksh . . . jñā . . . . . . 38c recto.

40b recto.

yaṁ śataṁ | sarve miśrāpi dṛishṭhaṁ cha śatāni . . . 39d recto.

. . dhanaṁ 1 . . . . . . 10 || esha ekaika bhāgā guṇitā jā . . . . 39b verso.

60 | 180 | 200 | 300 | evaṁ dhanaṁ 1200 pratyaya trairāsikena . 00 40c verso.

. . . . dhanaṁ 1200

.	pha° 144
.	pha° 16 .
.	pha° 180
20	pha° 200

38c verso.

39d verso.

40b verso.

A 2. [38-40.] The writing on the two sides differs (*recto* α_1, *verso* α_2) and there are other indications that the fragments consist of portions of two leaves at least.

A 3.

bdhāṁbupayaso ghaṭaḥ eka miśrikṛit . . . 40d recto.

karaṇam | havẏa tulyaṁ vinikshipyaḥ 40e recto.

4	5	6	kuru prakshepakaṁ tata praksh
4	5	6	
4	5	6	

A 3. [38-40 recto.] See the plate facing page 4. The meaning is not clear, but $x(x+y+z)=60$
$y(x+y+z)=75$
$z(x+y+z)=90$

whence $(x+y+z)^2=225$ and $x+y+z=15$. The answers are $x=4$, $y=5$, $z=6$.

The writing is classed as α_2.

A 3—*contd.*

sthāpya 39^{a} recto.

4 pa 15	5 aṁ 15	6 15
4 15	5 15	6 15
4 15	5 15	6 15

. . . . . kṛiyate ‖ chaturbhi paṁchabhish shaḍbhi g . . prathama

rāśi yoga $\begin{array}{|c|}60\\15\end{array}$ vartyaṁ 4 madhū ghaṭa . . . dvitīya paṅktyā yoga $\begin{array}{|c|}75\\15\end{array}$ 38^{a} recto.

vartyaṁ $\begin{array}{|c|}5\\1\end{array}$ pānīyaṁ ‖ tritīya paṅktya kṛiyate yogaṁ $\begin{array}{|c|}90\\15\end{array}$ vartyaṁ jātaṁ

$\begin{array}{|c|}6\\1\end{array}$ payasāṁ . . . .

. . . . kṛitvā . . . guṇetu ‖ eko . . . kṛitaṁ 40^{d} verso.

śatatrayaṁ paṁchabhi ϕ purushair labdhaṁ kiṁ ādyaṁ prathamaṁ dhanaṁ ‖

. . . . . 120 ‖ . 2257 . . ṇaṁ . . . . t śeshe kshepa

16 anenātra bhāga $\begin{array}{|c|}32\\16\end{array}$ labdha $\begin{array}{|c|}2\\ \end{array}$ $\begin{array}{|c|}40\\12\end{array}$ pha . . . / pha 120 39^{c} verso.

. . labdher bhāg . $\begin{array}{|c|}28\\2\end{array}$ jātā 14 labdha kshepaṁ . . . . . . dṛi°60

prakshepa yukti 30 vibhaktaṁ $\begin{array}{|c|}1\\30\end{array}$ . . . ṇitā jātā . . 14 | 18 38^{a} verso.

| 28 | evaṁ 60

A 3. [38-40 *verso.*] Five separate fragments, too mutilated to be intelligible.

A 4.

i. . . . . . . dviguṇaṁ cha tṛi-ūna cha tṛitīyasya dhanaṁ bhavet . . . . 54ᵃ verso.
saṁyutaṁ | eka-viṁshatibhi χ krīto dīnāraistu rai . . . . ya
tu daṁ sā pṛithag vachaḥ ||
karaṇaṁ || yasya padaṁ na jñāyate . . . . etat prathamasya
dhanaṁ |

ii. . . . . cha dattavān hastag . . . . yeshāṁ | 0 | 2 54ᵇ verso.
| 2+ . . . .

54ᵃ recto.

dhanaṁ . . .

1	2	4	8
1	1	1	1

yātā | tayor yogaviyo . . . . kṛitāṁ rāshayaḥ

2	1	2+	9+	dṛi° 82
1	1	1	1	
1	2	4	8	
1	1	1	1	

. . bhājyā hitveti | tatra uttara rāśi . . . . . . uttaraṁ riṇam jātaṁ
. . . .
(b) sūtraṁ || . . . . (c) jātaṁ 76 esha prathamasya . . . . . 54ᵇ, ᶜ recto.

A 4. [54.] Folio 54 possibly consists of two leaves, or rather fragments of them, for there are ten pieces. The writing on the two sides differs—that on 54 *recto* may be classed as α_1 and that on the left side as α_2 and in this respect the leaf resembles fol. 35ᵇ. There is a characteristic *ye* at the bottom of 54ᵇ *verso* which is also found on 29ᵇ *recto et verso.*

[54ᵃ *verso.*] Seems to contain portions of a *sūtra*, an example and solution. The phrase *dviguṇaṁ cha triūna* seems to be referred to on fol. 35ᵃ *recto* but there we have *tryūna* with a particularly noteworthy conjunct *tryū* (see table IV, 5 part ii). The term *hastag(ataṁ)* on 54ᵇ *recto* (not necessarily connected with 54ᵃ) occurs only once more on fol. 1 *recto.*

[54ᵃ *recto.*] The phrase *tayor yogaviyo* also occurs on fol. 35ᵇ *verso.*

A 5.

. . . . . kasmāt kāraṇā | tayor yogaviyogasy āviyogas . . . . 35ᵇ recto.
bhājitā purusha 15 anena bhaktvā dhanaṁ | 9 / 15 | . . . . padvaya .
sahitaṁ ||
. . . . . . mūleṇa | 1 / 1 / 2 | eta dviguṇaṁ | 3 | dviyuta . . . . . . yasya 35ᵇ verso.
dhanaṁ | tadeva svārdhaṁ | 3 | asyārdhaṁ | 1 / 1 / 2 | yutaṁ nyāsa

A 5. [35ᵇ.] The writing is different on the two sides (α_1 and α_2) and possibly the fragment is a portion of two leaves stuck together. The phrase *bhājitā purusha* occurs on 51ᵇ *recto.*

A 6.

bhājitā hitvā | tatrottarā 1 | 1 | yutaṁ | 2 | $\begin{matrix}1\\4\end{matrix}$ $\begin{matrix}3\\2\end{matrix}$ $\begin{matrix}3\\1\end{matrix}$ | . . . . 51[b] recto.

9 | . eshā ϕ . ītha bhājitā | purushaḥ | $\begin{matrix}1\\4\end{matrix}$ | $\begin{matrix}3\\2\end{matrix}$ | $\begin{matrix}3\\1\end{matrix}$ | eshāṁ sadṛiśe 35[a] recto.

| $\begin{matrix}4\\19\end{matrix}$ | dhanaṁ | $\begin{matrix}19\\1\end{matrix}$ | anena guṇitaṁ jātaṁ | 4 | esha prathamasya dhanaṁ

. . . . dviguṇaṁ | 12 | dvi-yutaṁ | 14 | eta dvitīyasya . . . . .

guṇaṁ | 21 | dvi-guṇaṁ | 42 | try-ūnaṁ | 39 | eshaḥ nyāsaḥ

pratya . . . . daśam agravṛindānāṁ chatur-daśa ekonachatvāriṁśa | tat

pād-ārdha tṛi-bhāgā . . . .

4 | $\begin{matrix}1\\4\end{matrix}$ | pha° 4 | evam dī° 21 | esha prashṇa etair . . .

A 6. [51[b] and 35[a] *recto.*] The position is uncertain but the writing is of the α_2 style and there are slight indications of connexion with folio 54. Fol. 35[b] is in α_1 writing. See the plate facing page 4. (Read 51 *recto* B, not *verso.*)

The fragmentary contents are not clear. We have $1+1=2$; $\frac{1}{4}+\frac{3}{2}+3=\frac{19}{4}$; $\frac{19}{19/4}=4$ and $\frac{2.3.\ (12+2)}{2}-3=39$

Apparently a fragment of the *sūtra* on which the solution depends is preserved on fol. 54[a] *verso*, but the evidence, consisting of the phrase *dviguṇaṁ cha tṛir-ūna*, is slender.

udā || . . . . | 6 | . . . . . yoga 111 śeshā ϕ 51[b] verso.

purusha bhājitā purushāḥ $\begin{matrix}1\\4\end{matrix}$ | $\begin{matrix}1\\6\end{matrix}$ | $\begin{matrix}1\\5\end{matrix}$ eshāṁ sadṛishe yutiṁ kṛitvā yutā | $\begin{matrix}37\\60\end{matrix}$ | 35[a] verso.

bhājitā | $\begin{matrix}60\\37\end{matrix}$ | . . . . . esha gavāśva mahishī pratyaika śāleshu bhāga . .

1	śā°	180 / 1	gā°	$\begin{matrix}1\\4\end{matrix}$	phalaṁ 45	. .	. .
$\begin{matrix}1\\1\end{matrix}$	śā°	180 / 1	asvā°	$\begin{matrix}1\\6\end{matrix}$	phalaṁ 30	4 +	26
$\begin{matrix}1\\1\end{matrix}$	śālā	180 / 1	mahi°	$\begin{matrix}1\\5\end{matrix}$	phalaṁ 36	5 +	9

[51[b] & 35[a] *verso.*] The writing is of the α_2 class. The 'find order' of folio 51 is 37 while that of 35 is not known. The position is very uncertain. What remains of the problem is

$\frac{1}{4}+\frac{1}{6}+\frac{1}{5}=\frac{37}{60}$ $111\times\frac{60}{37}=180$

1 enclosure : 180 cows :: $\frac{1}{4}$: 45 ? subtract 6=39.
1 ,, : 180 horses :: $\frac{1}{6}$: 30 subtract 4=26.
1 ,, : 180 buffaloes :: $\frac{1}{5}$: 36 subtract 5=29 ?

A 7.

51ª recto.

1	1	1	1	
1	1	1	1	

. . . . | 1 | 2 | 3 | eshāṁ yuta $\frac{6}{1}$ | . . . . . . | 48 | śeshā ϕ

purusha sa 4 || anena bhājitā-r-labdhā . . . . sya bhavati | 12 | 13

| 14 | 15 | ekatraṁ 54 ||

ii. udā° || · kaśchid rājā dade dānaṁ sapta – paṁchāśakaṁ budha |

paṁchā . . . . . pravakshyāmy = anupūrvaśah . . . . .

dvi-guṇa dvi-guṇaṁ chaiva rūpa rūpottare . . . .

. . . pratbame prāptaṁ kiṁ prāptaṁ apare jane ||

0	1	2	. . . .
1	1	1	. . . .

51ª verso.

					dṛi 329
					1
1	3	9	27	81	

karaṇaṁ | uttar . . . . . tatrottara rāśīnāṁ yoga 87 esha dhanā dṛishyā śodhanīyā jātā 242 . . . . . . | purusha | 1 | 3* | 9 | 27 | 81 | yoga 121 anena . . . . jātā | 2 | esha dvau prathamasya dhanaṁ ||

2 | 6 | 18 | 54 | 162 | uttara rāśī saṁyutaṁ jātaṁ

2	15	48	147	444	eshāṁ . . . .
1	2	2	2	2	

A 7. [51ª.] Either there are two leaves stuck together here or there is some over-lapping. The writing on both sides is α_2. The find order is 37.

[51ª *recto.*] i. There is not enough material for reconstruction but $x+(x+1)+(x+2)+(x+3)=54$ therefore $4x=54-6$ and $x=12$ is indicated.

ii. A certain Rāja makes presents to 57 wise men, etc. See 52 *recto.*

[51ª *verso.*] This apparently does not connect up with the other side. It exhibits the solution of an example which may be expressed by

$$\left.\begin{array}{l} t_1+3t_1+3^2t_1+3^3t_1+3^4t_1 \\ \frac{3}{4}t_1+\frac{3}{4}(t_1+t_2)+\frac{3}{4}(t_1+t_2+t_3)+\frac{3}{4}(t_1+t_2+t_3+t_4) \end{array}\right\} =329$$

Set $t_1=2$ then the first series becomes 242 and the second 87 and the combined series is

$$2+\frac{15}{2}+\frac{48}{2}+\frac{147}{2}+\frac{444}{2}=329.$$ See also page 47.

* Omitted in the MS.

A 8.

i. . . . . 57 . . . tvedaṁ jātaṁ 3 anena chālimśa guṇaye 52 recto.

jātā 120 . . . . vam śurāṇāṁ ||

pratyaya trai-rāśikena

1 1	vaṁ	120 1	1 10	12 1 …
1 1	vaṁ	120 1	1 8	15 1 …
1 1	vaṁ	120 1	1 4	30 1 …

ii. udā° || dhanā sva-m-ardho saṁśoddhya . . . chottarīyakaṁ |

tat seshā paṁchamo bhāgo . . . . śata dvayaṁ |

aśītyādhikaṁ dhanaṁ chaiva kiṁ ādyaṁ prathamaṁ dhanaṁ ||

. . . . asya dvayānāṁ śatānāṁ pāda . . . rdhaṁ 52 verso.

śataṁ bhavati 150 atrāpi paṁcha bhāga 30 || evam . . . . .

1 2	1 4 1 2 +	1 5 1 4 + 1 2 +	pha° piṇḍā 280 ||

paṁchamī jātī karaṇaṁ kṛita . . . 280 | aṁśa yuti | 28 / 40 | bhaktaṁ | 40 / 28 |

dhanu 280 *guṇitaṁ jātaṁ* 400 esha phalaṁ bhavati ||

A 8. [52 *recto.*] i. The writing on both sides is α_2 and exhibits examples of the 'sickle-shaped' medial ī and ǐ. The 'find order' is 57. It is possible that 52 *recto* gives parts of the solution of the example on 51 *recto* which would make that page the reverse, but I doubt the connexion. What is left of the solution means

$x(\frac{1}{10}+\frac{1}{8}+\frac{1}{4})=57$ or $x\frac{19}{40}=57$ and $x=120$. A proof by the 'rule of three'

$1 : 120 :: \frac{1}{10} : 12$
$1 : 120 :: \frac{1}{8} : 15$
$1 : 120 :: \frac{1}{4} : 30$

ii. The example, which is c ntinued ɔn 52 *verso*, may be expressed by $x(1-\frac{1}{2})(1-\frac{1}{4})(1-\frac{1}{5})=x-280$ whence $x=\frac{280}{28/40}=400$. A proof follows: $\frac{400}{2}=200$ and $400-200=200$; $\frac{200}{4}=50$ and $200-50=150$; $\frac{150}{5}=30$ and $150-30=120$; and $400-120=280$.

Again $\frac{1}{2}+\frac{1}{4}(1-\frac{1}{2})+\frac{1}{5}(1-\frac{1}{4})(1-\frac{1}{2})=\frac{1}{2}+\frac{1}{8}+\frac{3}{40}=\frac{28}{40}$ and $280\times\frac{40}{28}=40ᴦ$.

A 9.

. . . . vinirdiset || 29[d] recto.

udā[c] || dhana . . . .

ādya dvitīya yonmiśraṁ dhanaṁ tatra ttrayodashaḥ

dvitīya tṛitīya yonmi . . . . . *chatur*daśa

ādya tṛitīya yonmiśraṁ dhanaṁ paṁchadaśa smṛitaḥ

ekaikasya dhanam . . . chchhiche katthyatāṁ mamaḥ

13	14	15
1	1	1

. . . . . prathamaṇ yasya tatrechhā paṁchaḥ | 5 | tat prathama . . . 29[b] recto.

. . 13/1 14/1 15/1 | tadādīś† śodhayet kramāt † ādi . . . . . . eta

chatur-daśabhi śodhya śeshaṁ | 6 | etat paṁcha 29[c] recto.

. . . . dvitīya yonmiśraṁ dhanaṁ 29[d] verso.

dvitīya tṛitīya yonmiśraṁ dhanaṁ sapta-dasha smṛitaḥ

tṛitīyaś chaturthayo . . . .

chatu*ṗ* paṁchaka miśraṁ tu dhanaṁ ekona-viṁśati |

prathama tatra cha

ekaikasya dhanaṁ kiṁssyād vechchhi . . . . .

16	17	18	19	20
1	1	1	1	1

29[b] verso.

karaṇaṁ || ichchhā . . . . dani†sodhayet kramāt† tatrādi 16

śud . . . . . tṛitīyāyaṁ śoddhya 7 chaturthāyaṁ śoddhya 12 paṁ 29[c] verso.

. . .

A 9. [29.] Folio 29 consists of six fragments, of which only the four larger ones need be considered at present. The correct order is *d, b, c.* Fragment *b* fits under *d* and *c* under *b* while *a* goes with folio 27. See the plate facing page 4.

[29 *d, b, c recto.*] The problem and its solution here partly preserved may be represented by $x_1+x_2=13$, $x_2+x_3=14$, $x_3+x_1=15$. If $x_1=5$ then $x_2=8$, $x_3=6$ and $x_3+x_1=11$ and the correct values are found from $x_1=5+\frac{15-11}{2}=7$, $x_2=13-7=6$, $x_3=8$. The phrase "śodhayet kramāt" recurs in the next example and is a quotation from a lost *sūtra.*

[29 *d, b, c verso.*] The example here given (continued on folio 27 *verso*) is formulated with exactly the same phraseology as the previous one. It may be represented by

$x_1+x_2=16$, $x_2+x_3=17$, $x_3+x_4=18$, $x_4+x_5=19$, $x_5+x_1=20$. If $x_1=10$, $x_2=6$, $x_3=11$, $x_4=7$, $x_5=12$ and $x_5+x_1=22$

Therefore the correct value of x_1 is $10+\frac{20-22}{2}=9$, $x_2=7$, etc.

The phrases *ichchhā* . . . and *śodhayet kramāt* are quotations from a lost *sūtra.*

A 10.

. . . . masya dhanaṁ | eshaṁ anukkrameṇa 27 verso.

pūrvokt . . . .

9 pra°	7 dvi°	10 tṛi°	8 cha°	11 paṁ°
7 dvi°	10 tṛi°	8 cha°	11 paṁ°	9 pra°

yutaṁ jātaṁ pratyaik . . . . 16 | 17 | 18 | 19 | 20 . . . . . .

. . . evaṁ sarvatra kārayet || 29ᵃ verso.

karaṇaṁ | †pṛithak rūpaṁ vinikshipya† | pṛithak rūpaṁ kshiptaṁ jātaṁ . 27 recto.

†. . . bhyāso† tatra guṇa | 3 | 4 | abhyāsaṁ | 12 | † rūpahīnaṁ† 1

. . abhyāsā chatuϕ paṁchakā | atra kshiptaṁ jātaṁ | 15 | 16 |

eśa trigun . . . . *tā* mūla . . ni chatuϕ paṁcha | 5 | 4 | esha .

. . . .

sūtraṁ || guṇau . . . . ka dhanaṁ || 29 recto.

guṇ ābhyāso rūpa hīnaṁ labdhaṁ rū . . . . .

A 10. [27 *verso*] gives the answer of the problem given on fol. 29 *verso*, namely $x_1=9$, $x_2=7$, $x_3=10$, $x_4=8$, $x_5=11$, and the sums of the pairs are 16, 17, 18, 19, 20. (For general discussion see § 78, Part I.)

[27 *recto.*] Solution of a lost problem which may have been $xy-3x-4y\pm1=0$ of which solutions are : $x=\frac{3\cdot4-1}{7}+4=15$, $y=3+1=4$; $x=4+1=5$, $y=\frac{3\cdot4+1}{1}+3=16$. The quotations are from a *sūtra*, very much like the one that follows.

The phrase *pṛithak rūpaṁ vinikshipya* 'having added unity in each case' appears to be a quotation from a lost *sūtra*.

[29*a*] is wrongly placed on plate XX. It should come directly under 27, for of the letters —*vaṁ sarvatra kārayet*, the top portions are on 27 *verso* and the bottom on 29*a recto*.

The writing is classed as α_2.

A 11.

i. bhyasa | .. / 1 | rū . . . . chaturguṇaṁ paṁchaguṇaṁ hastagataṁ 1 recto.

dhanaṁ ja . . . . paṁchaguṇaṁ 25 ||

navama sūtraṁ 9

ii. sūtraṁ || guṇau pṛithag rūpayutau yāchanā yukti samguṇāḥ ||

guṇanena guṇe . . rūpahīnena bhājitau |

viparīta yāchanā kshiptau guṇaśāster ayaṁ vidhiḥ ||

evaṁ sūtraṁ || dvitīya patre vivaritāsti ||

daśama sūtraṁ 10 ||

iii. sūtraṁ || aṁśāṁ viśoddhya chchhedebhya kuryātat parivartanaṁ ||

. . . sāsyaṁ tata projjhya dhanānviśa vinirdiśet ||

iv. udā° || paṁchānāṁ vaṇijā madhye *maṇi* vikrīyate kilaḥ

tatroktā maṇi vikrīta maṇi mūlyam kiyad bhavet

. . . . daṁ

ardha tṛi-bhāga pādāṅśaṁ paṁcha-bhāga *śodaṁśa cha*

. . . . †tato projjhyaḥ† sadṛiśaṁ kṛiyate . . . . jātā 1 verso.

120	90	80	75	72
60	60	60	60	60

tatra projjhya† jātaṁ 120 | 90 | 80 | 75 | 72

eshāṁ yoga kṛite jāta 437 ato . . . śeshaṁ 377 eśa maṇi mūlyaṁ |

A 11. [1 *recto.*] The position is uncertain. The 'find order' (33) places this leaf next to the fragments of folios 27, 29, 38, 39, 40. The writing is α_2 (there is a 'sickle' *i*). The numbered *sūtras* seem to place the leaf fairly early but they are not a very safe criterion. Note (ii) below seems to connect folios 1 and 27.

(i) Nothing intelligible. It ends the earliest numbered *sūtra* preserved.

(ii) I have not yet made out the meaning of this *sūtra*. Compare the opening phrase with the quotation on 27 *recto*. The metre is irregular. The reference to the second leaf is possibly to folio 27.

(iii) The *sūtra* means change $\frac{a}{b}$ to $\frac{b}{b-a}$ and quotations from it are given on folios 1 *verso* and 2 *verso*.

(iv) The example is solved on 1 *verso* and 2 *recto* and appears to have been somewhat as follows:

The combined capitals of five merchants less one-half of that of the first, one-third that of the second, one-fourth that of the third, one-fifth that of the fourth, or one-sixth that of the fifth is equal to the cost of a jewel. Find the cost of the jewel and the capital of each merchant.

A 11. [1 *verso.*] This appears to give part of the solution and proofs of the question on 1 *recto*. Since < $\Sigma x-\frac{1}{2}x_1=\Sigma x-\frac{1}{3}x_2=\Sigma x-\frac{1}{4}x_3=\Sigma x-\frac{1}{5}x_2=\Sigma x-\frac{1}{6}x_5=c$, we have $\frac{1}{2}x_1=\frac{2}{3}x_2=\frac{3}{4}x_3=\frac{4}{5}x_4=\frac{5}{6}x_5=k$ > whence $\Sigma x=\frac{120+90+80+75+72}{60}=\frac{437k}{60}=\frac{437c}{437-60}$. If $k=60$ then $c=377$ and $x_1=120$, $x_2=90$, $x_3=80$, $x_4=75$, $x_5=72$.

Then follows a 'proof' which may be expressed by

$$90+80+75+72=317 \text{ and } 317+\tfrac{120}{2}=377$$
$$120+80+75+72=347 \quad ,, \quad 347+\tfrac{90}{3}=377$$
$$120+90+75+72=357 \quad ,, \quad 357+\tfrac{80}{4}=377$$
$$120+90+80+72=362 \quad ,, \quad 362+\tfrac{75}{5}=377$$
$$< 120+90+80+75=365 \quad ,, \quad 365+\tfrac{72}{6}=377 >$$

† Compare with *sūtra* 11 on 1 *recto*.

A 11—*contd.*

chaturthāṁ saṅka sarvasvam ||

prathamasya saṅka ardham . . . 90 | 80 | 75 | 72 chaturnāṁ yoga

317 prathamārdheṇa sashṭibhir yutaṁ 377 eśa *prathamasya dhanaṁ*

prathama dhanaṁ | tṛitīya chaturtha paṁchamasya dhanaṁ sarvasvaṁ 347

dvitīyā tṛi-bhāgaṁ 30 *eśa* yutaṁ 377 esa dvitīyasya dhanaṁ bhavati ||

puna prathama dvitīya chaturtha paṁchama . . sarvasvam 357 tṛitīyasya

pādam 20 eśa yutaṁ 377 esha tṛitīyasya *dhanaṁ bhavati* ||

punar api prathama dvitīya tṛitīya paṁchamasya 362 *chaturthasya paṁcho*

bhāga 15 *eśa yutaṁ* 377 esha *chaturthasya dhanaṁ* bhavati || . . . .

A 12.

i. Sya dhanaṁ bhavati ||

atha pratha . . . . tyaṁśashṭi śeshaṁ 377 || 2 recto.

atha dvitīyasya

120
30
80
75
72

evaṁ 377 dvitīyasya bhavati ||

atha tṛitīyasya kṛiyate

120
90
20
75
72

evaṁ 377 tṛitīyasya dhanaṁ bhavati ||

chaturthasya kṛiyate

120
90
80
15
72

evaṁ 377 chaturthasya dhanaṁ bhavati |

paṁchamasya kṛiyate | sthāpanaṁ

120
90
80
75
12

evaṁ paṁchamasya 377 |

A 12. [2 *recto.*] i. This appears to be another ' verification ' of the example on 1 *recto et verso* ; and means

<$\frac{120}{2}+90+80+75+72=377$>

$120+\frac{90}{2}+80+75+72=377$

$120+90+\frac{80}{4}+75+72=377$

$120+90+80+\frac{75}{5}+72=377$

$120+90+80+75+\frac{72}{6}=377$ and ' this is the measure of the price of the jewel.'

A 12—*contd.*

esha maṇi mūlyaṁ pra . . . . .

ii. udā° || anyonya vidita vibhavaṁ . . . . vaṇik*dvaya*ṁ |

tṛi . . . . dalaṁ tatha . . 2 verso.

7+	3+	5+
12	12	6
12	12	6

†aṁśāṁ viśoddhya† visodhayet riṇaṁ sthitaṁ | esha . . . kṛiyate

19	7	11
12	4	6

†kuryātat parivartanaṁ†

12	4	6
19	7	11

chchhede

. . . . jātaṁ asya

924	836	798
1463	1463	1463

projjhya jātā

924	836	798

eshāṁ yutiṁ kṛiyate . . . jātā | 2558 | chchheda projjhyaṁ 1095 etan maṇi mūlyaṁ

A 12. ii. This is possibly the question solved on 2 *verso*.

[2 *verso*.] The general meaning is: since $x_3+x_2-(\frac{1}{3}+\frac{1}{4})x_1=x_1+x_3-(\frac{1}{4}+\frac{1}{2})x_2=x_2+x_1-(\frac{1}{2}+\frac{1}{3})x_3=c$, or $\Sigma x-(1+\ _7\ )x_1=\Sigma x-(1+\frac{3}{4})x_2=\Sigma x-(1+\frac{5}{6})x_3=c$, whence $\frac{19}{12}x_1=\frac{7}{4}x_2=\frac{11}{6}x_3=\Sigma x-c$ and $\frac{\Sigma x}{\Sigma x-c}=\frac{12}{19}+\frac{4}{7}+\frac{6}{11}=\frac{924+836+798}{1463}$. Setting $\Sigma x-c=1463$ we have $x_1=924$, $x_2=836$, $x_3=798$; $\Sigma x=2558$ and $c=1095$ 'which is the price of the jewel.'

I do not, however, understand the form of the first statement; but see fol. 65 *verso* where $\begin{matrix}880\\40\\42\end{matrix}$ means $\frac{880+40}{42}$.

† *aṁśāṁ viśoddhya* and *kuryātat parivartanaṁ* are quotations from *sūtra* 11 on fol. 1 *recto*.

A 13.

udā° || . . . . dvitīyasya hayān navaḥ 3 verso.

ūshṭrā dasa tṛitīyasya . . . .

. . . . pradattaṁ cha parasparaṁ

pṛithag dhanaṁ tu vaṇijāṁ mūlyaṁ vā prāṇināṁ pṛithak

yadi . . . . vaktum tato me chchhindhi saṁsayaḥ ||

A 13. [3 *verso*.] Writing α_2. Note the looped medial *e* in the penultimate line. Possibly a double leaf. 'Find order' 49. The position is determined only by the writing and the numbered *sūtras* on the reverse. *Example:* One possesses 7 horses (*a°=asva*), another 9 horses (*ha°=haya*) and a third 10 camels (*ū°=ūshṭra*). Each gives one of his animals to both the others (and then their possessions are of equal value). It is required to find the capital of each merchant or the price of each animal. If thou art able, solve me this riddle.

We have $(7-2)x_1+x_2+x_3=(9-2)x_2+x_3+x_1=(10-2)x_3+x_1+x_2=c$ or $\Sigma x-(7-3)x_1=\Sigma x-(9-3)x_2=\Sigma x-(10-3)x_3$, whence $4x_1=6x_2=7x_3=k$ and $\Sigma x=\frac{42+28+24}{168}k$. If $k=168$ then $x_1=\frac{168}{4}=42$, $x_2=\frac{168}{6}=28$, $x_3=\frac{168}{7}=24$. Also $7x_1=294$, $9x_2=252$, $10x_3=242$ are the original capitals, and $c=262$.

Mahāvīra gives the following . example.

Rule.—The number of gems to be given away is multiplied by the total number of men. This product is subtracted from the number for sale: the continued product of the remainders gives rise to the value of the jewel provided the remainder relating to it is given up.

Example.—The first man had 6 sapphires, the second had 7 emeralds and the third 8 diamonds. Each by giving to each the value of a single stone became equal (in wealth to the others). *Answer* 20, 15, 12,

A 13—*contd.*

7 a°	9 ha°	ū° 10
1	1	1

vaṇijjakā 3 deyaṁ . . . . vaṇik piṇḍa hataṁ | piṇḍa 7 | 9 | 10 | deyaṁ 3 śuddha śesham 4 | 6 | 7 tata śeshaṁ paraspara kṛitaṁ guṇita jātaṁ | 168 | 168 | 168 | svaśeshena tu vibhaktaṁ

168	168	168
4	6	7

labdhaṁ 42 | 28 | 24 | esha pratyaika mūlyaṁ

ekaikasya . . . . guṇitā jātāni asvai hayai ūshṭrebhyaḥ 294 | 252 | 240 ekaikasya . . . . jātā 262 | 262 | 262 | etes sama dhanā

i. datvā ssamadhanā jātā prasta mūlyaṁ tad uchyatāṁ 3 recto.

4 ya°	5 go°	6 sa°
1	1	1

evaṁ prasta mūlyaṁ 2 | 3 | 6 dattais samadhanā jātā 17 | 17 | 17 . . . .

trayodasama sūtraṁ 13

ii. sūtraṁ || ekayutānāṁ saṁkhyā dvi . . . . hīnā cha ||
evaṁ tāvat kāryaṁ yāvat purushai samā bhavati ||

saptama patre bhilikhita sthita

chatur-daśama sūtraṁ 14

iii. sūtraṁ || gatisyaiva viśeshaṁ cha vibhaktaṁ pūrva gaṁtunāḥ
tenaiva kālaṁ bhavati stha . . . . kena tu ||

iv. udā° || addhyardha yojana gate śata . . . . . .
. . . .

A 13. [3 *recto.*] This is the reverse because *sūtra* 15 obviously begins a new section (B).

i. This appears to be a companion example to that on 3 *verso.* The abbreviations are possibly *ya°* for *yava* 'barley,' *go°* for *godhūma* 'wheat,' sā for *sāli* 'rice.' Here $(4-2)\ x_1+x_2+x_3=(5-2)\ x_2+x_3+x_1=(6-2)\ x_3+x_1+x_2=c$ whence $x_1=2x_2=3x_3$ and $x_1=6$, $x_2=3$, $x_3=2$ and $c=17$.

ii. Not understood. The reference to the seventh leaf is now only tantalising. No recognisable quotations from the *sūtra* are preserved. The phrase *tāvāt* ...... *yāvat* 'so much ...... as much' does not recur anywhere. In Bhāskara *tāvat* and *yāvat* (*tā°* and *yā°*) are used as algebraic quantities.

iii. The rule means $t=\frac{r_1 D}{r_2-r_1}$ where r_1 and r_2 are rates of progress and D is a given time. (See § 83, Part I.) The rule is quoted on 4 *recto* where *gatisyaiva viśesham cha* and *pūrva gata* occur.

B 1.

i. sūtram || dviguṇaṁ prabhavaṁ śuddhā dviguṇaṁ *niyathaṁ tathā* 8 recto.

uttareṇa bhajech chheshaṁ labdhaṁ rūpaṁ vinirdiśet ||

ii. udā° || var*tate* bhṛitakaχ kaschi tatraiko dasha māśakaṁ |

pratyahaṁ kurute tatra karmaṁ bhaṭṭika mānavah

dvitīyaṁ kṛiyate karmaṁ dvyādi tṛitayar uttaraṁ |

padaṁ tatra tu bhavati kena kālena sāsyatāṁ ||

ā°		u°		pa°		prati°	
	$\frac{2}{1}$		$\frac{3}{1}$		$\frac{0}{1}$		$\frac{10}{1}$

‡dviguṇaṁ prabhavaṁ śuddhā‡ prabhavaṁ | 2 | dviguṇaṁ | 4 | niyata puna dvi

. . . . . | 16 | [uttarārdheṇa bhājayet] uttaram . . . . .

i. sūtram || *hayor vibhajya gantavyaṁ ato bhāga* gantata 8 verso.

ekaś cha gamana jñeya yutās samguṇya . . .

udā° || niyo rathośvair daśabhir yujyate haya pamchakaṁ

gaṁtavyaṁ yojana *śatam* . . . . kim udbhavet

ha		haya lagna rathasya		gantavyo yojana	
	$\frac{10}{1}$		$\frac{5}{1}$		$\frac{100}{1}$

†hayor vibhajya gantavyaṁ† tatra hayā | 10 | gantavyaṁ yo° | 100 | †ato

B 1. [8 *recto.*] The position of folios 8, 9 and 7 is very doubtful. They fit in nowhere perfectly. Their find orders are 48, 43 and 45; but 7 *recto* indicates that this find order is not of much value here. See the notes on fol. 7 *verso*. The writing is α4.

i. The rule is another variation of that given on 7 *verso* and means $t=\frac{2A-2a}{d}+1$ where A is a fixed rate and t $A=((t-1)\frac{d}{2}+a)t$.

ii. The example is $A=10$, $a=2$, $d=3$ whence $t=\frac{2.10-2.2}{3}+1$ $<=6\frac{1}{3}$ and $s=\frac{190}{3}=63\frac{1}{3}>$.

The phrase *dviguṇaṁ prabhavaṁ śuddhā* is quoted from the *sūtra* above; while the phrase *uttarārdheṇa bhājayet* was wrongly quoted and was afterwards cancelled: Compare with the *uttarārdhena bhājitaṁ* quoted on 7 *verso*.

B 1. [8 *verso.*] i. The *sūtra* is partially reconstructed from the quotations in the solution below.

ii. The example is: There are ten horses of which five are yoked at a time to the chariot. How many changes should there be in a journey of one hundred yojanas and how much will each horse do?

The solution is $\frac{100}{10}=10$ stages and $10\times5=50$

Proof. 5×100 . . . . .

Mahāvīra gives a similar example (vi, 158).

ravi-ratha-turagās sapta hi chatvāro'śvā vahanti dhūryuktāḥ |
yojana-saptati-gatyah ko vyūḍhaḥ ko chaturyogaḥ | |

'It is well known that the horses of the Sun's chariot are seven. Four horses are yoked at a time. They have to perform a journey of 70 *yojanas*. How many times are they unyoked and how many times yoked.'

Mahāvīra's solution is expressed thus:

The number of the total *yojanas* divided by the total number of horses gives the *yojanas* in turn. These *yojanas* multiplied by the optionally chosen number of horses to be yoked gives the measure of the distance to be travelled over by each horse.

That is $\frac{70}{7}=10$ is the length of each stage; and $10\times4=40$ gives the distance each horse works.

The solution is rather cryptic, but the interesting point is that the problem was a traditional one. Probably something of its original quality has been lost.

B 1—*contd.*

bhāga† *hṛite l*abdha 10 tatra yuktāśva | 5 | etais saṁguṇya pariyoga jātaṁ

. . . . yojanānyaikośva rūḍha | pratyayaḥ paṁchabhis śata saṁguṇya

jātaṁ . . . *kṛi*yate || yadi da . . yojana paṁcha . . . . . .

B 2.

udā° || . . . . tat samāptaṁ dvijanmabhi | 9 recto.

tat punas te samaṁ bhaktvā daśa . . samāptavān |

saṁkhyāya χ kati māchakshu kati viprā χ kati prashṭaṁ ||

ā° 1 1	u° 1 1	pa° 0 1	labdhaṁ 10 1

karaṇaṁ || ‡labdhaṁ dviguṇitaṁ kṛitvā‡ tatra labdhaṁ | 10 | dviguṇaṁ

| 20 | tathādvyūnaṁ | 18
1 | ‡uttareṇa vibhājitaṁ‡ atrottaraṁ | 1 | anena

bhaktvā jātam tad e*sha rūpādh*ikaṁ | 19
1 | ayaṁ prashṇā brāhmaṇā ekona-

viṁśati

sthāpa . . .

ā° 1 1	u° 1 1	pa° 19 1	rūpoṇā karaṇena phalaṁ	190 1

. . . . 9 verso.

. . yo° 6 1	śa° yo° 1 1	yo° 70 1	gantavyaṁ

B 2. [9 *recto.*] See the notes on fol. 7 *verso*. The writing is of the same style, α4. Possibly there are two leaves stuck together. The example is a=1, d=1, A=10, and $10t=((t-1)\frac{1}{2}+1)t$ whence $t=\frac{2.10-2.1}{1}+1=19$ and by the *rūpoṇa* method s=190.

Dr. Hoernle gave the following restoration:

> "For a certain feast one Brāhman is invited on the first day, and on every succeeding day one more Brāhman is invited. For another feast 10 Brāhmans are invited on every day. In how many days will their numbers be equal; and how many Brāhmans were invited."

The use of the term *labdhaṁ* is here rather curious. The phrases *labdhaṁ dviguṇitaṁ kritvā, tathādvyūnaṁ, uttareṇa vibhājitaṁ* and *rūpādhikaṁ* are probably quotations from a *sūtra.*

B 2. [9 *verso.*] The example probably meant: 'A and B start for a place 70 *yojanas* distant. A travelled at the rate of 1 *yojana* a day and B at the rate of 6. At what point on his return journey did B meet A?'

Since $\frac{x}{1}=\frac{2.70-x}{6}$, where x is the distance traversed by A, we have $x=\frac{2.70}{6+1}=20$ as given in the text, and since A travels at the rate of one *yojana* a day, this is also the time.

Proof by the 'rule of three' 1 day : 6 yo° :: 20 days : 120 yo°, and 70−20=50 and 70+50=120. Also 1 day : 170° :: 20 days : 20 yo°.

The abbreviation *sa°* may be for *sanairga* 'slow goer.'

B 2—*contd.*

a(la)bdhe saṁyoga $\frac{7}{1}$ vibhaktaṁ $\frac{1}{7}$ *ganta*vyena guṇitā jātān labdha

$\frac{10}{}$ dviguṇaṁ $\frac{20}{}$ eshālpasyaḥ ||

atha . . . ayam kālo jñeyaḥ anena kālenash shat yojanāni gantavyaṁ |

. . . bhyām ekayojanikasya samāgamo bhavati ||

tadyathā trai-rāśikena *praty*aya | yady ekasya shaṭ yojanā tadā viṁśānāṁ kiṁ

$\frac{1}{1}$ $\frac{6}{1}$ yo° $\frac{20}{1}$ pha° $\frac{120}{1}$

atha saptati śoddhya śesha atra ssaptati $\frac{70}{}$ āgata paṁchāśa $\frac{50}{}$ *adhve*

. . . .

$\frac{1}{1}$ di° $\frac{1}{1}$ yo° $\frac{20}{1}$ di° pha° yo° $\frac{20}{1}$

B 3.

i. . . . . 7 verso.

ā° $\frac{3}{1}$ u° $\frac{4}{1}$ pa° $\frac{0}{1}$ nitya datta $\frac{7}{1}$

†ādiṁ viśoddhya† . . . . *ādi* $\frac{3}{}$ niyataṁ $\frac{7}{}$ viśoddhya $\frac{4}{}$

†uttarārdhena bhājitaṁ† | uttaraṁ $\frac{4}{}$ *anena bhāji*taṁ $\frac{4}{2}$ jātaṁ $\frac{2}{}$

†labdhaṁ sarūpa† | esha rūpādhikaṁ $\frac{3}{}$ eśa kāla . . . .

ā° $\frac{3}{1}$ u° $\frac{4}{1}$ pa° $\frac{3}{1}$ rūpoṇa karaṇena phalaṁ rū° 21 ||

dvitīyasya trai-rāśikena . . . $\frac{1}{1}$ di° $\frac{7}{1}$ $\frac{3}{1}$ di° pha° rū° 21

B 3. [7 *verso.*] Folio 7 is a very interesting sheet. The writing may be classed as α4. On examining the original I noted that it was a double sheet, but the reproduction (Plate vi) might lead one to conclude that it was a palimpsest. Probably, however, the writing underneath is showing through, or the faint writing marks have been impressed from the contiguous leaf. The two sides are definitely disconnected by their contents and the right side has now been definitely located between folios 6 and 65. Folios 7 (*verso*), 8 and 9 are difficult to place. Indeed there seems to be some duplication. Folio 5 certainly follows folio 4 and section C cannot very well include folios 7 (*verso*), 8 and 9.

i. The problem is $7t=((t-1)\frac{4}{2}+3)t$ whence $t=\frac{2(7-3)}{4}+1=3$. By the *rūpoṇa* method $s=[(3-1)\frac{4}{2}+3]t=21$ and by the 'rule-of-three' 1 : 7 : : 3 : 21.

The phrases *ādiṁ visoddhya, uttarārdhena bhājitaṁ* and *labdhaṁ sa rūpa* are quotations from a lost *sūtra.* Compare with fol. 8 *recto.*

B 3—*contd.*

esha ssamadhanā jātā ||

ii. udā° || ādyeka uttara dvayaṁ dvitīya paṁcha pratyahaṁ |

kena kālena samatām vada me gaṇakottama ||

ā°	u°	pa°	niyata nityaṁ
$\frac{1}{1}$	$\frac{2}{1}$	$\frac{0}{1}$	$\frac{5}{1}$

†ādiṁ viśoddhyā† . . . .

B 3. ii. The problem is $5t=((t-1)\frac{2}{2}+1)t$ whence $t= <2(\frac{5-1}{2})+1=5$ and $s=25>$.

B 4.

i. . . . . yojana paṁchakaṁ | sapta dinānī 4 recto.

tasyaiva gatasya | parataḥ dvitīya nava yojanaika gatake . . . taṁ

$\frac{1}{1}$	di°	$\frac{5}{1}$	yo°	dina	$\frac{7}{1}$	gatasya gata yojana	$\frac{35}{1}$	dvi°	$\frac{1}{1}$	di°	$\frac{9}{1}$ yo°

†gatisyaiva viseshaṁ cha† . . . yate | gati 5 | 9 | viśeshaṁ 4

vibhaktaṁ $\frac{1}{4}$ pūrva gata 35 esha pāder guṇitaṁ $\frac{35}{4}$ . . . bhir

dinai sama gatī bhavanti nava yojanaṁ ||

pratyaya trai-rāśikena

$\frac{1}{1}$	di°	$\frac{5}{1}$	yo°	$\frac{35}{4}$	. .	
$\frac{1}{1}$	di°	$\frac{9}{1}$	yo°	$\frac{35}{4}$	pha°	.

ii. udā° || ashṭā-daśa yojanā ekena dine yātī |

tasyāshṭa dinā gatasya |

dvitīya paṁcha-viṁśe yojanā dine yāti |

B 4. [4 *recto*.] The writing changes, due possibly to the use of a different pen, but it is different and may be termed α3. This leaf is closely connected with fol. 3 *recto* and with fol. 5.

i. The example may be restored : One goes at the rate of 5 *yojanas* for 7 days and then a second starts at the rate of 9 *yojanas* a day. When will they have traversed equal distances ?

The phrase *gatisyaiva visesham cha* is a quotation from *sūtra 15* (fol. 3 *recto*) and *pūrva gata* is a reference to the same rule.

The solution is $t=\frac{7\cdot 5}{9-5}=\frac{35}{4}$ days. 'Proof by the rule of three' $1:5::\frac{35}{4}:\frac{175}{4}$ and $1:9::\frac{35}{4}:\frac{315}{4}$ <and $\frac{315}{4}-\frac{175}{4}=35$>.

One travels at the rate of 18 *yojanas* in one day for a period of 8 days. A second goes at the rate of 25 *yojanas* in one day. Determine in what time.

The eleventh leaf must have been close by : indeed *pūrvepi* seems to indicate that it was 'just before.'

B 4—*contd.*

kena kālena sāsyatāṁ ||

evaṁ ekā-daśama pattre bhilikhita pūrvepi ||

paṁcha-daśama sūtraṁ 15

iii. sūtraṁ || *ād*yor viśesha kartavyaṁ uttarasya viseshataḥ

vibhaktaṁ muttare . . . .

i. . . . . 4 verso.

uttaraṁ $\frac{2}{1}$ vibhaktaṅ $\frac{1}{2}$ ādi śesha $\frac{2}{1}$ jātā $\frac{1}{1}$ dviguṇaṁ

2 rūpa saṁyutaṁ 3 esha . . . saṁkalite

pratyaya | *padam* iṇā ubhaye sthāpitavyā rūpoṇā karaṇe phalaṁ 21 . .
21 dvi

kiṁ prabhūtepi likhite || shodaśama sūtraṁ 17 sūtre bhrāntim asti

ii. sūtraṁ || ādyor viśesha dviguṇaṁ chaya suddhir vibhājitaṁ

rūpād*hikam tathā* kāl*aṁ* gati sāsyaṁ tadā bhavet ||

iii. udā° || dvayādi tṛi chayaś chaiva dvitīya tryādi-k-ottaraḥ

dvayo cha bhavate paṁthā kena kālena sāsyatām ||

sthāpanaṁ kṛiyate ||

ā° $\frac{2}{1}$	u° $\frac{3}{1}$	pa° $\frac{0}{1}$	dvi°	ā° $\frac{3}{1}$	u° $\frac{2}{1}$	pa° $\frac{0}{1}$

karaṇam | ‡ ādyor viśesha‡ . . . . . .

B 4. iii. The rule means (?) $t=2\frac{(a_1-a_2)}{d_2-d_1}+1$. Note that the next *sūtra*, on the reverse, commences with the same phrase *ādyor viśesha*.

[4 *verso*.] i. The example was $a_1=4$, $d_1=3$; $a_2=6$, $d_2=1$. Where a_1 and a_2 are the first terms and d_1 and d_2 are the increments of arithmetical progressions, the sums of which were equal. Therefore $(t-1)\frac{3}{2}+4=(t-1)\frac{1}{2}+6$ whence $t=2\frac{(6-4)}{3-1}+1=3$. The proof is by the *rūpona* method, namely, $s_1=((3-1)\frac{2}{3}+4)3=21$ and $s_2=((3-1)\frac{1}{2}+6)3=21$. But 'why should it be written out in full?' See Part I, § 73.

The remark that the *sūtra* is wrongly numbered was probably added later by some one other than the original scribe. The next *sūtra* is numbered 18 (fol. 5) and so on. This is not a copyist's error: it is one of an original MS.

ii. The rule is much the same as the previous one and means that $t=2\frac{(a_1-a_2)}{d_2-d_1}+1$ when $((t-1)\frac{d_1}{2}+a_1)t=((t-1)\frac{d_2}{2}+a_2)t$. The rule is quoted below and on fol. 5 *recto*.

iii. The example gives $a_1=2$, $d_1=3$; $a_2=3$, $d_2=2$ < whence $t=3$ and $s=15$ >.

C 1.

i. 5 recto.

ā°	$\frac{5}{1}$	u°	$\frac{6}{1}$	pa°	$\frac{0}{1}$	dha°	$\frac{0}{1}$
ā°	$\frac{10}{1}$	u°	$\frac{3}{1}$	pa°	$\frac{0}{1}$	dha°	$\frac{0}{1}$

karanaṁ | ‡ādyor viśeshaṁ‡ | ādi . . . . . ‡chaya śuddhi‡ chayaṁ 6 | 3 | śuddhi $\frac{3}{1}$ ādi śesha 5 dviguṇaṁ 10 uttara viśesha 3 vibhaktaṁ $\frac{10}{3}$ sa rūpaṁ $\frac{13}{3}$ anena ka . . . samadhanā bhavanti | pratyayaṁ | rūpoṇā karaneṇa phalaṁ $\frac{65}{\text{dvi}^\circ\ 65}$ esha padaṁ . . . . ashṭādaśama sūtraṁ 18

ii. (sūtram) || dina gamanam ādi rahitaṁ *dviguṇaṁ* tachchottarena saṁyutaṁ |
pratinihita ātmaguṇaṁ *jñeyaṁ kshepa saṁjñako* rāśi |
ashṭottara guṇite kshepa saṁjñako datvā mūlaṁ
pra*tinihita yutaṁ dviguṇottara bhājitaṁ*

. . . . hataṁ 30 . . . †dina gama- 5 verso.
nam ādirahitam‡ dina gamana yojanaϕ pañcha 5 ādi 3 rahitaṁ jātaṁ 2 †dviguṇaṁ† 4 †tachchottareṇa saṁyutaṁ† 8 . . . †ātmaguṇaṁ† 64 eśa †kshepa saṁjñako rāśi† ashṭottara saṁgu . . . labdha rāshi 30 ashṭa guṇaṁ 240 uttareṇa guṇaṁ uttaraṁ 4 guṇitaṁ jātaṁ 960 †kshepa saṁjñako datvā† | tatra kshepa saṁjñ . . . 64 yutaṁ jātaṁ 1024 asya mūlaṁ 32 †pratinihita† . . . . 8 yutaṁ jātaṁ 40 u . . . .

C 1. [5 *recto.*] The writing is the same as that on folio 4, namely α3, but it changes again in the middle of 5 *recto*.

i. The example is $a_1=5$, $d_1=6$; $a_2=10$, $d_2=3$, where $((t-1)\frac{6}{2}+5)t=((t-1)\frac{3}{2}+10)t$; and the solution is $t=2(a_2-a_1)/(d_1-d_2)+1$ or $2(10-5)/(6-3)+1=\frac{13}{3}$.

Proof by the *rūpoṇa* method $s_1=<((\frac{13}{3}-1)\frac{6}{2}+5)\frac{13}{3}=((\frac{13}{3}-1)\frac{3}{2}+10)\frac{13}{3}>=65$.

The *sūtra* number should probably be 17. See fol. 4 *verso*.

ii. The writing now changes to what may be termed the α4 style. The rule means that $<$ if $DT+Dt=((t-1)\frac{d}{2}+a)\,t>$ then

$$t=\frac{\sqrt{(d-2(a-D))^2+8dDT}+d-2(a-D)}{2d}$$

where *D* and *T* are fixed quantities and *a*, *d* and *t* are elements of an arithmetical progression of which *a* and *d* only are given. The quantity designated *pratinihita* 'set aside' is $d-2(a-D)$, while the *kshepa saṁjñako rāśi* 'the quantity known as *kshepa*' is $\{d-2\,(a-D)\}^2$

[5 *verso.*] Writing α4. Notːce a semi-looped medial *e* near the end.

i. The exꭤmple is $<D=5,\ T=6,\ a=3,\ d=4$; hence $t=\frac{\sqrt{(2(5-3)+4)^2+8.4.5.6}+2(5-3)+4}{2.4}>$. The solution proceeds step by step thus: $DT=5.6=30$, $D-a=5-3=2$, $2(D-a)=4$, $2(D-a)+d=4+4=8$; $(2(D-a)+d)^2=64$ and 'this is known as the *kshepa* quantity'; $8DT=240$, $8DTd= 960$; $8ĿTd+(2(D-a)+d)^2=1024$; $\sqrt{1024}=32$; $2(D-a)+d+32=40$; and $<\frac{40}{8}=5>$.

Almost the whole of the *sūtra* on 5 *recto* is quoted here and on the following pages.

C 2.

i. . śike pratyayaṁ [$\frac{1}{1}$ | $\frac{5}{1}$ | $\frac{5}{1}$] phalaṁ . . . . anenas saha 55 eśa 6 recto.

samābdhānaṁ ||

ii. udā° || ādi paṁchaṁ uttaraṁ *trīṇi* naro yojana gamyate |

dvitīya pratidinaṁs sapta gatasya dina paṁchakaṁ |

kena kālena samatāṁ katthyataṁ gaṇakottama ||

ā°	u°	pa°	prati° gati	dina
$\frac{5}{1}$	$\frac{3}{1}$	$\frac{0}{1}$	$\frac{7}{1}$	$\frac{5}{1}$

paṁcha dina ga . . . yojanikaṁ yojana [35]

karaṇaṁ | †dina gamanam ādi rahitaṁ† tatra dina gamanaṁ [7]

†ādi rahitaṁ† ādi 5 rahitaṁ . . . .

i. . . . anena guṇitaṁ jātaṁ [840] †samjñako datvā† tatra kshepa rāshi [49] 6 verso.

datvā jātaṁ [889] . . . . dāna dadāti samaṁ | karaṇī kriyate

ii. sūtraṁ || akṛite sli*shṭha krity ūnā* *śesha* chchhedo dvi-saṁguṇaḥ

tad vargaḥ dala saṁslishṭha hṛiti *śuddhi kṛiti ksha*yaḥ

anena sutreṇa slishṭha mūlaṁ ānaya svamatimā . . . . .

ii. . . labdhaṁ mūlaṁ [29 / 48 / 58] †pratinihitaṁ† [7] anena yutaṁ [36 / 48 / 58]

. . . [$\frac{2136}{58}$] †dviguṇottara bhājitaṁ† | tato . . . .

C 2. [6 *recto.*] i. Continues the example. 'Proof by the rule of three' 1 : D : : t : Dt or 1 : 5 : : 5 : 25 and DT+Dt=30+25=55.

ii. The next example is D=7, T=5, a=5, d=3 ; hence $t=\frac{\sqrt{((2)7-5)+3)^2+8.3.7.5}+2(7-5)+3}{2.3}$

Part of the solution is lost < DT=35, 2(D−a)+d=7, $7^2=49$ >. It is continued on 6 *verso.*

[6 *verso.*] i. Continues the solution: 8DTd=840 ; 8DTd+$(2(D-a)d)^2$=889. Here the solution breaks off in order to tackle the problem of obtaining the root of a surd quantity, and a subsidiary (un-numbered) *sūtra* is given.

ii. The rule recurs on folios 56 *recto* and on 57 *verso*, and with the help of these other versions it has been restored as above.

The rule means that an approximate root of $\sqrt{A^2+b}$ is $A+\frac{b}{2A}$ and that the difference between the squares of these two quantities is $(\frac{b}{2A})^2$; and that by continuing the process closer approximations can be obtained. For a discussion of this rule see Part I, §§ 68, 69, 85. The three versions as they now stand are—

akṛite śli . . . . chchhedo dvi-saṁguṇaḥ 6 Verso.
tad vargaḥ dala | saṁślishtha | hṛit . . . . . yaḥ

. . . kṛity ūnān śesha chchhedo dvi-saṁguṇaṁ | 56 Recto.
tad varga . . . . . ślishthaḥ hṛiti śuddhi kṛiti kshayaḥ ||

akṛite ślishṭha krity ūnā śesha chchhedo dvi . . . . . 57 Verso.
. *varga* dala saṁślishtha hṛiti śuddhi kṛiti kshayaḥ . . . .

iii. The solution is resumed: < since 889=841+48=29^2+48 > the first approximation to $\sqrt{889}$ is $29\frac{48}{58}$ and (terming this q_1) we have $q_1+2(D-a)+d=29\frac{48}{58}+7=36\frac{48}{58}=\frac{2136}{58}$ < and $t_1=\frac{2136}{58\times 6}=\frac{178}{29}$ > where t_1 is an approximate value depending on q_1.

C 3.

6
1
8
60*
16 cha°
60*
33 li°
60*
6 vi°
60*
śe° $\frac{6}{29}$

$\frac{447}{29}$ dalitā $\frac{447}{58}$ . . . . sāsye yutaṁ $\frac{737}{58}$ pada 7 recto.

ghnā | tatra padaṁ $\frac{178}{29}$ anena guṇitaṁ jātaṁ $\frac{65593}{841}$

śli . . . tya śeshaṁ kṛiyate | $\frac{65569}{841}$ bhāge hṛite

pratyayaṁ trai-rāśikena | $\frac{1}{1}$ | 7 yo° $\frac{}{1}$ | $\frac{178}{29}$ phalaṁ

yojana 42 śe $\frac{28}{29}$ niyataṁ tena . . . 77 . . . . . .

ekona-viṁśatima sūtraṁ 19 ||

C 3. [7 *recto*.] This continues the example started on fol. 6 *recto*. [The numbers marked with asterisks are change-ratios (see Part I, §§ 103—105).] The set of figures on the left expresses $\frac{178}{29}$ as a sexagesimal fraction (see Part I, § 58), *i.e.*, $\frac{178}{29}=6+8^{i}+16^{ii}+33^{iii}+6\frac{6}{29}^{iv}$. The portion of the statement above the 16 is missing but the restoration is certain. Of the abbreviations *cha°* has not yet been identified; *li°* stands for *liptā* (Gk. λεπτε); *vi°* for *viliptā*; *śe°* for *śeshaṁ* 'remainder.' In Hindu astronomical works *liptā* means a 'minute of arc,' and *viliptā* 'a second of arc.' This use of the sexagesimal notation for arithmetical purposes in an Indian work is unique. The solution proceeds to find the approximate value of s_1 which depends on t_1 and ultimately q_1. We have $s_1=((t_1-1)\frac{d}{2}+a)t_1$. Now $(t_1-1)d=(\frac{178}{29}-1)3=\frac{447}{29}$; $(t_1-1)\frac{d}{2}=\frac{447}{58}$; $(t_1-1)\frac{d}{2}+a=\frac{447}{58}+5=\frac{737}{58}$; and $((t_1-1)\frac{d}{2}+a)t_1=\frac{737}{58}$. $\frac{178}{29}=\frac{65593}{841}$.

But $DT+Dt_1=7(5+\frac{178}{29})=\frac{65569}{841}=77\frac{812}{841}$.

'Proof by the rule of three': $1 : 7yo° :: \frac{178}{29} : 42\frac{28}{29}$ and < $48\frac{28}{29}+35=77\frac{812}{841}$>.

[Note that $\frac{65593}{841}-\frac{65569}{841}=\frac{24}{841}=\left(\frac{48}{2.29}\right)^2\times\frac{1}{8.3}=\frac{e_1}{8d}$. This process of reconciliation is explained in Part I, § 85.]

The *sūtra* number should probably be 18. See fol. 4 *verso*.

C 4.

ā° $\frac{1}{1}$	u° $\frac{1}{1}$	pa° $\frac{0}{1}$	$\frac{60}{1}$

65 verso.

karaṇaṁ || †ashṭottaraghne guṇite† | ashṭa ghanaṁ | 480 | uttara ghana . . . dvi-ghnam ādi | ādi dvi-guṇa | 2 | chayojjhitam | cha . . . . uttaraṁ | ato uttaraṁ pātayitvā ekaṁ bhavati | 1 | va . . . nikshipya dhanasya | 481 | mūlaṁ ślishṭha karaṇyā $\begin{matrix}21\\40\\42\end{matrix}$ . . . . .

vaṁśaṁ $\begin{matrix}882\\40\\42\end{matrix}$ śeshaṁ chatvāriṁśa pṛithak sthāpya 40 . .

yojyaṁ $\frac{922}{42}$ tan mūla varjitaṁ || tan mūlaṁ 880

C 4. [65 *verso*.] Folio 65 consists of two leaves stuck together. The writing on both sides may be classed as α4. The left side has no direct connexion with fol. 7 *recto* but it belongs to the same section.

The *sūtra* here quoted from is lost, or hidden, for possibly when folios 7 and 65 are separated it may be discovered. It may be said to be one of the most important *sūtras* of the whole work judging by the care and elaboration with which it is illustrated. It must mean that < when $s=((t-1)\frac{d}{2}+a)t$ then $t=\frac{\sqrt{(2a-d)^2+8ds}-(2a-d)}{2d}$ > where a, d, t and s are respectively the first term, the common difference, the number of terms and the sum of an arithmetical progression.

The example is $a=1$, $d=1$, $s=60$; hence $t=\frac{\sqrt{(2-1)^2+8.1.60}-(2-1)}{2.1}=\frac{\sqrt{481}-1}{2}$

The solution proceeds $8ds=480$, $2a-d=1$, $(2a-d)^2+8ds=481$; by the square-root method (see fol. 6 *verso*) the first approximation is $21\frac{40}{42}=\frac{882+40}{42}=\frac{922}{42}$ and < $t_1=(\frac{922}{42}-1)\div2=\frac{880}{84}$.>

C 5.

$\begin{array}{c|c} 880 & 964 \\ 84 & 168 \end{array}$ guṇita jātaṁ $\begin{array}{c} 848320 \\ 14112 \end{array}$ chatvāriṅśa pṛithak sthānāṁ vargaṁ 56 verso.

$\begin{array}{c} 1600 \end{array}$ esha uparā pātya śeshaṁ $\begin{array}{c} 846720 \\ 14112 \end{array}$ vartya jātaṁ $\begin{array}{c} 60 \end{array}$

. . . . $\begin{array}{c} 21 \\ 20 \\ 21 \end{array}$ *teshā*m varggah tasthāt . . . . . 56 recto.

akṛite ślishṭha kṛityūnān śesha chchhedo dvi-saṁguṇaṁ |

tad varga dala saṁślishṭhaḥ hṛiti śuddhi kṛiti kshayaḥ ||

†śesha chchhedo dvi-saṁguṇa† kṛi . . . . .

$\begin{array}{c} 21 \\ 20 \\ 21 \end{array}$ $\begin{array}{c} 400 \\ 441 \end{array}$ dala $\begin{array}{c} 1 \\ 2 \end{array}$ saṁślishṭhaḥ $\begin{array}{l} 21 \text{ bha} \\ 20 \\ 21+ \end{array}$

śeshaṁ pātya . . . . . . dvā bhājita . . . †adhaṁ upare uparaṁ†

guṇitavyaṁ vargaṁ yāva marjaye

. . . $\begin{array}{c|c} 425042 & 400 \\ 19362 & 19362 \end{array}$ śeshaṁ $\begin{array}{c} 424642 \\ 19362 \end{array}$

C 5. [56 *verso*.] Continues the example. $s_1 = ((t_1 - 1)\frac{1}{2} + 1)\, t_1 = t_1 \frac{(t_1+1)}{2} = \frac{880}{84} \cdot \frac{964}{168} = \frac{848320}{14112}$, but $< \frac{e_1}{8d} = \left(\frac{40}{2.21}\right)^2/8. > = \frac{1600}{14112}$ and $\frac{848320-1600}{14112} = \frac{846720}{14112} = 60$.

The bottom half of fol. 56 *verso* is blank but the example is continued on 56 *recto*.

[56 *recto*.] This continues the example given on fol. 65 *verso*. The top part of the leaf is much broken up; but the square-root rule (see fol. 6 *verso*) is given. Why this rule is repeated is not quite understood nor is it understood why it comes between two approximations of the same surd. Anyhow the general aim is clear: since the first approximation is $21\frac{20}{21}$ the second is given by

$$q_2 = 21\tfrac{20}{21} - \tfrac{1}{2}\cdot(\tfrac{20}{21})^2/21\tfrac{20}{21} = 21\tfrac{20}{21} - {}^{1}\cdot\frac{400}{441} \times \frac{21}{461} = \frac{424{,}642}{19362}$$

C 6.

$\begin{array}{c|c} 405280 & 444004 \\ 38724 & 38724 \end{array}$ . . . . . . ardhaṁ kartavyaṁ . . . . . . 64 recto.

$\begin{array}{c|c} 405280 & 444004 \\ 38724 & 77448 \end{array}$ saṁguṇya jātaṁ | a . . . . hrarā hareshu guṇ

. . . $\begin{array}{c} 179945941120 \\ 2999096352 \end{array}$ asya ūrdhaṁ . . . . . . 160000+

C 6. [64 *recto*.] and $t_2 = (\frac{424642}{19362} - 1) \div 2 > = \frac{405280}{38724}$. Also $s_2 = \frac{t_2\,(t_2+1)}{2} = \frac{405280}{38724} \cdot \frac{444004}{77448} < = \frac{179{,}945{,}941{,}120}{2{,}999{,}096{,}352} >$ and $s_2 - \frac{160{,}000}{2{,}999{,}096{,}352} = \frac{179{,}945{,}781{,}120}{2{,}999{,}096{,}352} = 60$.

C 6—*contd.*

. . . †śesha chchhedo dvi-saṁguṇaṁ† $\begin{array}{c}6\\5\\12\end{array}$ śeshaṁ paṁchakaṁ pṛithak . **64 verso.**

. . . ansās vaṁsaṁ $\begin{array}{c}77\\12\end{array}$ tan mūla varjitaṁ **|** tan mūlam . . . . .

dvi-guṇottara saṁbhaktaṁ $\begin{array}{c}65\\24\end{array}$ esha padaṁ **||** . . . . yanaṁ

ā° $\begin{array}{c}1\\1\end{array}$ u° $\begin{array}{c}1\\1\end{array}$ pa° $\begin{array}{c}65\\24\end{array}$ rūpoṇā $\begin{array}{c}41\\24\end{array}$ . . . . . . dalita $\begin{array}{c}41\\48\end{array}$

ādi saṁyutaṁ $\begin{array}{c}89\\48\end{array}$ . . . .

C 7.

. . . . †ashtottara-ghne guṇite† 40 dvi-ghnam ādi cha . . **57 verso.**

. . . . . nikshipya 41 mūlaṁ $\begin{array}{c}6\\5\\6\end{array}$ †śesha chchhedo dvi-saṁguṇa†

. . . . śuddhaḥ tasmāt

akṛite ślishṭha kṛity ūnā śesha chchhedo dvi *saṁguṇaḥ* **|**

tad varga dala saṁślishthaḥ hṛiti śuddhi kṛiti kshayaḥ **||**

†akṛite slishtha† . . . . . tada dvi-saṁguṇa kṛita

$\begin{array}{c}6\\5\\12\end{array}$ tad vargatam $\begin{array}{c}\\\\12\end{array}$ $\begin{array}{c}6\\5\\12\end{array}$ $\begin{array}{c}\\25\\144\end{array}$ dala . . . . .

$\begin{array}{c}25\\1848\end{array}$ . . . . $\begin{array}{c}11833\\1848\end{array}$ hṛi 1848 kṛiti kshaya kritaṁ : eśa **57 recto.**

mūlaṁ **||** tan mūlaṁ . . . . mūlaṁ ekaṁ 1 esha sadṛiśe pātita jāta

$\begin{array}{c}9985\\1848\end{array}$ . . . sambhaktaṁ uttaram dvi-guṇaṁ 2 anena bhaktvā $\begin{array}{c}9985\\3696\end{array}$

esha paṁchakasya padaṁ **||** asya pra . . . . .

sūtraṁ **||** eko rāśi dvidhā sthāpyaś *chayase* . . . . .

C 6-7. 64 *verso* 57 *verso* and 57 *recto* are all (except the last line) concerned with one example the beginning of which is lost. The example is < a=1, d=1, s=5; therefore $t=\frac{\sqrt{41}-1}{2}$>. The first approximation to $\sqrt{41}$ is $q_1=6\frac{5}{12}=\frac{77}{12}$ and $t_1=\frac{65}{24}$. Therefore $s_1=((\frac{65}{24}-1)\frac{1}{2}+1)\frac{65}{24}=(\frac{1}{2}\cdot\frac{41}{24}+1)\frac{65}{24}=\frac{89}{48}\cdot\frac{65}{24}=<\frac{5785}{1152}=5\frac{25}{1152}=5+\frac{1}{8}(\frac{5}{12})^2>$. The second approximation is introduced by the square-root rule (as previously on fol 56 *recto*) and is given by $q_2=6\frac{5}{12}-\frac{1}{2}(\frac{5}{12})^2/6\frac{5}{12}=\frac{11858-25}{1848}=\frac{11833}{1848}$ and $t_2=\frac{1}{2}(\frac{11833}{1848}-1)=\frac{1}{2}\cdot\frac{9985}{1848}=\frac{9985}{3696}$ and 'this is the number of terms when the sum is five.'

Apparently a proof followed introduced by a *sūtra* of which, unfortunately, only a fragment remains.

C 8.

i. $\frac{10225}{32800}$ dalitā $\frac{10225}{65600}$ . . . . . ādi yutaḥ $\frac{108625}{65600}$ padaghnā 45 recto.

pada *saṁyutā* $\frac{6455040625}{3227520000}$ ato paṁcha-viṅśa . . . . uparāḥ

$\frac{6455040000}{3227520000}$ . . . . labdhaṁ 2 esha dhanaṁ ||

ii. ā° $\frac{1}{1}$ 2 | u° $\frac{1}{1}$ 2 | padu $\frac{0}{1}$ | dhanu $\frac{7000}{1}$

. . . . . 384 asya varga 147456 akṛi . . 21743271936 45 verso.

esha sarva guṇitā karaṇi . . . . kṛitvā bhājita jātaḥ $\frac{1158+}{671250}$ aṁśair

aṁśā guṇaye . . . . . . raśi varjya jātaḥ

579 $\frac{768}{1158}$ $\frac{294912}{777307500}$ | 579 $\frac{515520000}{777307500}$ | $\frac{294912+}{777307500}$

śeshaṁ 579 $\frac{515225088}{777307500}$ . . . . $\frac{450576267588}{777307500}$

dvayena mūle . . . .

C 8. [45 *recto.*] i. The greater portion of this example is lost, but can be restored. The example was < a$=1\frac{1}{2}$, d$=1\frac{1}{2}$, s$=2$; whence $t=\frac{\sqrt{10}-3}{6}$. The first approximation to $\sqrt{105}$ is $q_1=10\frac{1}{4}$ and the second is $q_2=10\frac{1}{4}-\frac{1}{2}(\frac{1}{4})^2/10\frac{1}{4}=10\frac{81}{328}$. This gives $t_2=\frac{10\frac{81}{328}-3}{6}=\frac{59425}{49200}$; and $s_2=\langle(\frac{59425}{49200}-1)\frac{3}{4}+\frac{3}{2})\frac{59425}{49200}>=(\frac{10225}{32800}\cdot\frac{1}{2}+\frac{3}{2})\frac{59425}{49200}=(\frac{10225}{65600}+\frac{3}{2})\frac{59425}{49200}=\frac{108625}{65600}\cdot\frac{59425}{49200}=\frac{6,455,040,625}{3,227,520,000}$. < Now $\frac{e_2}{8d}=\frac{625}{3,227,520,000}$ (see Part I § 86 (v)) > and $s=s_2-\frac{e_2}{8d}=\frac{6,455,040,625-625}{3,227,520,000}=2$.

ii. The statement without any formal question should be noted. The example is a$=1\frac{1}{2}$, d$=1\frac{1}{2}$, s$=7000$. The first part of the solution is lost but a good deal of the later working is preserved on folios 45 *verso* and 46 *recto.* We have $q_1=579\frac{584}{579}$. (See *part* I, § 86 (vi).)

45 *verso.* The second approximation is given by $q_2=579\frac{768}{1158}-\frac{1}{2}(\frac{384}{579})^2/579\frac{768}{1158}=579\frac{768}{1158}-(\frac{384}{579})^2\frac{1158}{671250}=579\frac{768}{1158}-\frac{294,912}{777,307,500}=579\frac{515,520,000-294,912}{777,307,500}=579\frac{515,225,088}{777,307,500}=\frac{450,576,267,588}{777,307,500}$. Continued on folio 46 *recto.*

C 9.

| 448244345088
4663845000 | | 443580500088
4663845000 | ------- | 221790250044
1554615000 | dalitā e . **46 recto.**

| 110895125022
1554615000 | ādi saṁyuta | 113227047522
1554615000 | pada-ghnā | 50753383762746743271936
7250483394675000000 |

. . . . . karaṇi pāta | 21743271936 | pātita jātā uparānyāsa sthāpa.

| 50753383762725000000000
7250483394675000000 | bhā 7000

C 9. [46 *recto*.] Continued from 45 *verso*. $t_2 = {}_1(\frac{450,576,26,588}{777,307,500} - 3) \div 6 = \frac{448,244,345,088}{4,663,845,000}$ and $t_2 - 1 = \frac{443,580,500,088}{4,663,845,000}$, $(t_2 - 1)\ d = \frac{221,790,250,044}{1,554,615,000}$, $(t_2 - 1)\frac{d}{2} = \frac{110,895,125,022}{1,554,615,000}$, $(t_2 - 1)\ \frac{d}{2} + a = \frac{113,227,047,522}{1,554,615,000}$ and finally $s_2 = ((t_2 - 1)\frac{3/2}{2} + \frac{3}{2})\ t_2 = \frac{50,753,383,762,746,743,271,936}{7,250,483,394,675,000,000}$. New $\frac{e_2}{8d} = \frac{21,743,271,936}{7,250,483,394,675,000,000}$ and $s = s_2 - \frac{e_2}{8d} = \frac{50,753,383,762,725,000,000,000}{7,250,483,394,675,000,000} = 7000$.

D 1.

| *maḍe* 8 | maḍe 6 | maḍe 3 | . . . . kā 20 apara prashṭaḥ pārā 46 verso.

a i e vīhujaṇa vī ha . . hai . . . ṇa | gore jā ma cha | uppaṇe

sā male a . . ḍha pa . ḍhale āpot diṇe āgaṇe vīhujaṇa ehu vī . . . . .

karaṇaṁ | trai-gore varehahipaṇehi sā . . . . .

D 1. [46 *verso*.] Writing α 4. Find order 9. This is quite unintelligible to me.

D 2.

tola 5 . . . . 70ᵃ recto.

. . . . . . $\frac{35}{2}$ | ete bhāgā . . . . . . 70ᵇ recto.

. . . . $\frac{17}{5}$ | $\frac{117}{70}$ . . . .

. . . . . 2 . 0 rītā 7 . . . . . pala 2 tola 1 . . . pala 6 || 70ᶜ recto.

udā° || samā . . . . ṇapeśī kṛitāni cha

dvecha tisraś . . . .

tisra samādāya] tulitāni trayo-daśe |

. . . . ekaikasya sārdhayaḥ . . .

| $\frac{1}{2}$ | $\frac{1}{3}$ | $\frac{1}{4}$ | d . . . . .

prakshepa yuktyā phalaṁ . . . . 70ᵃ verso.

. . . . ri . . . . ri 70ᵇ verso.

gunya phala rāśi . . . . .

. . . . katram pala 8 70ᶜ verso.

D 2. [70]. Folio 70 consists of 5 scraps not obviously connected. The writing may be classed as α_2. The 'find order' is 65 and this and the five following fragmentary leaves are placed in their 'find order,' for want of some more reliable basis of classification.

70 *recto* is mostly unintelligible but $x(\frac{1}{2}+\frac{1}{3}+\frac{1}{4})=13$ and $x=12$ is a solution.

70 *verso*. Here $x(\frac{1}{2}+\frac{1}{3}+\frac{1}{4})=65$ and $x=60$ is obviously connected in some way with the example on 70ᶜ *recto* but they are two separate examples.

D 2—*contd.*

udā° || ardha tṛi . . . . dāṅśā paṁcha śashṭi nṛipo dadau |

sevakānāṁ tu dī . . . .

$\begin{array}{c|c|c|c|c|}1 & 1 & 1 & \text{dṛishya}\ 65 \\ 2 & 3 & 4 & \ \ 1 \end{array}$ sadṛi . . . . . .

. . . .

D 3.

$\begin{array}{|c|c|c|}2 & 2 & 2 \\ 5 & 6 & 7\end{array}$ dṛishya . . . . ato sadṛisha . . . hakaṁ | 69 recto.

upari māṁsaṁ taṁḍulā bhavanti . . . . . chatvāliṁśa | dūnā chau

rāśi . . eta taṁḍulā | dvā-chatvāriṁ . . . . . vanti ete vrīhakā

sarvattraḥ sthāpanaṁ asya | . . . . .

pratyaya trai-rāśikena $\begin{array}{|c|c|c|c|}5\ \text{ā}^\circ & 2\ \text{taṁ}^\circ & 210 & \text{pha}^\circ\ \text{taṁ}^\circ\ 84 \\ 1 & 1 & 1 & \end{array}$

. . īyasya kṛiyate | $\begin{array}{|c}6 \\ 1\end{array}$ . . . .

. . . . yate *rāśiḥ* $\begin{array}{|c|c|c|c}7 & 2 & 210 & \text{phalaṁ} \\ 1 & 1 & 1 & \end{array}$

. . . . katraṁ 105 . . 69 verso.

udā° || . . . . tṛibhir dattai tṛiguṇā tṛiguṇena tu |

. . . . tad uchyatāṁ ||

$\begin{array}{|c|c|c|c|}1 & 3 & 9 & \text{dṛishya}\ 130 \\ 1 & 1 & 1 & \ \ 1\end{array}$ prakshepa $\begin{array}{|c|c|c|}10 & 30 & 90\end{array}$ ekatram $\begin{array}{|c|}130\end{array}$

. . . vān || taṁ śataṁ tṛibhir datyai paravaptrā pavaptṛi kai . . . .

$\begin{array}{|c|c|c|c|}4 & 6 & 9 & \text{dṛi}^\circ\ 190 \\ 1 & 1 & 1 & \end{array}$ . . . . . $\begin{array}{|c|c|c|}40 & 60 & 90\end{array}$ ekatram

---

D 3. Folio 69 consists of four pieces but is not quite so shabby as folio 70, for the two larger pieces fit together.

[69 *recto*]. The statement means $x(\frac{2}{5}+\frac{2}{6}+\frac{2}{7})=214$ whence $x=210$. The 'proof by the rule of three' is

5 *ā°* : 2 *tam°* : : 210 : 84 *taṁ°* < and 84+70+60=214 >
6 *ā°* : 2 *tam°* : : 210 : 70 *taṁ°*
7 *ā°* : 2 *tam°* : : 210 : 60 *taṁ°*

[69 *verso*]. Here $x(1+3+9)=130$ whence $x=10$ and the numbers are $10+30+90=130$. Again $x(4+6+9)=190$ and $40+60+90=190$.

D 4.

. . . . $\begin{array}{|c|}168\\\hline 4\end{array}$ . . . . deśa dvātya . . . . pātya jātā 68 recto.

śeshaṁ* 21 . . . . ekatraṁ 29 . . . . . . draṁ 2 . .

. . . . .bāri . . . .

D 4. [68 *recto.*] Consists of small fragments which probably belong to folio 67. Writing α_2. The phrase *pātya śesham* occurs on some six other occasions (on folios 31, 62, 63, 56).

D 5.

i. . . . . *trai-rāśikena* $\begin{array}{|c|c|c|}2 \text{ dine} & 3 \textit{ dram}^\circ & 168 \textit{ di}\\1 & 1 & 11\\ & 2 & \\\hline\end{array}$ . . . . 31 recto.

tīyasya kṛiyate . . . $\begin{array}{|c|c|c|}3 \text{ di}^\circ & 2 \text{ draṁ}^\circ & 168 \text{ dinā}\\1 & 1 & 11\\ & 2 & \\\hline\end{array}$ phalaṁ draṁ°

$\begin{array}{|c|}140\\11\\\hline\end{array}$ prathamena dattaṁ . . . . saptaḥ dattais samadhanā jātā |

. . . . sadṛiśaṁ $\begin{array}{|c|c|}77 & 294\\11 & 11\\\hline\end{array}$ pātya śeshaṁ† $\begin{array}{|c|}217\\11\\\hline\end{array}$ dvitīyasya

datta . . . . 77 eśas sama-dhanā jātā ||

ii. punānyaṁ sarva . . . . bhā $\begin{array}{c}4\\1+\\4\end{array}$ dine | draṁ° 15 jīvyā ||

dvitīyasya | bhā $\begin{array}{c}3\\1\\3\end{array}$ dine | draṁ° . . . . .

D 5. Folio 31 consists of two leaves stuck together and the writing on the two sides differs. The leaf is very ragged.

[31 *recto.*] The writing may be classed as α_2

i. The example may be restored with some uncertainty: A earns $3\frac{1}{2}$ *drammas* in 2 days, B earns $2\frac{1}{2}$ in 3 days. A gives B 7 *drammas* and this makes their possessions equal. How long had they been earning?

< Since $\frac{3\frac{1}{2}}{2}t-7=\frac{2\frac{1}{2}t}{3}+7$ we have $t=\frac{14}{7/4-5/6}=\frac{168}{11}=15\frac{3}{11}$ days. >

Proof by the rule of three 2 days : $3\frac{1}{2}$ drammas : : $\frac{168}{11}$ days : $\frac{294}{11}$ drammas

and 3 days : $2\frac{1}{2}$ drammas : $\frac{168}{11}$ days : $\frac{140}{11}$ drammas

and $\frac{294}{11}-\frac{77}{11}=\frac{140}{11}+\frac{77}{11}=\frac{217}{11}$.

ii. Another example of the same kind.

D 5—*contd.*

. . . . *kāraṇaṁ* . . . chchheda saṁ-guṇe 31 verso.

draṁ°	1	4	ya
	1	1	
		2	

draṁ°	1	6	mudgā
		1	
		2	

. . . rdha yutī hṛiti phalaṁ || asya guṇākāro dvayāna

2	1
9 2	13 2

. . . . . . †uparaṁ guṇaye† adau tāva dva . . . . . .

. . . guṇaya | 2 | . . . . . . dvi-nava-bhāgesu . .

sūtraṁ ||

[31 *verso.*] Some of the lower writing shows through and it is very difficult to differentiate. The word *guṇākāra* ? 'form of multiplication' occurs again on fol. 42 *verso.*

D 6.

i. . . . . chchheshaṁ ta dviguṇa . tā | 67 verso.

nirgachchha . . . praviśa māne chatvāri dattaḥ

puna dvi-gunaṁ

sūṇya hastaṁ gataṁ tasya kiṁ atra mūladhana syāt |

1	2	bhā°	1	1	2	bhā°	2	2	2	bhā°	3	3	2	bhā°	4
1	1		1	1	1		1	1	1		1	1	1		1

4	2	bhā°	5	1
1	1		1	4

. . . .

D 6. 67. The surface of the leaf is much worn and the writing is in some places rubbed off. The writing is $\alpha 2$.

[67 *recto.*] i. The example seems to relate to a game at which a certain quantity was staked and eventually all lost. The statement means $1+\frac{1}{2}(2+\frac{1}{2}(3+\frac{2}{2}(4+\frac{3}{2}(5+\frac{4}{2}\cdot\frac{1}{4}))))=<\frac{93}{16}>$

D 6—*contd.*

. . . . . . $\begin{array}{|cc|}49 & . \; 12\\ 8 & 8\end{array}$ jātā $\begin{array}{|c|}61\\8\end{array}$ . . sadṛiśaṁ $\begin{array}{|ccc|}8 & . & 11\\1 & 2 & 8\end{array}$ puna . . 67 recto.

$\begin{array}{|cc|}16 & 61\\8 & 8\end{array}$ jātā $\begin{array}{|c|}77\\8\end{array}$ . . . . sadṛiśaṁ ekasya $\begin{array}{|c|}16\\16\end{array}$ yutaṁ $\begin{array}{c}77\\16\end{array}$

jātaṁ $\begin{array}{|c|}93\\16\end{array}$ esha phalaṁ bhavati **|** . . . . . .

pratyayaḥ $\begin{array}{|c|ccc|ccc|ccc|c}93 & 1+ & 2 & 2+ & 2+ & 2 & 3+ & 3+ & 2 & 4+ & 4\\16 & 1 & 1 & 1 & 1 & 1 & 1 & 1 & 1 & 1 & 1\end{array}$ . . . .

ii. huṇḍikā samānayana sūtraṁ **||**

dina bhakta viśeshaṁ cha . . . . dvi-guṇaṁ kriyatc chaiva

kālam eshāṁ vinirdiśet trai-rāśika vidhānena

. . dattaṁ cha pātavyaṁ‡ sūkshme dattaṁ cha tatsamaṁ **||**

udāharaṇaṁ **||** dvi-guṇa . . . .

[67 *verso.*] Worked out by steps < $\frac{3}{2}(5+\frac{4}{2}.\frac{1}{4})=\frac{33}{4}$, $\frac{33}{4}+4=\frac{49}{4}$ > $\frac{1}{2}(\frac{49}{4}+3)=\frac{61}{8}$, $\frac{1}{2}(\frac{61}{8}+2)=\frac{77}{16}$, and $\frac{77}{16}+1=\frac{93}{16}$ which is the answer.

Proof. $((((\frac{93}{16}-1)\,2-2)\,2-3)-4)\,\frac{2}{3}-(5+\frac{4}{2}\,.\,\frac{1}{4})=0$.

ii. This *huṇḍikā sūtra* should be intelligible but it is not yet clear to me.

D 7.

. . . . dviguṇaṁ dviguṇaṁ bhāraṁ . . . . labdhaṁ 28 recto.

14 **||** puna kṛiya . . . . . .

. . . . vet **||** . . . . . . guṇaye $\begin{array}{|c|c|}1 & 1\\16 & 8\end{array}$ guṇi . jātā . . . . 28 recto.

āhuṭva aḍho . . . . guṇa . . . . bhāgasya divardhā χ kim

$\begin{array}{|c|c|}1 & 1\\96 & 1\\ & 2\end{array}$ phalaṁ . . . . phalaṁ 5 . . . . . .

D 7. [28]. Find order 60. Writing ? α4. Not intelligible.

E 1.

. . . . . ekārghaṁ tu paṇyānāṁ eka-dvi-tṛi-chatush-shaṭ 66 recto.

. . . . . . paṇyān imānayaḥ

sthāpanaṁ kṛiyate

1 1	1 1	dram°	1 1	2 1	dram°	1 1	3 1	dram°	1 1	4 1	dram°	1 1	6 1	

. . . .

pratyaya trai-rāśikena 66 verso.

1 1	dram°	1 1	rū°	12 1	dram°	phalaṁ rūpa	12
1 1	dram°	2 1	rūpa	6 1	dram°	phalaṁ rūpa	12
1 1	dram°	3 1	rūpa	4 1	dram°	phalaṁ rūpa	12
1 1	*dram°*	4 1	rūpa	3 1	dram°	phalaṁ rūpa	12
1 1	*dram°*	6 1	*rūpa*	2 1	*dram°*	*phalaṁ rūpa*	12

E 1. Folio 66 consists of a bad piece of birch-bark containing a large knot. The knot is repeated on folio 53. The find order is 58. Writing is probably α4.

The problem may have been something like this: The rates of purchase are one, two, three, four and six articles for one dramma. What will be the cost of twelve of each?

The cost of one of each would be $1+\frac{1}{2}+\frac{1}{3}+\frac{1}{4}+\frac{1}{6}=\frac{27}{12}$ therefore the cost of 12 of each is 27 drammas, and the numbers of articles are 12, 6, 4, 3 and 2.

'Proof by the rule of three'

1 *dram°* : 1 ru° : : 12 dram° : 12 rū.
1 ,, : 2 ,, : : 6 ,, : 12 ,,
1 ,, : 3 ,, : : 4 ,, : 12 ,,
1 ,, : 4 ,, : : 3 ,, : 12 ,,
1 ,, : 6 ,, : : 2 ,, : 12 ,,

E 2.

. . . .

1 1 2	yo°	1 1	di°	6 1

53 recto.

viśeshaṁ tu tatra gatiṁ

3 2	2 1

viśeshaṁ

1 2

. . . sarva gatī

E 2. Folio 53 resembles fol. 66 in appearance and has the same large knot. Its find order is not known. Writing ? α4. The problems are, however, similar to that on fol. 9 *recto*.

[53 *recto.*] The following conjectural restoration of the problem is offered:

One goes $1\frac{1}{2}$ yojana in a day and another 6 in 3 days. If the first had a start of 9 yojanas when would the second overtake him?

$$\text{Since } \tfrac{3}{2}t+9=\tfrac{6}{3}t \text{ we have } t=\frac{9}{\frac{6}{3}-\frac{3}{2}}=18 \text{ days.}$$

'Proof by the rule of three': 1 day : $\frac{3}{2}$ yo° : : 18 days : 27 yo° and 27+9=36
1 ,, : 2 yo° : : 18 ,, : 36 yo°

E 2—*contd.*

yojana | 9 | anena guṇaye | 18 | anena . . . . bhavishyati

pratyaya trai-rāśikena	1 di° 1		yo° 27 dina 6 ādau yojana 9 1

. .	1 1	1 di° 1	. . yo°	phalaṁ yojana

. . . .	18 yojana 1	20 dina 20 ghaṭike 35* gha° dina	phalaṁ yo° 27/7 53 verso
. . .	27 yo	20 20 ghaṭike 35* ghaṭike dina	pha° yo 36/7

53 verso.

[53 *verso.*] The following is merely a guess at the problem: One goes 18 yojanas in 96 days and another 27 yojanas in 108 days. The first starts from A and the second from B and the distance AB is 9 yojanas. When will they meet if they go only for $\frac{7}{12}$ or 35 *ghaṭikas* of each day ? (60 *ghaṭikas*=24 hours).

In one day they together go $\frac{18}{96}+\frac{27}{108}=\frac{7}{16}$, that is they meet at the rate of 1 yojana in $\frac{16}{7}$ days and actually meet each other in $\frac{9\times16}{7}$ days $=20\frac{4}{7}$ or 20 days 20 ghaṭikas.

Proof 96 days : 18 yo° : : 20d. 20 gha° : $\frac{27}{7}$ yo° and $\frac{27}{7}+\frac{36}{7}=9$.

108 days : 27 yo° : : 20d. 20 gha° : $\frac{36}{7}$ yo.

E 3.

udā° || śaḍ-vimśas cha tṛi-paṁchāśa ekona-tṛimśe vacha | 58 recto.

dvā-śa . . . śaḍ-viṁśa chatuś-chatvāliṁśa saptati |

chatush-shashṭi na*va* . . . . . ṁśa naṁtaraṁ |

trir-āśīti ekaviṁśa ashṭa . . . . . pakaṁ |

. . . 296226447064994 . . . .

E 3. [Folio 58.] Find order not known. Writing ? α4. Possibly two leaves stuck together.

[58 *recto.*] This gives pairs of numbers, first in words and then in figures, thus:

Twenty-six and fifty-three and one less thirty

. . . . . twenty-six, forty-four, seventy

sixty-four . . . .

eighty-three, twenty-one, . . eight . .

and in figures . . . . 29 62 26 44 70 64 99 4 . . .

E 3—*contd.*

. . . . 58 verso.

sthāpanaṁ kṛiyate . . .			
1 1	yuvī 1 1	sūḍha 1 1	dṛishya 20
. . . 3	maṁ 1 1 2	maṁḍa 1 2	maṁḍe 20

. . . . ta datta jātaṁ maṁḍa 2 yu 5 sūḍhe 1 . . . .

[58 *verso.*] There is basis for the following restoration—

A man earns 3 in one day, a young woman $1\frac{1}{2}$ in 1 day and . . . $\frac{1}{2}$ in one day. If 20 earn 20 *maṇḍas* in one day, how many of each will there be ?

Let x, y, z be the numbers of each class, then $x+y+z \quad =20$ individuals

$3x+\frac{3}{2}y+\frac{3}{8}=20$ *maṇḍas*

of which the only solution in positive integers is that given in the text, namely $x=2$, $y=5$, $z=13$. This problem known as the 'Hundred Hens' problem in China, and as the *Regula Virginum*, etc., in Europe is noted upon in Part I, §80 (a).

E 4.

. . . . . tṛi-bhāga . . dine tatha | tṛi rūpa paṁchabhi dinai | 21 recto.

eshāṁ da . . . .

rū° 1 1	rū° 1 1	rū° 3 1	dṛishya 100 1
1 di° 3	1 di° 2	5 di 1	

karaṇaṁ || . . . . kṛitvā

3 1	2 1	3 5	dṛi° 100 1

E 4. Folio 21 consists of 7 scraps of which the largest piece is partly intelligible. The find order is 55 and the writing α1, 4.

[21 *recto.*] Apparently this means : 1 *rū°* is given or obtained in $\frac{1}{3}$ days, 1 in $\frac{1}{2}$ day and 3 in 5 days by three separate individuals (or classes) and the total amount given or obtained is 100.

In one day $\frac{1}{\frac{1}{3}}+\frac{1}{\frac{1}{2}}+\frac{3}{5}=3+2+\frac{3}{5}=5\frac{3}{5}$ is given, so that one is given in $\frac{5}{28}$ days and 100 in $\frac{500}{28}=17\frac{6}{7}$ days.

E 4—*contd.*

. . . . vārdhaṁ   tritīyasya   21 verso.

jīva-lokāt eshāṁ dīnār . . . . . . kasya kiṁ bhavati ||

2 dī° 1 2	3 dī° 1 2	4 dī° 1 2
1 dī° 1 2	1 dī° 1 3	1 dī° 1 4

. . . . . parivartanam kṛiyate

10 6	21 8	36 10	dṛi 500 1

prakshe . . . .

[21 *verso.*] Here the main elements of a problem are preserved and the problem is continued on folio 22. The problem probably was to the effect that: A gave $2\frac{1}{2}$ *dīnāras* in $1\frac{1}{2}$ days, B gave $3\frac{1}{2}$ in $1\frac{1}{3}$ days and C $4\frac{1}{2}$ in $1\frac{1}{4}$ days. In what time would they have given 500 *dīnāras* ?

In one day $\frac{2\frac{1}{2}}{1\frac{1}{2}}+\frac{3\frac{1}{2}}{1\frac{1}{3}}+\frac{4\frac{1}{2}}{1\frac{1}{4}}=\frac{10}{6}+\frac{21}{8}+\frac{36}{10}<=\frac{947}{120}$ is given. Therefore 500 is given in $\frac{500\times 120}{947}=\frac{60000}{947}=63\frac{239}{947}$ days>

Continued on fol. 22 *recto*.

E 5.

. . . .

473500 947

vartita jātā phalaṁ dī 500 || 22 recto.

asya *pratyaya trai-rāśikena*

2 dī° 1 2	1 dī° 1 2	100000 dī° 947	phalaṁ di 60000 947
3 dī° 1 2	1 dī° 1 3	157500 dī° 947	phalaṁ di 60000 947
4 dī° 1 2	1 dī° 1 4	216000 dī° 947	phalaṁ di 60000 947

E 5. [22 *recto*] continues the solution of the example on fol. 21 *verso*.

<The gifts are therefore $\frac{100,000}{947}+\frac{157,500}{947}+\frac{216,000}{947}>=\frac{473,500}{947}=500$ *dīnāras*.

'Proof of this by the rule of three' $2\frac{1}{2}$ *dī°* : $1\frac{1}{2}$ days : : $\frac{100,000}{947}$ dī : $\frac{60,000}{947}$ days.

$3\frac{1}{2}$ „ : $1\frac{1}{3}$ „ : : $\frac{157,500}{947}$ „ : $\frac{60,000}{947}$ „

$4\frac{1}{2}$ „ : $1\frac{1}{4}$ „ : : $\frac{216,000}{947}$ „ : $\frac{60,000}{947}$ „

F 1.

i. . . . . . . dvi-guṇaṁ dvitīyasya prathama . . . . . tīya . . | prathamā 22 verso.

chaturguṇaṁ chaiva chaturthe chaiva dattavān cha . . . . śatam ekaṁ

dvayānvayaṁ || vadasva prathame dattaṁ kiṁ pramāṇāṁ . . sya .

0 1	2 1	3 1	4 1	dṛishya 200 1

†śūnyam eka-yutaṁ kṛitvā† 1 | 2 | 3 | 4 | . . . †kshepa yuktyā†

phalaṁ || 20 | 40 | 60 | 80 | evaṁ 200 || eshām . . . .

ā° 20 1	u° 20 1	pa° 4 1

rūpoṇā karaṇena phalaṁ 200

ii. sūtraṁ || yadṛichchha pinyase sūṇye tadā vargaṁ tu kārayet

---

F 1. [22 *verso.*] i. This appears to be the beginning of a new section. The *sūtra* is lost. Find order 54, writing α4.

The problem was something like this: A certain amount was given to the first, twice that to the second, thrice it to the third, and four times to the fourth. State the amount given to the first and the shares of the others, if the total amount given was 200.

The shares are represented by 0, 2, 3, 4. 'Having added one to the nought' the sum is 1+2+3+4=10. <Then the proper share of the first is $\frac{200}{10}=20$ > . Having added in this value the series becomes 20+40+60+80=200.

The proof by the *rūpoṇa* method gives $<((4-1)\frac{20}{2}+20)\ 4>=200$.

For the method of solution, the *regula falsi*, see Part I, §§71 and 72, and for the *rūpoṇa* method see §73. The whole section is dealt with in §87, and the use of the symbol for 'nought' in §60.

ii. The *sūtra* begins "Put what number you please in the empty place (or for the nought)." This is quoted on fol. 23 *recto* and so is *tadā vargaṁ tu kārayet*, etc.

F 2.

i. . . . . cha tṛi-guṇaṁ . . . . 23 recto.

. . . . prathamasya tu kiṁ bhavet

0 1	tadā 2 1	tadā

†yadṛichchhā vinyase sūṇye† . . chchhā | 1 | †tadā vargaṁ tu kārayet†

1 1	2 1	2 3 1 1	6 1 . . .

prakshipe guṇitaṁ 1 | 2 | 6 | 24 |

. . prakshiptaṁ 33 || dṛishyaṁ vibhajet

132 33

vartyaṁ jātaṁ

4 1

F 2. [23 *recto.*] The find order is 52.

i. The example may be represented by $x+2T_1+3T_2+4T_3=132$. Where T_1, T_2, etc., represent the values of the first, second, etc. terms. Make x=1 then the terms are 1+2+6+24=33 and the proper value of x is $\frac{132}{33}=4$ and the series becomes 4+8+24+96 =132.

All the technical terms here employed are of interest and will be dealt with in due course: *ichchhā* 'an assumed number'; *varga* 'a series'; *prakshepa* 'something thrown in' or 'an interpolation'; *vartya* 'cancelled'; *dṛishya* 'the given number': etc.

F 2—*contd.*

. . . . . dattaṁ || ato nyāsaḥ | 4 | 8 | 24 | 96 | . . . .

esha varga krama gaṇitam || atha yuti vargam kṛi . . .

ii. sūtraṁ || kāmikaṁ śūṇye vinyastaṁ tadā chaiva krame guṇam

i. . . . . kṛitvā chaturtha . . . . 23 verso.

. . . . prathamasya tu kiṁ bhavet

0	2	1	3	3	12	4	dṛi°	300
1	1	1	1	1	1	1		1

‡kāmikaṁ śūṇye piṇyastaṁ‡ kāmikaṁ 1 || esha ṇyastam . . . .

‡tadā chaiva krameṇa guṇitaṁ‡ | 1 | 2 | 9 | 48 | eshām yu . | 60 |

anena dṛishyaṁ bhājitaṁ | 1/60 | 300/1 | jātā | 5 | e . . . . .

anena kshepaṁ guṇaye | 5 | 10 | 45 | 240 | . . . . . . yuti varga gaṇitaṁ ||

ii. udā° || prathamasya na . . . . dattaṁ chaivā dhanaṁ |

sa cha dvyārdha yuta dhanaṁ . . . .

---

ii. The term *kāmika* is practically synonymous with *ichchhā* or *yadṛichchhā* 'what you please'; 'an assumed number.' Bhāskara uses *ishṭa* much in the same way. A good deal of the *sūtra* is quoted on fol. 23 *verso*.

[23 *verso*.] i. The example may be represented by $x+2T_1+3(T_1+T_2)+4(T_1+T_2+T_3)=300$. Put $x=1$ then the series becomes $1+2+9+48=60$ and the proper value of x is $\frac{300}{60}=5$ and we have $T_1=5$, $T_2=10$, $T_3=45$, $T_4=240$ and $\Sigma T=300$.

ii. The example is solved on fol. 24 *recto*.

F 3.

śataṁ chatuś-chatvaliṁsā . . . . . **dattaṁ chaiva chaturguṇam** 24 recto.

kiṁ prathamasya . . . . . .

0	1	2	2	3	3	4	4	dṛi°	144
1	1	1	1	1	1	1	1		1
	2		2		2		2		2

F 3. [24 *recto*.] The example may be represented by

$$[x(1+1\tfrac{1}{2})]+[2T_1+2\tfrac{1}{2}x]+[3T_2+3\tfrac{1}{2}x]+[4T_3+4\tfrac{1}{2}x]=144\tfrac{1}{2}$$

Set $x=1$ and the series becomes $\frac{5}{2}+\frac{16}{2}+\frac{52}{2}+\frac{217}{2}=\frac{289}{2}=144\frac{1}{2}$ which is the same as the given sum and therefore $x=1$ is correct.

The phrase marked ** is deleted in the original. The expression "*upare uparaṁ adhe adhaṁ guṇaye*" is obviously quoted from a well known rule relating to fractions: 'numerator should be multiplied by numerator and denominator by denominator.' See also C5, D5.

F 3—*contd.*

. . . . śūṇyeśu $\begin{array}{c}1\\1\end{array}$ . . . . . . †yutaṁ ḉhaiva guṇaṁ† tataḥ

yutaṁ chaiva guṇaṁ kṛitvā kāraye gaṇa . . . . $\begin{array}{|c|}5\\2\end{array}$ guṇaṁ | †upare

uparaṁ adhe adhaṁ guṇaye† $\begin{array}{|c|}10\\2\end{array}$ sārdha dv . . yutaṁ . tīya rāśyā guṛaṇaṁ |

sārdhais saptabhi trīṇi $\begin{array}{|c|}45\\2\end{array}$ sārdha traya yutaṁ . . chaturtha rāśi

guṇayesh shaḍviṁśatibhi | jātā $\begin{array}{|c|}208\\2\end{array}$ sārdha chatvāri yu . . . . . .

$\begin{array}{|c|}289\\2\end{array}$ evaṁ dṛiśyaṁ | sarvaṁ tadeva jātaṁ

i. . . . . tri-sārdha yu . . . . . — 24 verso.

. . chatur-guṇaṁ chaturthena navārdha yutaṁ dattaṁ |

. . *dviśa*tā dvāviṁśādhikā kim atra prathamasya dattāsit

0 3 1 2	2 5 1 2	3 7 1 2	4 9 1 2	ekatraṁ dattaṁ 222

‡śūnya . . . . datvā‡ | 1 | yuta guṇita yuta krameṇa jātaṁ ||

sthāpā . . . . .

5 2	15 2	67 2	357 2	dṛishya 222

prakshepena

jātaṁ 222 || . . dṛiśyāḥ 222 ||

ii. udā° || prathamaṁ na jānāmi | divardha yutaṁ . . . . .

[24 *verso*]. i. The example may be represented by

$[x\,(1+1\tfrac{1}{2})]+[2T_1+\tfrac{5}{2}x]+[3\,(T_1+T_2)+\tfrac{7}{2}x]+[4\,(T_1+T_2+T_3)+\tfrac{9}{2}x]=222.$

Set x=1 and the series becomes $\frac{5}{2}+\frac{15}{2}+\frac{67}{2}+\frac{357}{2}=222$.

The same quotation *śūnya sthāne* . . . *rūpaṁ datvā* occurs on fol. 25 *verso*. See also at the bottom of fol. 26 *recto*.

F 4.

| $\begin{matrix}0 & 3\\ 1 & 2\end{matrix}$ | $\begin{matrix}2 & 5+\\ 1 & 2\end{matrix}$ | $\begin{matrix}3 & 7+\\ 1 & 2\end{matrix}$ | $\begin{matrix}4 & 9+\\ 1 & 2\end{matrix}$ | dṛi° $\begin{matrix}78\\ 1\end{matrix}$ | 25 recto.

. . . . . . yutaṁ jātaṁ | $\frac{5}{2}$ | dvitīya guṇaṁ | $\frac{10}{2}$ | . . tṛitīya ekatre guṇitaṁ **|** yutena **|** yutaṁ | $\frac{10}{2}$ | $\frac{23}{2}$ | . . yutaṁ | $\frac{33}{2}$ | guṇitaṁ | $\frac{132}{2}$ | riṇaṁ jātaṁ **|** pārya . . eśa *nyāsa* | $\frac{5}{2}$ | $\frac{5}{2}$ | $\frac{23}{2}$ | $\frac{123}{2}$ | dṛishya $\frac{78}{1}$ |

. . . . | $\frac{156}{2}$ | vibhaktavyaṁ | $\frac{2}{156}$ | $\frac{78}{1}$ | . . . . .

i. karaṇaṁ **|** †śūṇya sthāne†...........†rūpaṁ datvā† | 1 | yutā jātā | $\frac{5}{2}$ | 25 verso.

.... | $\frac{15}{2}$ | prathamā tṛitīyasya tṛi-guṇaṁ yutaṁ jātaṁ...........

chaturguṇaṁ navārdha yuṭam jātaṁ | $\frac{29}{2}$ | ekatra *nyāsa*...... | $\frac{5}{2}$ | $\frac{15}{2}$ | $\frac{22}{2}$ | $\frac{29}{2}$ | dṛi° $\frac{71}{2}$ | prakshiptaṁ | $\frac{71}{2}$ | bhaktaṁ dṛishyaṁ jātaṁ 1........anena sarvaṁ guṇitaṁ tadeva | $\frac{5}{2}$ | $\frac{15}{2}$ | $\frac{22}{2}$ | $\frac{29}{2}$ | ekatraṁ **||**

eshāṁ aparo vidhiḥ **||**

ii. udā° **||** prathama dhanaṁ dattam najātaṁ kiṁ tu divardha yutaṁ **|**
tadā dvitīyena dvi-guṇaṁ dattaṁ paṁchārdha hīnaṁ **|**
tadā tṛitīyena tṛiguṇaṁ dattaṁ saptārdha . .
chaturtheṇa chatur-guṇaṁ navārdha hīnam . . .
dattaṁ ekatraṁ ta . . . .

| $\begin{matrix}2 & 5\\ 1 & 2+\end{matrix}$ | $\begin{matrix}3 & \text{gu}^\circ & 7\\ 1 & & 2+\end{matrix}$ | $\begin{matrix}4 & \text{gu}^\circ & 9\\ 1 & & 2+\end{matrix}$ | *dṛi* $\begin{matrix}29\\ 2\end{matrix}$ |

F 4. [25 *recto.*] The example may be represented by

$$[x(1+\tfrac{1}{2})]+[2T_1-\tfrac{5}{2}x]+[3(T_1+T_2)-\tfrac{7}{2}]+[4(T_1+T_2+T_3)-\tfrac{9}{2}x]=78.$$

Set x=1 and the series becomes $\frac{5}{2}+\frac{5}{2}+\frac{23}{2}+\frac{123}{2}=\frac{156}{2}$ and $\frac{78}{156/2}=1$.

[25 *verso.*] i. The example, of which only the solution remains, is

$$[x(1+\tfrac{3}{2})]+[2T_1+\tfrac{5}{2}x]+[3T_1+\tfrac{7}{2}x]+[4T_1+\tfrac{9}{2}x]=\tfrac{71}{2},$$ which, when x=1, becomes $\frac{5}{2}+\frac{15}{2}+\frac{22}{2}+\frac{29}{2}=\frac{71}{2}$.

ii. The example is $[x(1+\frac{3}{2})]+[2T_1-\frac{5}{2}x]+[3T_1-\frac{7}{2}x]+[4T_1-\frac{9}{2}x]=\frac{29}{2}$.
The solution is given on fol. 26 *recto*.

F 5.

i. karaṇaṁ || †śūnya . . . . †rūpaṁ datvāḥ† yutam jātaṁ | $\frac{5}{2}$ | 26 recto.

. . . . | $\frac{5}{2}$ | prathama tṛitīyaṁ tṛi-guṇam . . . . . prathamā

chaturthaṁ chatur-guṇaṁ navārdha rahitaṁ | śesham | $\frac{11}{2}$ | e . . . .

| $\frac{5}{2}$ | $\frac{5}{2}$ | $\frac{8}{2}$ | $\frac{11}{2}$ | dṛi° $\frac{29}{2}$ | prakshepa yuktiḥ | $\frac{29}{2}$ |

. . . . bhaktaṁ | $\frac{2}{29}$ | $\frac{29}{2}$ | jātaṁ | 1 | . . . gunitaṁ tad eva |

evaṁ riṇa rāśī bhavanti |

ii. tṛi-prakāraṁ . . samāptaṁ || śūṇya sthāne rūpaṁ datvā | tadanu

yuktaṁ | guṇita . . . . .

F 5. [29 *recto.*] i. This is the solution of the example given at the bottom of fol. 25 *verso.* Let x=1, then the series becomes $\frac{5}{2}+\frac{5}{2}+\frac{8}{2}+\frac{11}{2}=\frac{29}{2}$ and the correct value of x is $\frac{29}{2}\div\frac{29}{2}=1$.
ii. "The three-fold method is completed," namely, "having put unity in the nought (empty) place; then having added . . .
The *śūnya sthāne rūpam datvā* is quoted on folios 24 *verso,* 25 *recto* and at the beginning of 26 *recto.*

F 6.

atha dvau 4 | 36 asya dalaṁ pha° 26 verso.

athāshṭa 8 | 32 dalaṁ pha° . . . .

16 | 28 dalaṁ pha° . . . .

4 bhu° 36 24 | 16 dalaṁ pha° . . . .

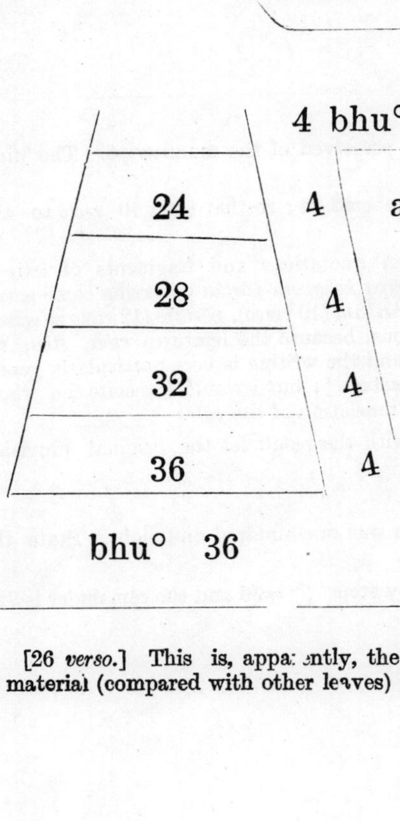

24 4 atha *trīṇi usārā* da . . . .

28 4 36 | 20 | 4 asya tri . . . .

32 4 32 | 20 | 8 a . . . .

36 4 28 | 20 | 12 puna . . . . .

bhu° 36 24 | 20 | 16 . . . . .

F 6. [26 *verso.*] This is, apparently, the beginning of another section, but it is isolated and although there seems to be abundance of material (compared with other leaves) I can make nothing of the problem.

G 1.

i. sūtraṁ 24 10 recto.

ii. sūtraṁ || *kritvā rūpa kshayaṁ pārtha* dhānta samguṇcnaṁ *tataḥ*

pravṛ*ittir guṇanaṁ tatah* . . . . vinirdiset ||

iii. udā° || tṛi-bhāga maladagdhasya tṛi-dhāntasy aiva . . .

ashṭottara-śatāni dattaṁ kiṁ śeshaṁ vada paṇḍita ||

108	1	1	1
1	1	1	1
	3+	3+	3+

kṛitvā ṛūpa kshayaṁ pārtha† jātā | 32 | śesha || prathamab† dhānte

kshayaṁ 36 | śeshaṁ | 72 | dvitīyab dhānte kshayaṁ | 24 | śeshaṁ 48

tṛitīyab dhānte kshayaṁ | 16 | śeshaṁ | 32

pratyayaṁ kṛiyate | sthāpanaṁ

0	1	1	1	bhā°	śeshaṁ	32
1	1	1	1			1
	3+	3+	3			

phalaṁ mūlā 108 || atha . .

sajāti kṛiyā

G 1. Folios 10 to 15 form a fairly well defined section and the leaves are among the best preserved of the manuscript. The 'find order' is 42, 41, 40, 39, ?, 29 and the writing α2. The *sūtra* numbers 24 and 25 occur.

[10 *recto*.] i. The end of the *sūtra* is marked with the usual design and the *sūtra* is numbered 24; so that from 10 *recto* to the end of 15 *recto* consists of one *sūtra* (25) and its illustrative examples.

ii. Of *sūtra* 25 the only complete word preserved is *vinirdeset*. It is reconstructed from quotations and fragments of letters. The *sūtra* is the most quoted one in what remains of the original text, the phrase *kṛitvā rūpa kshayaṁ pārtha* occurring some seven times. The last word of this phrase is, however, variously written *pārtha* (fol. 10 *recto*), *pāstham* (10 *verso*), *pāstam* (12 *recto et verso*), *pāstha* (14 *verso*) and is rather curiously omitted on fol. 11 *recto*. This variation is very curious, because the ligatures *rtha*, *stha*, *sta* are so very unlike that the differentiation can hardly be one of carelessness in writing (and the writing is here particularly good). The meaning of the term is still obscure. Dr. Hoernle suggested *prāsta* 'thrown out' or 'wastage'; but I would translate the whole phrase by 'Having calculated for unity the loss per term.' The following is Dr. Hoernle's translation of the *sūtra*—

'Calculate the loss in one; let the instalments of wastage be multiplied together; with the result let the original provision be multiplied; take the result to be the required remainder.'

iii. The example may be rendered:

The third part of the burnt bronze in three instalments (is lost). The amount given was one-hundred and eight. State the remainder, O Pandit.

The solution according to the rule gives $108\left(1-\frac{1}{3}\right)\left(1-\frac{1}{3}\right)\left(1-\frac{1}{3}\right)=32$. But proceeding by steps $\frac{108}{3}=36$ and the remainder is 72; $\frac{72}{3}=24$ and the remainder is 48; $\frac{48}{3}=16$ and the remainder is 32.

The proof may be represented by $x^1=\frac{32}{\left(1-\frac{1}{3}\right)\left(1-\frac{1}{3}\right)\left(1-\frac{1}{3}\right)}$

Continued on the reverse.

G 1—*contd.*

i. . . . . . . | 0 / 1 / 1 / 3+ / 1 / 3+ / 1 / 3+ | . . . . . tṛibhi tryashṭa-bhāga saṁyutaṁ . . . 10 verso.

tadāshṭottara-śatā$_{\chi}$ kiṁ | $\frac{27}{8}$ | $\frac{1}{1}$ | $\frac{108}{1}$ | pha° śe° 32 |

. . . . yadyekasya trayas traya ashṭa bhāga tadā dvā-

tṛiṅśānāṁ kim iti | $\frac{1}{1}$ | $3\frac{3}{8}$ | 32 | phalaṁ $\frac{108}{1}$ |

ii. udā° || sakṛid dhāntasya lohasya daśāṁshā kshīyate-s-trayaṁ |

saptate dviguṇā . cha kiṁ śeshaṁ vada paṇḍitaḥ || | $\frac{3}{10}$ | 140 |

†kṛitvā rūpa kshayaṁ pāsthaṁ† iti . . rūpaṁ | 1 | $\frac{3}{10}$ | kshayaṁ kṛitvā

jātaṁ śesha | $\frac{7}{10}$ | mūlaṁ | $\frac{140}{1}$ anena guṇitaṁ jātaṁ | 98 | kshayaṁ | 42 |

evaṁ | 140 |

. . . . | $\frac{7}{10}$ | $\frac{1}{1}$ | $\frac{98}{1}$ | phalaṁ 140 |

G 1. [10 *verso.*] i. Gives further proofs of the example on the obverse, namely:

$x^1 (1-\frac{1}{3}) (1-\frac{1}{3}) (1-\frac{1}{3}) =$ <32, hence x=108>;

then two proportions in words and figures $\frac{27}{8} : 1 :: 108 : 32$ and $1 : 3\frac{3}{8} :: 32 : 108$.

ii. *Example.*—Of iron once refined three-tenths is lost. What is the remainder of twice seventy, tell me Pandit?

The loss on unity is $\frac{3}{10}$ and the remainder is $\frac{7}{10}$. The original quantity is 140 and $\frac{7}{10}$ of 140=98. The loss is therefore 42 and 98+42=140.

Proof. $\frac{7}{10} : 1 :: 98 : 140$

Continued on fol. 11 *recto.*

G 2.

i. pratyayaḥ | 0 / 1 / 1 / 3+ / 10 | . . . . . . 11 recto.

ii. udā° || . . palā krīte pala tṛi-bhāgaṁ kshya vrajati |

ashṭā-daśa . . . . thatāṁ brūhi ||

G 2. [11 *recto.*] i. Continued from fol. 10 *verso.* 'Proof $x(1-\frac{3}{10})$=<98, therefore x=140>

ii. *Example.*—In purchasing one and a half *palas* the loss is one-third. State what would be the loss on eighteen.

Since $\frac{1}{3}/\frac{3}{2}=\frac{2}{9}$, the loss on unity, the remainder is $\frac{7}{9}$. Now $\frac{7}{9}$ of 18 =14 and the loss is 4.

Proof by the rule of three :— $1\frac{1}{2} : \frac{1}{3} :: 18 : 4$ and $\frac{1}{3} : 1\frac{1}{2} :: 4 : 18$.

G 2—*contd.*

1	3 bhā	18
3	2	1

karanaṁ | addhyardha palaṁ-s-chhedebhya idam

2
9

†kṛitvā rūpa kshayaṁ† rūpaṁ | 1 | kshayaṁ kṛitvā jātaṁ

7	18
9	1

guṇitaṁ jātaṁ | 14 | kshayaṁ | 4 |

pratyaya trai-rāśikena ||

addhyardha pala krīte tṛi-bhāgaṁ kshaya gachchhati |
ashṭā-daśa pala krīta kiṁ kshayaṁ vada paṇḍita ||

1	1	18	phalaṁ 4
1	3	1	1
2			

puna tṛi-bhāga divardhaṁ tadā chatubhi χ kiṁ iti

1	1	4	phalaṁ 18
3	1	1	1
	2		

iii. udā° || chatur-bhāga mala *dagdha* *suvar*ṇa śata-paṁ*chakaṁ* |

. . . . atha pratyay . . . . 11 verso.

0	158 su°
1	1 to°
1	5*
4+	1
1	64
4+	
1	
4+	
1	
4+	

phalaṁ mūla 500 || punar eva prastāra kramaṁ

500	1	1	1	1
1	1	1	1	1
	4+	4+	4+	4+

phalaṁ śesha tad iti

śesha 158 to° 1 śe° $\frac{1}{64}$

---

iii. *Example.*—In refining bronze there is a loss of one-fourth. What would be the loss on 500 *suvarṇas* four times refined ?

The solution is lost. It amounted to < 500 $(1-\frac{1}{4})$ $(1-\frac{1}{4})$ $(1-\frac{1}{4})$ $(1-\frac{1}{4})$ = $158\frac{13}{64}$ = 158 *suvarṇas*+$1\frac{1}{64}$ *tolās,* since 5 tolās = 1 *suvarṇa*>.

Continued on the reverse.

G 2. [11 *verso.*] This appears to have contained five proofs of the example on the obverse, for the present third proof is designated 'the fourth.' The proofs are—

i. Missing.
ii. $x^1(1-\frac{1}{4})(1-\frac{1}{4})(1-\frac{1}{4})(1-\frac{1}{4})=158$ *su*°+$1\frac{1}{64}$ *to*° therefore $x^1=500$.
iii. $500(1-\frac{1}{4})(1-\frac{1}{4})(1-\frac{1}{4})(1-\frac{1}{4})=x^1$ and x=158 *su*°+$1\frac{1}{64}$ *to*°.
iv. x^1=(158° *su*°+$1\frac{1}{64}$ *to*°)$\div(1-\frac{1}{4})(1-\frac{1}{4})(1-\frac{1}{4})(1-\frac{1}{4})$ and $x^1=500$.
v. The first loss is $\frac{500}{4}=125$ and the remainder is 375.
The second loss is $\frac{375}{4}=93\frac{3}{4}$=93 *su*°+3 *to*°+9 *māsha.* (Since 12 *mā*°=1 *to*°) and the remainder is $281\frac{1}{4}$.
The third loss is $281\frac{1}{4}\div4=7\ \frac{5}{16}$ add the remainder is $210\frac{15}{16}$.
The fourth loss is $210\frac{15}{16}\div4=52$ and the remainder is $158\frac{1}{6}$.

G 2—*concld.*

anyaṁ chaturtha pratyayaṁ kṛiyate

0	1	1	1	1	bhā°	śesha	158	phalaṁ	500 ‖
1	1	1	1	1			1		
	4+	4+	4+	4			5*		
							1		
							64		

ādyaṁ kshayaṁ 125 śeshaṁ 375 | dvitīye kshayaṁ 93 to° 3 māśa 9

śeshaṁ $\begin{array}{c}281\\1\\4\end{array}$ kshayaṁ $\begin{array}{c}70\\5\\16\end{array}$ śeshaṁ $\begin{array}{c}210\\15\\16\end{array}$ kshayaṁ $\begin{array}{c}52\\47\\64\end{array}$

śeshaṁ $\begin{array}{c}158\\13\\64\end{array}$ eśa sarvatra kartavyā ‖

G 3.

. . . . . . prastha . . . . madhunās tathāh 12 recto.

aṁbhasa . . . .

†kṛitvā rūpa kshayaṁ pāstam† iti : tatra kshayaṁ : pāstam : iti : tatra kshaya :

rūpaṁ guṇya śeshaṁ $\begin{array}{c}3\\4\end{array}$ $\begin{array}{c}3\\4\end{array}$ $\begin{array}{c}3\\4\end{array}$ $\begin{array}{c}3\\4\end{array}$ $\begin{array}{c}4\\1\end{array}$ gadyūti gadyūti gatvā-

t-prasthaṁ pivet . . . . gadyūti yojanaṁ | chatu prasthai

ādhakaṁ | tadā dhāntaśor gu . . tataḥ $\begin{array}{c}81\\256\end{array}$ āvṛitti pravṛittir-guṇanaṁ tataḥ

| 4 | anena guṇitaṁ jātaṁ $\begin{array}{c}81\\64\end{array}$ eśa maddhva bhāgā bhāge hṛite labdhaṁ

| madhu prastha 1 ku° 1 śe° $\begin{array}{c}1\\16\end{array}$ aṁbha bhāgā prastha 2 kuḍava 2

śe $\begin{array}{c}15\\16\end{array}$ evaṁ | 4 | . . . kuḍavokti prakshepake āḍhakā śoḍasha kuḍavā

bhavanti | 16 | ato ma . . . . śeshaṁ 12

G 3. [12 *recto.*] This is not directly connected with folio 11 but is probably correctly placed here. The find order places it between folios 11 and 13 and it is definitely connected with folio 13. Also it quotes from *sūtra* 25 on folio 10 *recto.* It has the same knot as folio 13.

The example may be conjecturally restored : A traveller goes a journey of 4 *gavyūtis* and takes with him 4 *prasthas* of wine. After each *gavyūti* he drinks 1 *prastha* and then fills up his bottle with water. How much wine and how much water will there be at the end of his journey ?

The preliminary part of the solution is rather confused. Possibly the *visarga* marks denote deletion. The general solution is $4.\frac{3}{4}.\frac{3}{4}.\frac{3}{4}.\frac{3}{4}=4.\frac{81}{256}=\frac{81}{64}=1\frac{17}{64}$ *prasthas* of wine remain and $2\frac{47}{64}$ *prasthas* of water. The number of *gavyūtis* in a *yojana* are mentioned (?2), and the number of *prasthas* in an *āḍhaka* are said to be 4 and the number of *kuḍavas* in an *āḍhaka* are given as 16. Therefore the wine left over $\frac{81}{64}=1$ *prastha* $+1\frac{1}{16}$ *kuḍavas* and the water $=2\frac{47}{64}=2$ *prasthas* $+2\frac{15}{16}$ *kuḍavas* and the sum of these is 4 *prasthas.*

Continued on the reverse.

G 3—*contd.*

i. . . . . prastha kuḍavā | 4 | 3 | śesha chatvāra . . . 12 verso.

kuḍavaḥ $\begin{array}{|c|c|}2 & 2\\ 1 & 1\\ 4 & 4\end{array}$ śeshā cha kuḍavā pītā | ma° $\begin{array}{|c|c|}7 & 9\\ 1 & 1\\ 4+ & 4\end{array}$ puna

chatvāri kuḍavā bhuktaṁ śeshaṁ $\begin{array}{|c|c|}81 & 175\\ 16 & 16\end{array}$ jala bhāgaṁ | madhu kudava

$\begin{array}{|c|}5\\ 1\end{array}$ śe° $\begin{array}{c|}1\\ 16\end{array}$ jala kuḍava $\begin{array}{|c|}10\\ 1\end{array}$ śe° $\begin{array}{c|}15\\ 16\end{array}$ evaṁ kuḍava 16 ||

ii. udā° || datvā śulkaṁ chatur bhāgaṁ ashṭau āṇīta kuṁkumā |

chatu śulka śālais tu kiṁ śeshaṁ vada paṇḍita ||

$\begin{array}{|c|c|}8 & 1\\ 1 & 1\\ & 4+\end{array}$

karanaṁ | †kritvā rūpa kshayaṁ pāstaṁ† pāstaṁ $\begin{array}{|cc|}8 & 3\\ 1 & 4\end{array}$ guṇitaṁ

jātaṁ $\begin{array}{|c|}6\end{array}$ śulke $\begin{array}{|c|}2\end{array}$ śeshaṁ $\begin{array}{|cc|}6 & 1\\ 1 & 1\\ & 4+\end{array}$ anena guṇitaṁ jātaṁ

$\begin{array}{|c|}4\\ 1\\ 2\end{array}$ kshayaṁ $\begin{array}{c|}1\\ 1\\ 2\end{array}$ śesheṇa $\begin{array}{|cc|}4 & 1\\ 1 & 1\\ 2 & 4+\end{array}$ datvā guṇita jātā $\begin{array}{|c|}27\\ 8\end{array}$

. . . . . .

G 3. [12 *verso.*] i. The solution of the example on the obverse is now done by steps. The original amount of 4 *prasthas* is expressed in *kuḍavas*, namely 16.

Of these 16 *kuḍavas* of wine he drinks $\frac{16}{4}$ and 12 are left and he adds 4 of water. He then drinks $\frac{12}{4}$ of wine and there are 9 *kuḍavas* left and the water is made up to 7 *kuḍavas*. Then he consumes $\frac{9}{4}=2\frac{1}{4}$ of wine and there are $9-2\frac{1}{4}=7-\frac{1}{4}$ and the water is made up to $9\frac{1}{4}$. He then drinks $\frac{6\frac{3}{4}}{4}=\frac{27}{16}$ and there is left $6\frac{3}{4}-\frac{27}{16}=\frac{81}{16}$ and the water is made up to $\frac{175}{16}$. There is, therefore, finally $\frac{81}{16}=5\frac{1}{16}$ *kuḍavas* of wine and $\frac{175}{16}=10\frac{15}{16}$ *kuḍavas* of water and these added together give 16 *kuḍavas*. See part I, § 89.

ii. *Example.*—Having given one-quarter as toll at four toll-houses eight of saffron is brought in. State, O Pandit, what is left.

Solution. $8\times\frac{3}{4}=6$ and 2 is paid in toll ; $6\,(1-\frac{1}{4})=4\frac{1}{2}$ and the loss is $1\frac{1}{2}$: $4\frac{1}{2}\,(1-\frac{1}{4})=\frac{27}{8}$ < $=3\frac{3}{8}$ and the toll is $1\frac{1}{8}$; $3\frac{3}{8}\,(1-\frac{1}{4})$ $=\frac{81}{32}=2\frac{17}{32}$ and the last toll is $\frac{27}{32}$; and the total toll paid is $2+1\frac{1}{2}+1\frac{1}{8}+\frac{27}{32}=5\frac{15}{32}$ which leaves $8-5\frac{1}{32}=2\frac{17}{32}=\frac{81}{32}$>.

Continued on fol. 13 *recto.*

G 4.

i. $\begin{array}{|ccccc|} 8 & 1 & 1 & 1 & 1 \\ 1 & 1 & 1 & 1 & 1 \\ & 4+ & 4+ & 4+ & 4+ \end{array}$ guṇitaṁ jātaṁ $\begin{array}{|c|} 81 \\ 32 \end{array}$ *punānyaṁ* 13 recto.

$\begin{array}{|ccccc|} 8 & 3 & 3 & 3 & 3 \\ & 4 & 4 & 4 & 4 \end{array}$ phalaṁ $\begin{array}{|c|} 81 \\ 32 \end{array}$ punānyaṁ $\begin{array}{|c|} 8 \\ 1 \\ 1 \\ 4+ \\ 1 \\ 4+ \\ 1 \\ 4+ \\ 1 \\ 4+ \end{array}$ ph..laṁ

$\begin{array}{|c|} 81 \\ 32 \end{array}$ puna pratyayaṁ $\begin{array}{|c|} 0 \\ 1 \\ 1 \\ 4+ \\ 1 \\ 4+ \\ 1 \\ 4+ \\ 1 \\ 4+ \end{array}$ bhā śe $\begin{array}{c|} 81 \\ 32 \end{array}$ phalaṁ kuṁkuma 8 ||

ii. udā° || ṭri-bhāga shaḍ-bhāga paṁchāṁśam guḍapiṇḍ āshṭabhārakaṁ |
kiṁ śeshaṁ dattabhir bhavet || . . .

$\begin{array}{|c|c|c|c|} 8 & 2 & 5 & 4 \\ 1 & 3 & 6 & 5 \end{array}$ guṇitaṁ jātaṁ $\begin{array}{|c|} 32 \\ 9 \end{array}$ etat phalaṁ ||

iii. udā° || chatu ϕ paṁchaka lābhena daśa droṇāt prayojita |
tad vai ṭribhis tu kiṁ lābhaṁ katthyatāṁ gaṇakottama ||

$\begin{array}{|c|c|c|c|} 10 & 5 & 5 & 5 \\ 1 & 4 & 4 & 4 \end{array}$ guṇitaṁ jātaṁ $\begin{array}{|c|} 1250 \\ 64 \end{array}$

G 4. [13 *recto.*] i. Here are four 'proofs' of the example given on folio 12 *verso.*

(*a*) $8\,(1-\frac{1}{4})\,(1-\frac{1}{4})\,(1-\frac{1}{4})\,(1-\frac{1}{4})=\frac{81}{32}$.

(*b*) $8.\frac{3}{4}.\frac{3}{4}.\frac{3}{4}.\frac{3}{4}=\frac{81}{32}$.

(*c*) $8\,(1-\frac{1}{4})\,(1-\frac{1}{4})\,(1-\frac{1}{4})\,(1-\frac{1}{4})=\frac{81}{32}$

(*d*) $x^1=\frac{81/32}{(1-\frac{1}{4})\,(1-\frac{1}{4})\,(1-\frac{1}{4})\,(1-\frac{1}{4})}$, whence $x^1=8$.

ii. *Example.*—There is a quantity of molasses weighing eight *bhārakas.* What will be left after giving away one-third, one-sixth and one-fifth ?|

$8.\frac{2}{3}.\frac{5}{6}.\frac{4}{5}=\frac{32}{9}$ and this is the answer.

iii. *Example.*—By a gain of five-fourths ten *droṇas* are obtained. Let it be said, O best of calculators, what will be the gain by three transactions.

(Here the term *lābha* seems to have meaning 'capital + profit,' what is termed the 'mixed quantity' *miśraka* on folio 62:

$10 \cdot \frac{5}{4} \cdot \frac{5}{4} \cdot \frac{5}{4}=\frac{1250}{64}$ < $=19\frac{17}{32}=19$ *dro°* +2 *āḍ°* +0 *pra°* +2 *ku°*>

For these measures see part I, §109.

Continued on the reverse.

G 4—*contd.*

i.

0	*bhā°*	śe° 19		phalaṁ 10 ‖	0	phalaṁ dro° 19 ā°	13 verso.
1		1			1		
1		ā° 2			1	2 pra° 0 ku° 2 .	
4		4*	dro°		4		
1		pra° 0			1	. . .	
4		4*	ā° pra°		4		
1		ku° 2			1		
4		ku° 4*	prasthi		4		

ii. udā° ‖ kasyāpyarjjakasya shashṭhi sva-dalena kshayaṁ gata |

puna vṛiddhyā tṛi-bhāgena sva-pādena tatojjhitaṁ

vṛiddhyā tu paṁcha-bhāgenas tathā vṛiddhi dvayo gataṁ |

kā vṛiddhi . . . syā kiṁ vā śeshaṁ tad uchyatāṁ ‖

60	1	1	1	1	rūpa lā . . . jātā 36 ‖
1	1	1	1	1	
	2+	3	4+	5	

pratyayaṁ punasyaiva

0	1	1	1	1	bhā°	36	phalaṁ 60 ‖
1	1	1	1	1		1	
	2+	3	4+	5			

punānyaṁ pratyayaṁ

60	phalaṁ 36 ‖ . . . . mūlam na jñāyate
1	
1	
2+	
1	
3	
1	
4+	
1	
5	

0	1	1	1	1
1	1	1	1	1
	2+	3	4+	5

phalaṁ .

G 4. [13 *verso.*] i. Continued from the obverse.

(*a*) $x^1 = \frac{19 \text{ dro}^\circ + 2\, \bar{a}^\circ + 0 \text{ pra}^\circ + 2 \text{ ku}^\circ}{(1+\frac{1}{4})\,(1+\frac{1}{4})\,(1+\frac{1}{4})} = 10.$

(*b*) $x^1 (1+\frac{1}{4})(1+\frac{1}{4})(1+\frac{1}{4}) = 19$ *dro°* $+2$ *ā°* $+0$ *pra°* $+2$ *ku°* <whence $x^1=10$>. See Part I, p. 62.

ii. *Example.*—The capital of a certain banker is sixty. One half of it goes in loss and then he gains by one-third; next he loses one-fourth of it and finally gains one-fifth; so that he has two gains. What is his gain and what is his loss and what the remainder and let that be stated.

Solution: $60\,(1-\frac{1}{2})\,(1+\frac{1}{3})\,(1-\frac{1}{4})\,(1+\frac{1}{5}) = 36.$

Proofs. (*a*) $x^1 = \frac{36}{(1-\frac{1}{2})\,(1+\frac{1}{3})\,(1-\frac{1}{4})\,(1+\frac{1}{5})}$, whence $x^1 = 60.$

(*b*) $60\,(1-\frac{1}{2})\,(1+\frac{1}{3})\,(1-\frac{1}{4})\,(1+\frac{1}{5}) = 36$

(*c*) $x^1\,(1-\frac{1}{2})\,(1+\frac{1}{3})\,(1-\frac{1}{4})\,(1+\frac{1}{5}) = 36$ < whence $x^1 = 60$ >

. . . . 14 recto.

yasya tanmayatā chakshu . . . .

$\frac{1}{3}$ | $\frac{1}{4}$ | $\frac{1}{5}$ | apahṛita śulka piṇḍaṁ 24 ||

karaṇaṁ || †kritvā rūpā kshayaṁ pāsta† | $\frac{2}{3}$ | $\frac{3}{4}$ | $\frac{4}{5}$ | jāṭu saṁgunya

jātaṁ | $\frac{2}{5}$ | etāvad api rūpa saṁśudhā jātaṁ | $\frac{3}{5}$ | anena bhaktvā śulka

piṇḍaṁ guṇitaṁ jātaṁ | 40 | eśa piṇḍaṁ

pratyayaṁ | $\frac{2}{5}$ $\frac{40}{1}$ | guṇita jātaṁ 16 śeshaṁ | 24 | evaṁ | 40 |

anyaṁ asya pratyayaṁ | 40 $\frac{1}{1}$ $\frac{1}{3}+$ $\frac{1}{4}+$ $\frac{1}{5}+$ | phalaṁ 16 kshayaṁ 24 evaṁ 40 ||

ii. udā° || guḍa piṇḍa jñāta tulyoś chatu . . . avye guḍaṁ |

tṛi-chatu ϕ -paṁcha-shaḍ vṛiddhyā chatvāriṁśa (bha*) ve kshaya

---

G 5. [14 *recto.*] i. The find order of folio 14 is unknown. It introduces a variation of the problems given on folios 10 to 13, but it still quotes from the same *sūtra* or a very similar one. The first example can be represented by $x(1-\frac{1}{3})(1-\frac{1}{4})(1-\frac{1}{5})=x-24$.

Solution: $\frac{2}{3}\cdot\frac{3}{4}\cdot\frac{4}{5}=\frac{2}{5}$, $1-\frac{2}{5}=\frac{3}{5}$, $\frac{24}{3/5}=40$ and this is the quantity (*piṇḍaṁ*).

Proof: $\frac{2}{5}$ of 40=16 and 40—16=24.

Another proof of this: $40(1-\frac{1}{3})(1-\frac{1}{4})(1-\frac{1}{5})=16$ and 40—16=24.

ii. *Example.*—A known amount of molasses equal to . . four is increased by one-third, one-fourth, one-fifth, one-sixth and then forty is lost . . . . .

No solution is preserved.

i. udā° || ajñātārambha-lohasya tṛi-chatu ϕ-paṁchakā kshaye | 14 verso.

sapta-viṁśati piṇḍasya tṛi-dhānta śeshya dṛishyate |

kiṁ sarvaṁ vada tatvajña kshayaṁ cha mama katthyatāṁ ||

| $\frac{1}{3}$ | $\frac{1}{4}$ | $\frac{1}{5}$ | śe° $\frac{27}{1}$ |

G 5. [14 *verso.*] (i) *Example.*—An unknown quantity of lapis-lazuli loses one-third, one-fourth, and one-fifth; and the remainder after the three-fold operation on the original quantity is twenty-seven. State what the total was, O wise one, and also tell me the loss.

Solution $\frac{2}{3}\cdot\frac{3}{4}\cdot\frac{4}{5}=\frac{2}{5}$; $1-\frac{2}{5}=\frac{3}{5}$; $27\div\frac{3}{5}=45$ and 45—27=18 and this is the loss.

The meaning of *ambha-loha*=lapis-lazuli was suggested by Dr. Hoernle.

G 5—*contd.*

karaṇaṁ | †kritvā rūpa kshayaṁ pāstha†

2/3	3/4	4/5

guṇitaṁ

jātaṁ $\frac{2}{5}$ rūpa kshayaṁ $\frac{3}{5}$ anena śeshaṁ bhaktaṁ śeshaṁ 27

bhaktaṁ jātaṁ 45 asya saptā-viṁśa | pātya seshaṁ 18 | eta kshayaṁ ||

iii. udā° || parikshīṇasya lohasya tṛi-dhāntaṁ paṁcha māśakaṁ |

na jñāyatet pravṛittkāṁ na tu śesha pradṛisyate |

pravṛitti śeshaṁ yo piṇḍaṁ kevalaṁ viṁśati sthitaṁ |

ajñāta kāṁ pravṛittī syā kiṁ vā śeshaṁ vadaśva me ||

1/3	1/4	1/5

kritvā

ii. *Example.*—Of the loss of iron the third is one-fifth of a māsha. The original quantity is not known and neither is the remainder given; but only the original remainder which quantity stands at twenty. Tell me what is the unknown original quantity and what is the remainder.

This interpretation, however, is by no means certain. The solution is lost.

G 6.

. . . . pravṛitti bhavet *sakhe* || 15 verso.

1/3	1/3	1/3	1/3

śe 16/1

karaṇaṁ || dhāntaśo ghātitaṁ tena | †rūpa kshayaṁ kṛitvā† jātaṁ

G 6. [15 *verso.*] There is a suspicion that this is a double leaf. The lenticels on the left side are well-marked but hardly any trace of them appears on the right side. Also the contents are to some extent incongruous.

The example may be represented by $x\,(1-\frac{1}{3})\,(1-\frac{1}{3})\,(1-\frac{1}{3})\,(1-\frac{1}{3})=16$. Now $\frac{2}{3}.\frac{2}{3}.\frac{2}{3}.\frac{2}{3}=\frac{16}{81}$ and $16 \div \frac{16}{81}=81$ and this is the original quantity.

Another method by *kala-savarṇa.* (This term laterally means 'parts resembling one-sixteenth,' but by Mahāvira it is used to denote fractions generally iii. 1). The question is inverted: 'Of iron (refined) four times eighty-onė is given. What is the remainder, state, O expert, which is solved by working hard in calculating.'

$$81\,(1-\tfrac{1}{3})\,(1-\tfrac{1}{3})\,(1-\tfrac{1}{3})\,(1-\tfrac{1}{3})=16.$$

" Another proof is made and the original amount is not known."

$$x^1=\frac{16}{(1-\frac{1}{3})\,(1-\frac{1}{3})\,(1-\frac{1}{3})\,(1-\frac{1}{3})}=81 \ \textit{pala}.$$

G 6—*contd.*

2	2	2	2	guṇitaṁ	16	bhaktaṁ	81	śeshena guṇaye ।
3	3	3	3		81		16	

śeshaṁ	16	guṇita jātā	81	. . . . pravṛittir ity arthaḥ ॥ athānya
	1		1	

vidhi kalā savarṇe

chatur dhānta lohasya ekāśītis-cha dattavān

kiṁ śeshaṁ vada dharmajña ya gaṇite kritaṁ śramaṁ ॥

81	1	1	1	1	phalaṁ śe° 16 ॥
1	1	1	1	1	
	3+	3+	3+	3+	

puna pratyayaṁ kṛiyate mūlaṁ na jñāyate

0	1	1	1	1	bhā°	śe°	16	phalaṁ loha pala 81 ॥
1	1	1	1	1			1	
	3+	3+	3+	3+				

udā° . . . . 15 recto.

. . . . . kaśchi yadi śakya tad uchyatām ॥

etan me saṁśayaṁ prājñad dhānta ksh*ayaṁ* vichāraṇāḥ

2	3	4	ksha°	śe°	32
3	4	5			1

karaṇaṁ ॥ dhānta saṁguṇya guṇi*taṁ* jātaṁ $\frac{3}{5}$ rūpaṁ dadyā $\frac{8}{5}$

bhāge hṛite labdhaṁ bhak......... $\frac{5}{8}$ $\frac{32}{1}$ phalaṁ 20 eśa sā pravṛitti ।

śesham 12............32 ॥ paṁcha-viṁśatima sūtraṁ ॥ 25

[15 *recto*.] Only the end of the formal question is preserved—If thou canst state . . . this is my doubt, O wise man, by examination . . . . . .

The example may have been :— $(1-\frac{1}{3})(1-\frac{1}{4})(1-\frac{1}{5})=x-r$ and $x+r=32$. From this $\frac{2}{5}x=x-r$, $(1-\frac{2}{5})x=r$, $\frac{3}{5}x+x=32$ and $x=32\div\frac{8}{5}=20$, and $r=\frac{3}{5}.20=12$.

G 7.

i. vibhaktaṁ jātaṁ

2	śe°	10	. . . .	.9
9		1		7

16 recto.

anena guṇitaṁ jātaṁ

90
7

bhāge hṛite labdhaṁ 12 ||

asya pratyaya trai-rāśikena

7	1	10	pha°	12
6	1	1		6
	2			7

ii. udā° || mākshikag-ghaṭakasyaiva dvi-tṛi-bhāga pravardhitaṁ

dvitīye dvi-paṁchamo-bhāgo tritīye dvi-saptakod*bhavaṁ*

chaturthe dvi-navaṁ-bhāgaṁ evaṁ jāta pala trayaṁ |

babhūvā saulkikai hṛitvā *kiṁ* sarvaṁ vada paṇḍita ||

2	2	2	2	śe°	3
3	5	7	9		1

dhāntaso . . . . . iti | kṛitvā

G 7. [16 *recto*.] i. The find order is 30 and the writing is α2,4. Only the remnants of a problem: Loss on $1\frac{1}{2}$ is 7/6; what is the original when the remainder is 10? Loss on 1 is $\frac{7}{6}\div 1\frac{1}{2}=\frac{7}{9}$ therefore x $\frac{7}{9}=10$ and $x=\frac{90}{7}=12\frac{6}{7}$.

Proof by the rule of three : 7 : $1\frac{1}{2}$: : 10 : $12\frac{6}{7}$.

ii. *Example.*—Of a *ghaṭaka* of honey two-thirds is given, to the second two-fifths, to the third tw -sevenths, to the fo rth two-ninths, till only three *palas* (are left). O Pandit, state how much altogether was taken away by the tax collector.

H 1.

. . . . sūtraṁ . . . . . 13 verso.

i. idāni suvarṇa kshayaṁ vakshyāmi . . syedaṁ

ii. sūtraṁ ‖ kshayaṁ saṁguṇya kanakās tadyutir bhājayet tataḥ

saṁyutair eva kanakair ekaikasya kshayo hi saḥ

iii. udā° ‖ *eka-dvi-tṛi-chatus* saṁkhyā suvarṇā māshakai riṇai |

eka-dvi-tṛi-chatus saṁkhyā rahitā sama-bhāgatāṁ ‖

sthāpanaṁ kṛiyate | eshāṁ

1+	2+	3+	4+
1	2	3	4

karaṇaṁ ‖ †kshayaṁ saṁguṇya kanakādibhi† kshayena saṁguṇya jātaṁ

| 1 | 4 | 9 | 16 | ... | esha yuti | 30 | kanakā yuti 10 anena

bhaktvā labdhaṁ . . . . .

H 1. [16 *verso.*] i. The end of a *sūtra* is marked but the number is not preserved (probably 26) and then a new section is introduced by the remark—" Now I shall speak about *suvarṇa kshaya.*" It should be noted that Mahāvira uses the term *kshaya* as synonymous with *varṇa* in his section (vi, 169ff) on *suvarṇa kuṭṭīkāra.* In our text there seems to be some confusion about the meaning of *kshaya* which here really means *varṇa* or ' quality,' although the author obviously thought it denoted a loss. Mahāvira's rule is—

Kanaka kshaya saṁvargo miśrasvarṇāhṛitaḥ kshaya jñeyaḥ |
paravarṇa pravibhaktaṁ suvarṇa guṇitaṁ phalaṁ hemnaḥ ‖ 169 ‖

" It should be known that the products of gold *kshaya*, when divided by the mixed gold gives rise to the *kshaya.* When divided by the last *varṇa* (=*kshaya*) and multiplied by the gold gives the corresponding quantity of gold."

ii. *Rule.*—Having multiplied the parts of gold with the *kshaya* let this sum be divided by the sum of the parts of gold. The result is the average *kshaya.* This means $f=\frac{f_1 g_1+f_2 g_2 \ldots . f_n g_n}{g_1+g_2 \ldots . g_n}$ where f denotes *kshaya* and g gold.

iii. *Example.*—$f_1=1$, $f_2=2$, $f_3=3$, $f_4=4$ and $g_1=1$, $g_2=2$, $g_3=3$, $g_4=4$ therefore $f=\frac{1.1+2.2+3.3+4.4}{1+2+3+4}=\frac{30}{10}=3$.

Continued on fol. 17 *recto.*

H 2.

i.

1	1					
10	30	4	pha°	mā°	śe°	12
1	1	1				1

17 recto.

ii. udā° ‖ eka-dvi-tṛi-chatus *saṁkhyā* suvarna projjhitā ime

māśakā dvi tṛitām chaiva chatu *saṁkhyā* paṁchakarāṁśakaṁ

kiṁ kshayaṁ

1	2	3	4
1	1	1	1
2	3	4	5

H 2. [17 *recto.*] i. The remnant of a proof of the example given on 16 *verso:*

. . . . :

10 : 30 : . 4 : 12, *i.e.,* $\Sigma g : \Sigma fg :: g_r : g_r F$.

ii. *Example.*—Gold one, two, three, four ; ' abandoned ' the following *māshakas* one-half, one-third, one-fourth and one-fifth.

$$F=\frac{\frac{1}{2}\cdot 1+\frac{1}{3}\cdot 2+\frac{1}{4}\cdot 3+\frac{1}{5}\cdot 4}{1+2+3+4}=\frac{163}{60}\div 10=\frac{163}{600}.$$

' Proof by the rule of three' $\Sigma g \cdot \Sigma fg :: g_r : g_r F$.

H 2—*contd.*

karaṇaṁ || †kshayaṁ saṁguṇya kanakā† eśa sthāpayate |

1 2	2 3	3 4	4 5

†taa yutir bhājayet tataḥ† hara sāsye kṛite yutaṁ | 163
60 | † saṁyutai χ

kanakair† bhaktvā tadā kanaka | 10 | anena bhaktaṁ jātaṁ | 163
600 | eśa

ekaika suvarṇasya kshayaṁ ||

pratyaya trai-rāśikena . . .

10 1	163 60	1 1	*pha*° 163 600
10 1	163 60	2 1	*pha*° 163 300
10 1	163 60	3 1	*pha*° 163 200
10 1	163 60	4 1	*pha*° 163 150

17 verso.

---

kramena dvaya māshādi uttare eka hīnatāṁ |

suvarṇaṁ me tu sammiśrya katthyatāṁ gaṇakottama ||

sthāpaṇaṁ

4+ 5	5+ 6	6+ 7	7+ 8	8+ 9	9+ 10	1+ 2	2+ 3	3+ 4

†kshayaṁ samguṇya† jātaṁ | 20 | 30 | 42 | 56 | 72 | 90 | 2 | 6 | 12* | eśāṁ yuti | 330 || *kanakānāṁ* yuti 45 | anena bhaktvā labdhaṁ | 330
45 | paṁcha-daśa bhāge chchheda kṛiyate | phalaṁ | 7 śe° 1
1 3 |

eśaa ekaika māśaka kshayaṁ |

pratyaya trai-rāśikena | 45
1 | 330
1 | 1
1 | phalaṁ 22
3 |

evaṁ sarveshāṁ *pratyayaṁ*

H 2. [*17 verso.*] I do not understand the problem but it is explained by Dr. Hoernle in the *Indian Antiquary* of 1888 (Vol. XVII, p. 43).

The solution is $F=\frac{5.4+6.5+7.6+8.7+9.8+10.9+2.1+3.2+4.3}{5+6+7+8+9+10}=\frac{330}{45}=7\frac{1}{3}$.

Proof by the rule of three— 45 : 330 : : 1 : $\frac{22}{3}$ and 'so for all of them.'

* Inadvertently omitted in the manuscript.

H 3.

i. (sūtraṁ) || *aprāpta saṁguṇā kati* *kaṁchanāni tatojjhitaṁ* 18 recto.

. . . .

kāṁchanai yad bhave labdha sa kshaya jñāta māśaka ||

ii. udā° || eka-dvi māshako prāpto dvau cha prāptaṁ cha paṁchabhi |

trayaś cha katibhi ϕ prāpta shaḍ eva . ni kevalaṁ |

chaturbhi māshakair hīṇaṁ kaṭi dṛishṭvā mayā sakhe |

trayaś cha katibhi ϕ prāptā suvarṇāṁ maśako vadaḥ |

1	2	3	6
2	5	0	4+

karaṇaṁ || †aprāpta saṁguṇā kaṭīd† iti $\begin{array}{c}6\\1\end{array}$ *aprāpta kaṭi* chatvāra 4

saṁguṇya jātaṁ 24 †kāṁchanāni tatojjhitaṁ† dvābhyā*ṁ e*ka paṁchabhi

dvayaṁ saṁguṇya jātaṁ 2 $\begin{array}{c}10\\1\end{array}$ tad yuti 12 | hitvā 2 . . . .

hitvā jātaṁ śeshaṁ 12 || aprāpta gaṇḍikai . . . .

H 3. [18 *recto.*] i. The *sūtra* is largely restored from the quotations given in the solution below. The application of the terms *aprāpta* and *kaṭi* are not at all clear ; but given that

$F=\frac{f_1 g_1+f_2 g_2+f_3 x}{g_1+g_2+x}$ then the *sūtra* states that $x=\frac{F.\Sigma g-(f_1 g_1+f_2 g_2)}{f_3}$.

ii. *Example.*—Māshakas of one and two, gold of two and five, māshakas of three and gold unknown. All that is known is the sum of *māshakas,* six ; and the average *māshaka* four. State the *māshaka* of the unknown gold.

Statement $f_1=1,\ f_2=2,\ f_3=3\ ;\ g_1=2,\ g_2=5,\ g_3=x\ ;\ F=4.$

Solution $x=\frac{4.6-(2.1+5.2)}{3}=\frac{24-2-10}{3}=\frac{24-12}{3}=\frac{12}{3}=4$........

. . . . ashṭa-viṁśatima sūtraṁ 18 verso.

i. sūtraṁ || ūnais saṁguṇya kanakā tat piṇḍaṁ cha viśodhayet

suvarṇa kanakābhyastā rāśi sheshaṁ vibhājayet

aprāpta gaṇḍika śeśa śuddhena kanakena tu |

yal labdhaṁ tat pramāṇaṁ tu gaṇḍikā yā vinirdiset ||

H 3. [18 *verso.*] The end of the 28th *sūtra* is marked.

i. *Rule.*—Having multiplied together the (known) gold pieces and their *varṇas* determine the sum of that. Divide the remainder of that quantity and the sum of the product of the average *varṇa* and known gold by the difference between the average *varṇa* and the *varṇa* of the unknown gold. That which results consider to be the measure of the unknown gold.

This may mean, for example, that if $F=\frac{f_1 g_1+f_2 g_2+f_3 x}{g_1+g_2+x}$ then $x=\frac{(f_1 g_1+f_2 g_2)-F\ (g_1+g_2)}{F-f_3}$

H 3—*contd.*

ii. **udā° || eka-dvi-tṛi-chatus saṁkhyā aprāpta māśakāni tu**

eka-dvi-tṛi-chatus saṁkhyā ekatrāvartitā kilaḥ

gaṇḍikā jñāta kanakā ūnaikā daśa māshakai |

aprāpta jñāta kanakai pra . . . . . . yah

1	2	3	4	0
1	2	.	.	.

karaṇaṁ . . . .

ii. **The example is not understood.**

J 1.

. . . . 30 recto.

sūtraṁ | eka yuta nara . . . . sarvash shaḍbhi pa . . .

. . . . anena labdhaṁ . . . . hītā prathaṁ

36	42	48	54	6
. .	. .	78	7 .	

. . . . sadṛiśa kṛi . . . . bhāga hāraṁ kṛiyate

234
70

. . 30 verso.

. . . tulāḍhe

3	mudgāḍhe	1	
24		47	
70		. .	

. . . . kṛiyate

J 2. [Folio 30.] Find order 32. Writing α4. By appearance this fragment and fol. 28 perhaps belong to the same leaf. See also fol. 31.

[30 *recto.*] A restoration is suggested in *part* I, §78, vii, but I doubt its being correct.

[30 *verso.*] We have $\frac{234}{70}=3\frac{24}{70}$ and $3\frac{24}{70}\div 2=1\frac{47}{70}$. The term *mudga* ? 'a kidney bean' occurs also on folio 31. See also *Līlāvatī*, §97.

J 2.

etat-kāla . . . . timanushyā ya . . . . lagyanti . . . 65 recto.

apara prashṇaḥ

yady eka purushasya drammāsh-shaṭ . *triṁś*abhir dinai jīva-lokā | tat kāryaṁ prastutaṁ . ssaptatīnāṁ . . . . pāka rākshakānāṁ drammaish-shaḍbhi kati dinā jīva-lokaṁ bhav*ati* . .

karaṇaṁ | ādau tāva yady ekapurushasya drammāsh-shaṭ triṇśabhi - - - jīvyāḥ | tat saptatīnāṁ kiṁ

1 pu° dram° 6	30 di°	70 pu°
1 1		1

phalaṁ . . . .

drammā . . trīṇi śata-sā . . . .

J 2. [65 *recto.*] Folio 65 consists of two leaves stuck together. The *verso* side has been definitely placed as C 4. The writing is here α4. The find order is unknown.

[Example :—If a man requires six drammas for his livelihood for 30 days, for how many days will 70 men (guards of a fort ?) live on six drammas ? The details are, however, uncertain.—K. N. D.]

J 3.

. . . . drammā ashṭa dvā-chatvāliṁśabhir dinai | tat saptati . . . . 41 recto.

ya 42/1 dine | draṁ° 8/1 jīvyā | 70/1 purushā 42/1 . .

drammā 560 || yadi paṁcha-śata-śashṭyādhika . . . . dva-chatvāliṁśabhi

tad drammai ashṭabhi kati dinā . . .

. . . . 2 adhe dāpaye dattaḥ | 17 8* 2 3 | adhenopari saṁ . . . uparima 41 verso.

rāśī dvaya guṇaye | 51 6* 2 3 | . . . . ; . upari yukta kṛiyate eka-

paṁchāśānāṁ 51/6

sthāpanaṁ | 1 53 . | phalaṁ ā 17 ṭri . 2

J 3. Folio 41 is much damaged and the illustration (Plate xxviii) suggests a double leaf; but the illustration is deceptive, for the cause of the uneven colour is the presence of gum on the original leaf. The find order is unknown: writing α4.

[41 *recto*.] This is undoubtedly closely connected with fol. 65 *recto* and the repair of fol. 41 and the separation of the two parts of fol. 65 would possibly make both intelligible.

[41 *verso*.] Not understood. *Possibly the 8 and 6 are change-ratios.

K

i. *udā°* || ko rāsi paṁcha yutā *mūladaḥ* sā rāśis sapta hīna mūlada ko so rāśir iti prashṇaḥ **59 recto.**

0	5	yu°	mū°	0	sā	0	7+	mū°	0
1	1			1		1	1		1

karaṇaṁ | †yuta hīnaṁ cha-m-ekatvaṁ† | 12 | tad dalaṁ | 6 | dvi hṛiṇaṁ | 4 | dalaṁ | 2 | ṿargaṁ | 4 | ṿ† hīne yutiṁ cha kartavyā† | hīnaṁ | 7+ | anena yuti | 11 | eśa sā rāśi || asya pratyānayaṁ *kriyate*

11	yu°	5	mū°	4	11	7+	mū°	2		
1		1		1	1	1		1		

paṁchāśama sūtraṁ 50

ii. sūtraṁ | gavāṁ viśesha kartavyaṁ dhanaṁ chaiva puna . .

. . . .

K [59]. The find order is unknown but the *sūtra* number is 50 and it probably originally preceded fol. 60. The reverse is blank, which possibly means that there are portions of two leaves stuck together.

(i) *Example.*—What number with five added is a square and that same number with seven subtracted also being a square ? What is that number ? is the question.

Statement $x+5=s^2$, $x-7=t^2$.

Solution <$x=[\frac{1}{2}(\frac{5+7}{2}-2)]^2+7=11$ by steps thus> : having combined the added and subtracted numbers $5+7=12$: that halved $=6$; two subtracted 4 ; halved 2 ; squared 4 ; then the subtractive number (7) is to be added and by the addition of this $4+7=11$ and this is the required quantity.

Proof : $11+5=4^2$, $11-7=2^2$. See Part I, §81.

(ii) There appears to be a reference to this fragment on fol. 60 *recto* where *sūtra* 51 is closed.

L 1.

. . . . 60 recto.

i. . . ekona-vimśatima

gāvo 10 1	rūpa 8 1	. . . ,

. . vivaritāsti ||

eka paṁchāsama sūtraṁ 51 ||

ii. sūtraṁ || āya vyaya viśeshaṁ tu vibha*jya* dṛishya saṁguṇaṁ |

yal labdhaṁ sā bhavet kālam ayam prashne . ya vidhi ||

iii. udā° || dvi-dine ārjaye paṁcha tṛi-dine nava bhakshaye

bhāṇḍāgāraṁ tasya tṛiṅśā kiṁ kālaṁ ārja bhakshaṇam ||

dī° 5 di° 2	dīnāra 9 dina 3	*dṛi*° 30

karaṇaṁ | †āyā vyaya viśeshan tu† | tatrāyaṁ

5 2

. . . . . .

. . . . . .

L 1. [60.] Writing α2. Notice the 'sickle' *i.* Find order unknown. Connected with fol. 59 on one side and folios 61—63 on the other. Folios 60—63 form a fairly definite section (L) relating to earning and spending.

[60 *recto.*] (i) This fragment is connected with the *sūtra* at the bottom of fol. 59, but very vaguely.

(ii) *Rule.*—The known quantity is divided by the difference between the expenditure and earning. This result is the time .... This means $t=\frac{s}{f-e}$

(iii) *Example.*—In two days one earns five ; in three days he consumes nine. His store is thirty. In what time will his earnings be consumed ?

Solution : $t=\frac{30}{\frac{9}{3}-5/2}=$ <60 and the amount earned in this time is $\frac{5}{2}$ of 60=150 dīnaras.>

i. . . . . bodi | phalaṁ 180 | dvāpaṁchāśama sūtraṁ 52 || 60 verso.

ii. sūtraṁ | aha dravya harāśauta tad viśeshaṁ vi*bhājayet*

yal-labdhaṁ dviguṇaṁ kālam° dattā sama-dhanā prati ||

[60 *verso.*] (i) Remnant of proof of the example on the obverse. The complete proof probably was :—

2 days : 5 *dīnāra* : : 60 days : 150 *dīnāra*

3 days : 9 *dīnāra* : : 60 days : 180 *dīnāra*

and 180—150=30.

(ii) *Rule.*—(If one earns e_1 in d_1 days and another e_2 in d_2 days and the first gives g to the second then $\frac{e_1}{d_1}t-g=\frac{e_2}{d_2}t+g$ and) $t=\frac{2g}{\frac{e_1}{d_1}-\frac{e_2}{d_2}}$

L 1—*contd.*

iii. udā° | tṛi-dine ārjaye paṁcha bhṛitako-m-eka paṇḍitaḥ

dvitīyaṁ paṁcha divase rasam ārjayate budhaḥ

prathamena dvitīyasya sapta dattā *nidhānataḥ*

datvā sama-dhanā jāṭā kena kālena katthyatāṁ

5	rū	6	. . . .
3	di	5	. . . .

See *Indian Antiquary*, XLII (1888), pp. 41, 44 ; but in 1915 Dr. Hoernle sent me the following note :—" The textual difficulty was not fully understood by me : the text is badly corrupted ; a portion (the 2nd *pāda*) has dropped out, and another (the 1st *pāda*) has been mixed up with the commentary. The real text of the first *pāda* is quoted in obverse line 8 of the next folio, in the commentary of the second example of the *sūtra*, and the missing part of the second *pāda* must be supplied from obverse lb. 4 and 5 of *sūtra* 52 ; which is merely a variant of *sūtra* 53. The latter *sūtra* should really run as follows :—

ahadravya viśeshaṁ cha *vibhajya datta saṁguṇaṁ* |
yal-labdhaṁ dviguṇam kālaṁ *dattā sama-dhanā prati* ||

i.e., " the difference of the daily earnings, having divided (invested), is multiplied with the given amount : the result being doubled is the time ; the given amount goes towards making the possessions equal."

(iii) *Example.*—In three days one pandit earns a wage of five and a second wise man earns six (*rasa*) in five days. The second is given by the first seven from his store and by this giving their possessions become equal. Let it be stated in what time.

Solution : $t=\frac{2\times 7}{5/3-\frac{6}{5}}=30.$

L 2.

i. . . . . anena kālena sama-dhanā bhavanti || 61 recto.

pratyayaṁ trai-rāśikena kṛi*yate*

3/1	5/1	30/1	pha° 50
5/1	6/1	30	36

prathame dvitīyasya (s) sapta dattā | 7 |

śeshaṁ 43 | . . . . 43 |

43 | ete sama-dhanā jātā ||

L 2. [61 *recto.*] i. The end of the solution of the example given on 60 *verso.*
Proof by the rule of three : 3 : 5 : : 30 : 50 and 5 : 6 : : 30 : 36 and 50—7=43=36+7.

L 2—*contd.*

ii. udā° || rājaputro dvayo kechi nṛipatis sevya santi vaiḥ

mekāsyāhne dvayash shaḍ *bhāgā* dvitīyasya divardhakam |

prathamena dvitīyasya daśa dīnāra dattavān

kena kālena samatāṁ gaṇayitvā vadāśū me ||

13	3	dattaṁ	10
6	2		1

karaṇaṁ || †aha-dravya viśeshaṁ cha† | tatva

---

ii. *Example.*—Two Rājputs are the servants of a king. The wages of one are two and one-sixth a day, of the second one and one-half. The first gives to the second ten *dīnāras*. Calculate and tell me quickly in what time there will be equality. (***Indian Antiquary***, 1888, p. 44).

Statement : $\frac{13}{6}$, $\frac{3}{2}$, given 10.

Solution : The difference of the daily earnings, ...

Continued on the reverse.

i. . . . . 61 verso.

1	13	30	pha 65	prathamena dvitīyasya .
1	6	1		
1	3	30	pha 45	. . . . r dattā jātā |
1	2	1		

55 | 56 || sama dhanā jātā ||

ii. sūtraṁ tri-paṁchāsamaḥ sūtraṁ 53 ||

sūtraṁ || *vikra*yena krayaṁ bhājyaṁ rūpa hīnaṁ punar bhajet

lābhena guṇaye tatra nīvī bhavati tatra cha ||

iii. udā° || dvibhi X krīṇāti *yas sa*pta vikṛiṇāti tṛibhish shaṭ

ashṭā-daśa bhaved lābhā kā nīvī *tatra* katthyatāṁ ||

7	6	18	lābhā
2	3	1	

karanaṁ | †vi . . . .

L 2. [61 *verso.*] i. Proof of example on the obverse—

1 : $\frac{13}{6}$: : 30 : 65

1 : $\frac{3}{2}$: : 30 : 45 and 65—10=45+10.

ii. The rule means $C=\frac{p}{c/s-1}$ where C is the capital, *p* the profit, *c* the rate of purchase and *s* the rate of sale.

iii. *Example.*—One buys 7 for 2 and sells 6 for 3 and 18 is his profit. What was his capital ?

Solution.— $C=\frac{18}{\frac{7}{2}\div\frac{6}{3}-1}=24$. The proof is given on folio 62 *recto*.

L 3.

nīvī jātā | sya pratyaya *trairāśik*ena || 62 recto.

yadi dvibhis sapta labhyate | tadā chaturviṁśatibhi χ kiṁ |

2	7	24
1	1	1

phalaṁ rū° 84 ||

i. asya vikrayaṁ kṛiyate | yadi shaḍbhi traya . . labhyate tadā chaturāśītibhi χ kiṁ |

6	3	84
1	1	1

phalaṁ 42 | mūlaṁ 24 | pātya śeshaṁ 18 eśa lābhāḥ

chau-paṁchāsama sūtraṁ 54.

ii. sūtraṁ || vikrayaṁ bhājaye chaiva guṇayet kraya piṇḍatāṁ |

rūp*one* mūla guṇaye labdha lābhaṁ cha prāpyate ||

iii. udā° || dvibhi krīṇāti yas sapta vikriṇāti tribhish shat . . . .

mūlā chā . . . . . .

L 3. [62 *recto.*] i. Continued from folio 61 *verso.*
"If for two 7 are obtained, then what for twenty-four ?"
2 : 7 : : 24 : 84 articles.
Again "If by six three are obtained then what for eighty-four ?"
6 : 3 : : 84 : 42
and the original quantity was 24 and the difference 42—24=18.
ii. The rule means $p=C\,(c/s-1)$.
iii. *Example.*—Articles are bought at 7 for 2 and sold at 6 for 3.

---

i. . . . . .

2	7	24	pha° 84
1	1	1	1

atha vikrayaṁ

6	3	84
1	1	1

62 verso.

pha° 42
1

. . . 24 | pātya śeshaṁ 18 | esha lābhaṁ ||

paṁcha-paṁchāsama sūtraṁ 55

ii. sūtraṁ || vikrayaṁ bhājaye chaiva guṇayet kraya piṇḍavat

vibhaktaṁ sa cha kartavyaṁ guṇaye miśrakaṁ budhaḥ

yal labdhaṁ sā bhaven mūlaṁ *yat*ch . chheshaṁ lābha piṇḍatāṁ ||

L 3. [62 *verso.*] i. *Solution.*—Continued from the obverse; $p=24\,(\frac{7}{2}\div\frac{6}{3}-1)=18$.
Proof.—2 : 7 : : 24 : 84 and 6 : 3 : : 84 : 42 and 42—24=18 is the profit.
ii. *Rule.*—$C=\frac{M}{C\div S}$ where M=C+p is called the 'mixed' quantity.

L 3—*contd.*

iii udā° ‖ tṛibhiś cha labhater ashṭau chaturbhiś cha vikrayaṁsh shaṭ

sa mūla lābham utpaṇṇa śataṁ śashṭi vimiśritaṁ |

kīṁ mūlaṁ kaścha lābhaṁ cha kathayed gaṇakottamaḥ ‖

8	6	miśra 160
3	4	1

karaṇaṁ | †vikrayaṁ bhājaye chaiva guṇayet†

. . . .

---

iii. *Example.*—Eight articles are obtained for three and six are sold for four. The sum of the capital and profit is one-hundred and sixty. State, O best of calculators, what was the capital and what is the profit.

The solution is lost except for the first quotation, but part of a proof is given on folio 63 *recto*. The solution was <$C=\frac{160}{\frac{8}{3}+\frac{4}{6}}=90$ and the number of articles bought was $\frac{8}{3}$ of $90=240$>.

L 4.

i. . . . . .

8	*3*	*240*	*phalaṁ 90*	. . . . 63 recto.
1	*1*	*1*	*1*	

6	4	240	phalaṁ 160	mūlaṁ	90	pātya śeshaṁ	70
1	1	1	1				

ii. *shat* paṁchāśama sūtraṁ 56

‖ vikrayaṁ cha vibhaktavyaṁ guṇitaṁ kraya rāsivat

kṛitvā rūpa kshayaṁ chaiva vibhaktaṁ mūlam *āpn*uyāt

iii. udā° ‖ paṁchabhiś chatu vargaṁ tu gṛihitaṁ kena mānava

. . kenash shaṭ vikrītamśh shaṭ paṁchaśa ṛiṇaṁ kṛitaṁ |

kraya vikraya saṁguṇya nīvis tasyaiva kathyatāṁ ‖

16	6	riṇam 56+
5	1	1

. . . . . bhājaye chaiva $\frac{1}{6}$ . . . .

L 4. [63 *recto.*] i. Proof of example given on folio 62 *verso.*

8 : 3 : : 240 : 90 and 6 : 4 : : 240 : 160 and 160=90+70.

ii. The rule means $C=\frac{l}{1-c/s}$ where l is the loss sustained, *i.e.*, having investigated the selling rate multiply with the purchase rate and having subtracted from unity divide—and the capital is obtained.

iii. *Example.*—With five four-squared are obtained by some man. For one six are sold and fifty-six is the loss. Calculating purchase and sale let his capital be stated.

The solution is <$C=\frac{56}{1-\frac{16}{5}\cdot\frac{1}{6}}=120$ and the number of articles is $\frac{16}{5}$ of $120=384$>.

L 4—*contd.*

i. punāsya vikraya $\frac{6}{1}$ | $\frac{1}{1}$ | $\frac{384}{1}$ phalaṁ | 64 | mūlaṁ | 120 | 63 verso.

chatush shashṭi pātya śeshaṁ | 56 | eśa riṇaṁ kṛi.

saptā-paṁchāśama sūtraṁ 57.

ii. **sūtraṁ** || vastra śulkaṁ yad bhavati tada . hṛita vastrataṁ |

trai-rāśika vidhānena śulka vikraya tatvataḥ ||

iii. udā° || paṭasya śulka viṁśāṅśaṁ ka . . . . tṛis-śataṁ |

paṭa-kānāṁ paṇa kṛite dvau paṭau hṛita śaulkikau |

. . *mūlyaṁ* paṇa daśas teshāḥ kiṁ mulyaṁ . . . . .

L 4. [63 *ver.o.*] Proof of example on the obverse : $<\frac{16}{5} : 1 :: 384 : 120>$, then with the selling rate 6 : 1 : 384 : 64 and 120—64=56.

ii. *Rule.*—That which is the tax on cloth is taken in cloth : by the method of the rule of three tax and sale alike.

iii. The example is not understood but reads something like this : The tax on a piece of cloth is one-twentieth part. Some one sells three-hundred. O ι the pieces being brought to market, two pieces are taken by way of tax : ten is (?) the selling price. What is the value ?

M 1.

i. 1 20 rakti dhā° 1 su° 1 chhe° 80* rakti-su° rakti 1 | pha° dha° 4 20 recto.
1 1 1 1 1 1
4 a° 0
ya° 4 ya° 1 *pā*° 3 *mū* 1
1
4

puna tṛitīyasyaiva 2 20 1 1
1 1 3 1
4 1 1 chhe° $\frac{1}{2}$*
2 2
1 chhe° 1*
4 4

. . . . chhedaṁ 6 dhā°-dra° pha° dhā° 4 ya° 1 . . . pā° 2 mū° 1 ||

suvarṇasya māṇaṁ samā . . .

ii. udā° || sa paṁcha nava bhāgāṇi dināni trayo-daśaḥ

. . . . ṇāṁ kiṁ ||

M 1. [20 *recto.*] Section 'M' begins. Writing β.

i. A fragment of a solution or 'proof'. There were at least three statements, of which the second is $1\frac{1}{4}$ of 20 *rakti* : 1 *dhā*+$4\frac{1}{4}$ *yā* : : 1 *su*°+1 *ra*° : 4 *dhā*°+1 *yā*°+3 *pā*°+1 *mū*° <or 25 *ra*° : 2025 *mū*° : : 81 *ra*° : 6561 *mū*°>.

Then a similar statement of the third (restored) $2\frac{1}{4}$ of 20 *rakti* : $\frac{1}{3}$ *dra*°+$\frac{1}{2}.\frac{1}{2}$ *dhā*°+$\frac{1}{4}$ *ya*° : : 1 *su*°+1 *ra*°: 4 *dhā*°+1 *ya*°+1 *ka*° +2 *pā*°+1 *mū*° <or 45 *ra*° : 3625 *mū*° : : 81 *ra*° : 6525 *mū*°>.

The numbers marked with asterisks are change-ratios. See Part I, §§ 103-104 ; and § 110 for the measures employed.

ii. *Example.*—Too mutilated to restore.

i. mū 12000 . . . . 20 verso.

udāharaṇaṁ | sarposhṭā-daśa hasto prāviśaty ārdhāṁgulaṁ
sa nava bhāga . . ti ekaviṁśati bhāgaṁ mapaharaṁti |
pratidinenaḥ kiṁ kālena vilaṁ saṁprāpyate ||

1 1 1 1 18 chhe° 24* aṁ° ha° | *phalaṁ va*° 2 *mā*° 4 *di* $10\frac{1}{2}$
2 21+ 1 360 1 1
1
9

udāharaṇaṁ | kīṇa x kilārdhaṁgulaṁ divase divase . . . . .

. . . .

[20 *verso.*] i. A mere fragment : 12,000 *mūdṛikas* ....

ii. *Example.*—A snake eighteen *hastas* long enters its hole at the rate of one-half *plus* one-ninth of that *minus* one-twenty-first part of an *aṅguia* a day. In what time will it have completely entered its hole ?

$(\frac{1}{2}+\frac{1}{18}-\frac{1}{21})$ *aṅ*° : $\frac{1}{360}$ years : : 18×24 *aṅ*°: 2 years 4 months $10\frac{1}{2}$ days.

iii. *Example.*—A worm......(see MAHAVIRA, V, 5).

M 2.

udā° **||** sumeru pṛithivi śaṁku surānāṁ parimāśrayaṁ **||** 33 verso.

āga . . . χ kaśchi tarasā suramadiraṁ **||**

satataṁ sapta-sārdhāṇāṁ sa pāmadhya . . . **||**

sa tṛi-bhāgā tṛi-paṁchāṁśa nityam evaṁ cha gachchhati **|**

yojanānāṁ sahasrāṇichatur-āśitir uchchhṛitam **|**

kena kālena sau gachchhe vada me ta śuniśchitaṁ **||**

. 7	di° 1	yo° 84000	adha chchhedaṁ 360* di
1	1	1	
2			

. . . .

M 2. [33 *verso*.] *Example*.—From the home of the gods a certain person desires to ascend swiftly SUMERU, the pole of the Earth and the dwelling place of the gods. He goes constantly at the rate of seven times one and a half and its quarter with one-third and one-fifth. The height of Sumeru is eighty-four thousand *yojanas*. In what time will he reach the summit ? Give me well considered answer.

There is some doubt about the rate of going and the only clear parts of the statement are the second and third terms (1 day and 84,000 yojanas), but possibly the complete statement was

$7\ (1+\frac{1}{2})\ (1+\frac{1}{4})\ (\frac{1}{3}+\frac{1}{5})$ *yo* : 1 day : : 84,000 *yo* : $\frac{12000}{360}$ years$=33\frac{1}{3}$ years.

i. udā° **||** dīnāra ko nāma viśā . . ttī du χ khārjanīyaṁ sukha-bhojane cha **|** 33 recto.

tasyārdham ardhaṁ cha yad ardham ardhaṁ ta ke . *deva* guru prasādaṁ

kṛipaṇa dhana bhuktaṁ **||**

1	1	1	1	1	1	1	108	pha° dī° 1 dhā 8 d .
1	2	2	2	2	2	2	1	

ii. uda° **||** ardhaṁ . . . . stāraṁ nava roma śatāni cha **|**

dvādaśa stīti charmāni kati romā . . . .

		12 24		
		1 1		
900	12 24		12 24	pha° roma
	1 1		1 1	

[33 *recto*.] i. *Example*.—The earning of *dinăras* is difficult but consuming them is easy. One gives one-half increased by ration of one-half (six times) for food for the poor. What is the amount consumed in 108 days ?

$1 : \frac{1}{2}.\ \frac{1}{2}.\ \frac{1}{2}.\ \frac{1}{2}.\ \frac{1}{2}.\ \frac{1}{2}$: : 103 : 1 *dī*° 8 *dhā*° 1 *aṁ*°

<*i.e.*, $\frac{108}{64}=1\frac{44}{64}=1+\frac{8}{12}+\frac{1}{48}$ and 4 *aṁ*°=1 *dhā*° and 12 *dhā*°=1 *di*°>.

See Part I, § 110.

ii. *Example*.—(This is n t underrstood, but appears to refer to the number of hairs on the skin of an animal.)

M 3.

i. . chandraanibhāṇa 32 recto

. . . . tu gaganaṁ nīta rāvaṇe | ra . . . . . . yaṁ

tyakta sutaya śetayā | sā kai kena parāvartaṁ dhanur bhāga śa

pa . . . vane patamāṇasau daśa bhāgaṁ nidhāryate | evaṁ tat

parimāṇa . hīya mānaṁ tu nityaśaḥ kiyatas tu parāvartaī bhūmiṁ

prāpyayate ja .

dha° 1 1 / 1 10 + / 5	parā° 8	yoja° 30 / 1 chhe° 8000 / 1 yo°-ja ⊙

phalaṁ parā° 218181 śe $\frac{9}{11}$

ii. udā° || nāga śva chchharma gāmi *dratama daśa*

. . . .

M 3. [32 *recto.*] Folios 32 and 36 have the same knot.

i. A mutilated example about Rāvaṇa and (?) Sītā. When Sītā had been carried up 30 *yojanas* into the air she dropped something to earth, which turned over 8 times in $1\frac{1}{10}$ *dhanus.* How many revolutions did it make before reaching the earth ?

Solution.—$(1\frac{1}{5}-\frac{1}{10})$ *dha°* : 8 revolutions : : 30×8,000 *dhanus* : : 218,181$\frac{9}{11}$ revolutions. (There is a fair amount of conjecture here. See Part I, § 47).

ii. *Example (?).*—A snake which is 100 *yojanas*, 6 *krośas*, 3 *hastas* and 5 *aṅgulas* long sheds its skin at the rate of 1 *aṅgula* in 2 days. In what time will it be free ?

(The solution is given (?) on the reverse.)

i. 32 verso.

100 / 6 / 8* / 3 / 4000* / 5 / 24*	ūrdha chchhe° 768000 a°-yo° / 1 — va° 429867 mā° 1 di° 4	1 / adha chchhedaṁ 768000 phalaṁ

ii. udā° || vraja . . charīśvāktā patitaṁ bhūmi tale paṭaṁ |

tṛi-śatāṁsya . . nāṁ tu . sapta yojana hīyate |

chatur daśas tu koṭṭī . *h*ūyata paṁcha-śashṭi cha |

kai dinai bhūtale prāpya va*da me gaṇa*kottama ||

nyāsa sthāponaṁ kṛiyate |

[32 *verso.*] i. 1 *aṅ°* : 2 d : : 100 *yo°*+6 *kro°*+3 *ha°*+5 *aṅ* : 429,867 years 1 month and 4 days.

<or 1 : 2 : : 77,376,077 *aṅ°* : $\frac{154752154}{360}$ years>. See Part I, § 108, for the measures employed.

ii. An example about some garment falling to the earth. The elements are uncertain. Compare with the problem on the obverse (i).

M 4.

i. hyā paṁcha tṛiguṇita sakhē 36 recto.

. . . . eṣha deśa pramāṇaṁ samāptaṁ ||

ii. udā || sa . . lavanasya rāshe koshṭhatāṁ vā kṛitāṁ rharai |

eshāṁ chaikāṁ rāśi punar e . . . . dhā nītā |

saptāṇāṁ m api chaikā rāśis tulitāṇi |

paṁcha saptatyā . . sahasraṁ bhavet saptāshṭa guṇaṁ kim

rā 1 1	1075 1	56 1

adha chchhedaṁ 2000* pa°-bhā° | pha- bhā 30 pa° 200

esha rāśi lavaṇa pramāṇaṁ

iii. kākinī daśa bhāgasya dadyād ashṭādaś*īti* . . |

tasyāṁ viṁśati bhāgas cha śata bhāgaṁ prayachchhati |

naro vakshaśa . . . .

M 4. [36 *recto.*] i. 'This land measurement is completed' may refer to the fragmentary example at the bottom of folio 32 *verso*, but I doubt it.

ii. The example appears to refer to heaps of salt. If one heap or quantity weighs 1,075 *palas* how much will 56 heaps weigh ?

1 : 1075 : : 56 : 30 *bhā°*+200 *pa°*

$<$ or $\frac{1075\times 56}{2000}=\frac{60200}{2000}$ *bhāra*=30 *bhā°*+200 *pa°* $>$.

iii. One tenth of a cowry is given in eighty-eight.... Of this one-twentieth and one-hundredth.........

i.

ya° 3 1	1 1* 1 360	yo° 5 chhe° 4608000* ya°-yo° 1	pha° va° 21333 mā° 4

36 verso.

ii. yojanasya tṛibhāgārdhaṁ sa tṛibhāga padonakaṁ |

yā nau dinat tṛibhāgena . . . gena gachchhati |

śā puna ϕ paṁcha bhāgārdhaṁ yojanasya tathāshṭamam

. . . . ti nivartaṁte vāyu vega valāhatā |

yojanānāmshṭau tara śataṁ *kena* kālena gachchhati ||

di 1 bhā 1 1 gu 1 1 1 3 bhā 3 3 2 5 2+ 8

[36 *verso.*] i. The statement means 3 *ya°* : 1 day : : 5 *yo°* : : 21,333 years 4 months or 3 *ya°* : $\frac{1}{360}$ years : : 5×4,608,000 *ya°* : 21,333 years 4 months $<\frac{5\times 4,608,000}{5\times 360}=21,333\frac{1}{3}>$. For the measures see Part I, § 108.

$$\frac{108}{(\frac{1}{3}\text{ of }\frac{1}{2})\div 3-(\frac{1}{5}\text{ of }\frac{1}{2})+\frac{3}{8}}=<1\text{ year, 3 months, }12\tfrac{6}{7}\text{ days}>.$$

ii. A boat travels $\frac{1}{2}$ of $\frac{1}{3}+\frac{1}{3}-\frac{1}{4}$ *yojanas* in $\frac{1}{3}$ of a day, but is driven back by the wind $\frac{1}{2}$ of $\frac{1}{5}$ of a *yojana* in $\frac{1}{8}$ of 3 days. In what time will it travel 108 *yojanas* ?

The problem is something like this but the details are not clear and the lower part of the statement has disappeared. See Part I, p. 51.

M 5.

i. *khagā ekādaśā* bhuktā prasṛitiṁ chaiva meva cha ‖ 34 recto.

. . shṭau vada sakhe   kiṁ khagaṁ vada sundari ‖

pra° 1 1	kha° 11 1	khā 5760 1	phalaṁ khaga* 63360.

esha bāhu pramāṇaṁ ‖

ii. kaśchit pumāṁ suvarṇas tu   kalā pāda yutaṁ yavaṁ ‖

pratyahaṁ sūline śuddhi . . . . kila dattavāṁ |

paṁchābdai māśam evaṁ tu   dinai paṁchadaśas tathāḥ

datvā . . sya sarvāya jñātum ichchhāmi tatvata ‖

di 1 1	1 6* bhā 1 1 1 4 4*	5	chchhedaṁ 192* yava-tola

M 5. [34 *recto.*] i. The problem is: Eleven birds feed on a *prasṛiti* (handful) of corn; how many can feed on 8 Kharis of corn? It ends "Say, O friend, say what are the *Khagas*, O SUNDARĪ."

If this is correct, the name Sundarī, 'beautiful one,' is used in exactly the same way as Līlāvatī is used by Bhāskara.

The solution is 1 *pra°* : 11 *kha°* :: 8 *khā°* : 63,360 *khagas* which would make 720 *prasṛiti*=1 *khāri*; but there are many elements of doubt and the application of *esha bāhu pramāṇaṁ* to this particular problem is not clear.

ii. By certain persons one *kalā* plus one *pāda* and one *yava* are given in gold daily at the shrine of ŚŪLIN. What would be the amount of the gift in five years, five months and fifteen days........I desire to know that.

Solution.—1 day : 1 *ya*+1 *ka*+1 *pa* : : 5 y. 5 m. 15 d. : x <or 1 d. : 30 *pa°* : : 1,965 d. : $\frac{1965\times30}{192\times25}$ *tola*=12 *to°*+3 *dhā°*+$1\frac{1}{2}$ *aṁ°*> See Part I, § 111.

i. . . . . 34 verso.

chittṛitāṁgai | tānī yata *śara-paramparay* ārjunena gṛiddhra . . . . tayā

spṛiśaṁti	1 8 1 9 1 5	śa° 1 1	yoja° 777 1 8 222 7 chchhe	phalaṁ . . . 940

ii. māśakārdha yuto dhyanta . . . vista paṁchapaṁchāśa satereṇa vajra maṇai labdham . . tra kathayaśva mūlyaṁ śāṇa chaturbhāgasya siddhārtha paṁcha bhāgasya.

ku° 1 1 2 chhe° 128* 1 mā°-ku° 1 2 mā° chhe 40* si°-mā°	sa° 55 1

[34 *verso.*] i. The fragment *chitritāṁgai* .... *ārjunena griddhra* is extremely interesting although it throws no light on the problem. See Part I, § 47.

The statement is puzzling: it may mean

$$\frac{1}{8}+\frac{1}{8.9}+\frac{1}{8.9.5} : 1\ sa : : 777\ yo°+222\frac{1}{7}\ kro° : .....40$$

But all the terms except the second are ambiguous.

ii. The problem is about a diamond weighing $1\frac{1}{2}$ *māshaka*, and obtained for ? 55 *satera*.

The statement means $1\frac{1}{2}$ *ku°*+$\frac{1}{2}$ *mā°*; 55 *sa°*, and indicates that 128 *mā°*=1 *ku°* and that 40 *si°*=1 *mā°*. See Part I, § 111.

The whole page is an interesting puzzle. (Is the leaf a double one? Neither side shows any clear lenticle.)

M 6.

i. . . . . . sūrya māṇasya 37 recto.

divākarasya ghaṭiḳai χ kiṁ prayātasya vada niśchitaṁ

30 mu° chhe° 2* gha°-mu°	500,000,000	gha° 1	pha° yo° 83·333333$\frac{1}{3}$
1 1	1	1	

ii. bhāṇo rathaṁ sura mahoraga siddhasaṁ(g)hai vidyādharai ϕ parivritaṁ . . .

ahorātru | koṭī śatārdhaṁ sa rathaṁ pryāsyāt tad brūhi śastra

kuśalo . . . vaktuṁ || muhūrtam ekena kiṁ gachchhe brūhi me

ganakottamā ||

500000000	gha° 2	pha° yo° 166,666,666$\frac{2}{3}$

M 6. [37 *recto.*] i. The question may be roughly restored: The Sun (*sūrya*) traverses 500,000,000 *yojanas* in a day. State with certainty the amount of the journey of the sun (DIVĀKARA) in a *ghaṭikā*.

The statement means 30 *mu*° : 500,000,000 : 1 *gha*° : 83,333,333$\frac{1}{3}$ *yo*° and it indicates that 2 *ghaṭika*=1 *muhūrta* (=$\frac{1}{30}$ of a day). The origin of the length of the daily journey of the sun, namely 500,000,000 *yojanas*, is not known. See Part I, § 100.

ii. The chariot of the sun (BHĀNU) is surrounded by the groups of gods, great snakes, SIDDHAS and VIDYĀDHARAS. In a day and night its journey is said to be half a hundred *koṭis*. Tell me, O best of calculators, how much in one *muhūrta* ?

30 *mu*° : 500,000,000 :: 2 *gha*° : 16,666,666$\frac{2}{3}$ *yo*°.

i. . . . . bhāge bhaved rāśi | 37 verso.

ūrdha chhhedaṁ 108000 viliptāṇam . . . . liptā 5

ii. paṁchārdha saṁvatsare bhukte rāśaikā yadi bhānujaḥ brūhi . . . ka tatvajña

samaśve vāsareṇa kiṁ

2	rā° 1	1 aṁ°	1 / 360
1	1	1	
2			

ūrdha chhhedaṁ 108000 viliptāṇāṁ rāśi | adha chchhedaṁ $\frac{1}{6}$ viliptā lipta ||

phalaṁ viliptā 2 | esha graha gatiṁ ||

iii. udā° || rāja yudhisṭhiro nāma ϕ pāṇḍu-vaṁśa . . . .

[57 *verso.*] i. The remnant of a problem possibly related to the daily motion of Jupiter, which according to the *Sūrya Siddhānta*, amounted to very nearly 5 minutes of arc (*liptā*).

ii. If BHĀNUJA (Saturn) moves through a sign in two and a half years, state, O knower of the truth, what will its motion in a solar day be equal to.

The solution is 2$\frac{1}{2}$ years : 1 sign :: 1 degree : x

and $x=\frac{1 \text{ sign} \times \frac{1}{360} \text{ degrees}}{2\frac{1}{2} \text{ years}}=\frac{30\times60\times60\times2}{5\times360}=\frac{108,000}{900}=120''$ = 2 minutes of arc (not 2 seconds as stated in the text, where *viliptā* appears to have been written by mistake for *liptā*). The terms employed are all orthodox except perhaps *vāsara* for 'solar day', but its special use is quite intelligible.

See Part I, § 100 ; and also my *Hindu Astronomy*, p. 57.

iii. This fragment is of interest because of the reference to YUDHISṬHIRA. See Part I, § 48.

M 7.

i. . . . . vyūha pārthaṁ hehayakī . . . . ghnata 47 verso.

sāyakaiś chaiva ϕ patti sva-pāda dala śodaśai |

a . . nyā chatasrā vai hatā tena mahātma vāṁ ||

śarāṇāṁ cha parīmāṇam . . . . viśārada ||

śi 1	16	4 a° chhe° 21870	phalaṁ śarā 2624400
1	1	1 1	
	1		
	4		
	1		
	2		

anyā . . . . . ī pramāṇaṁ

ii. sūtraṁ || eko ratho gaja . . . .

M 7. [47 *recto.*] i. This appears to relate to PĀRTHA the Mahābhārata hero, who pierced each soldier with 16 $(1+\frac{1}{2})$ $(1+\frac{1}{4})$ arrows and slew four divisions of the army. How many arrows did he use ?

1 *śi*° : 16 $(1+\frac{1}{4})$ $(1+\frac{1}{2})$ :: 4 × 21,870 : 2,624,400.

The abbreviation *śi*°=? ; *a*°=*anīkinī*. See Part I, § 52.

There is a very similar example about Pārtha in the *Līlāvatī* (§ 67) which has already been quoted (Part I, § 47).

ii. *Rule.*—There is little doubt that this rule relates to the constitution of an army and is exemplified on the reverse (fol. 47 *recto.*)

i. . . . . 47 recto.

. . . . vichakshaṇaḥ

chamūs tu pṛitanās tisras tisraś cha . . . .

anīkini daśaguṇām āhu arakshohanī buddaḥ ||

[47 *recto.*] i. Apparently 3 *chamūs*=1 *pṛitanā*, 3 *pṛitanās*=1 *anīkinī* and 10 *anīkinīs*=1 *akshauhiṇī*. The statement mean : a *patti* consists of 1 *ratha*+1 *gaja*+5 *nara*+3 *turaga* (*i.e.*, 1 chariot+1 elephant+5 foot soldiers+3 horsemen) and that an *akshauhiṇī* contains $3.^{7}10$ of each of these, namely—

$3.^{7}10.1$ chariots	=21,870 chariots.
$3.^{7}10.1$ elephants	=21,870 elephants.
$3.^{7}10.5$ foot-men	=109,350 foot-men.
$3.^{7}10.3$ horsemen	=65,610 horsemen.
TOTAL	=218,700.

Albīrūnī (Chap. xlviii) gives the following scheme :—

Each akshauhiṇī	has	10 anīkinī.
„ anīkinī	„	3 chamū.
„ chamū	„	3 pṛitanā.
„ pṛitanā	„	3 vāhinī.
„ vāhinī	„	3 gaṇa.
„ gaṇa	„	3 gulma.
„ gulma	„	3 senāmukha.
„ senāmukha	„	3 patti.
„ patti	„	1 ratha

and " a *ratha* comprehends besides, one elephant, three riders and five footmen."

Possibly all these terms were included in the example but *vāhinī*, *gaṇa*, *gulma* and *senāmukha* are now missing. Numerically Albīrūnī's scheme is identical with that given in our text.

The abbreviation *tu*° in the text is probably for *turaga* 'a horse.'

See Part I, §§ 51 and 94.

M 7—*contd.*

akshohi . . .

ra° 1	esha pati	$\frac{3}{1}$ $\frac{3}{1}$ $\frac{3}{1}$ $\frac{3}{1}$ $\frac{3}{1}$ $\frac{3}{1}$ $\frac{3}{1}$ $\frac{10}{1}$ gu°	
ga° 1		guṇitā jātā	ratha 21870
na° 5			gaja 21870
tu° 3			nara 109350
			haya 65610
			(218700)

esha akshohiṇī pramāṇaṁ ||

ii. udā° || kaśchid rāja kumāra śatrudama | . . . .

---

ii. *Example.*—A certain prince SATRUDAMA [The phrase may as well mean: 'a certain prince (engaged in) curbing (his) enemies, (employed or fought so many soldiers)—K. N. D.]

M 8.

ki	di°	ra° 1		va° $\frac{3}{1}$	. chhe . . 48 recto.
		ya° 1	$\frac{3^*}{1}{5}$ bhā°	ksha-ya $\frac{80^*}{1}{3}$	1
		ka° 1	$\frac{6^*}{1}{4}$ bhā°		mā° $\frac{3}{1}{12^*}$
		pā° $\frac{1}{4^*}$			di° $\frac{1}{30^*}$
		śe° $\frac{1}{3}$			

*chhe*daṁ 480* rakti-pala . . . guṇitaṁ jātaṁ $\frac{419942}{115200}$ . . . . . . 36 pala

to° 8* pale -to° 3 tolen āsti dhā° 12* dhā° 7 dhāṇe nāsti aṁ° 4* aṁ° 2

M 8. [48 *recto.*] This is a statement belonging to some lost problem and, omitting the change-ratios (marked with asterisks), it means

5 days : 1 *ra*°+1 *ya*°+1 *ka*°+1 *pā*+$\frac{1}{3}$—$\frac{1}{3}$? :: ? years+$\frac{1}{3}$ month+1 day : 36 *pa*°+3 *to*°+7 *dhā*°+2 *aṁ*° ....

or *5* : 110$\frac{1}{3}$ *pā*°—... :: ? years, etc. : 36 *pa*+3 *to*+7 *dhā*+2 *aṁ*°...

or *5* : $\frac{331\text{--}...}{115200}$ *palas* :: ? : $\frac{419942}{115200}$.

(Therefore the third term must be of the order $\frac{5\times4199420}{115200}\times\frac{115200}{330\times360}$ or nearly 180 years.) The abbreviations employed, the change-ratios, and the measures are explained in Part I, §§ 103 and 111.)

M 8—*contd.*

phalaṁ bhā° 2 . . . . enāsti 48 verso.

i. ... pala 2000 bhā° | pa° 270 || . . . . to° 8

tola-pala to° 6 tole nāsti dhāṇe 12 dhā° 8 ||

chhe°	8*		
dhā°	2		
chhe°	12*		
guṁ°	3		
chhe°	5*		
ya°	2	3*	bhā
	1	1	
		5	

ii. yadi dinam ekena esha dattaṁ tad dvādaśa varsheṇa . . . .

di 1	216	bhā°	varshe 12	3 . . .	phalaṁ bhāra 93 . . .
1	270	pa°	1	1	
	2000*	chhe°			
	6	to°			
	8*	chhe°			
	8	dhā°			
	12*	chhe°			

M 8. [48 *verso.*] This exhibits two mutilated statements of proportion that evidently belong to the same problem.

i. The first is ? : ? +2 *dhā°*+3 *guṁ°*+2 *ya°* :: ? : 216 *bhā°*+270 *pa°*+6 *to°*+8 *dhā*+......

ii. If this is given in one day what is that in twelve years..... 1 day : 216 *bhā°*+270 *pa°*+6 *to°*+8 *dhā°*+ ... :: 12 years+ ... : 93...... *bhāra* or <1 : 216[1] *bhāras* (nearly) :: 4320 days (nearly) : x, and x=216$\frac{1}{7}$×4320=933700 *bhāra* (roughly)>.

M 9.

. . . . rakti kshaya . . . . . paṁcha guṇaṁ . . . 49 verso.

divasā viṁśatikaṁ kiṁ *śuṁdya*ti mah . . . . vada niśchayaṁ

1	to°	3		kshaya 4+	va° 25	chhe° 360
1	mā°	2		60*	ma° 5	1
		12*		śi° 4	12*	
	aṁ°	3		8*	di° 20	
		4*			30*	
	ya°	3				
		4*				
	ka°	1	6*			
	pā°	1	1			
		4*	4			
	mū°	1				
		4*				

62321 / 19200 kshayaṁ śodhya . . 60881 / 19200 adha chchhedaṁ 2000 . .

sarva guṇitam 558278770 / 192·00 . . . . 7 tola palaṁ . . . .

M 9. [49 *verso.*] The statement means (omitting the change-ratios which are marked with asterisks) 1 day : 3 *to°*+2 *mā°*+3 *aṁ°*+3 *ya°*+1 *ka°*+1 *pa°*+1 *mū°*—(+4 *ra°*+4 *si°*) :: 25 years +5 months+20 days : x or 1 day : $\frac{62321}{19200}$ *tolas*—$\frac{1440}{19200}$ *tolas* :: 9170 days : x, and x=$\frac{558,278,770}{19,200}$ *tolas*= *bhā°*+1634 *pa°*+5 *to°*+0 *mā°*+0 *aṁ*+3 *ya°*+3 *ka°*+3 *p°* 1$\frac{2}{3}$ *mū.*

M 9—*contd.*

. . . . **49 recto.**

i. ya° 3 yava*nā*sti ka° 6

1 . . . . ka° 4 kalānāsti pā

4

. . *pā*danāsti mūdṛi° 4 *pāmu* mū° 2 ||

ii. udāharaṇaṁ ||

. . . śūkhyair yajaṁti devī pratimahni kechit dadāmi devyā . . kamchaḥ

krītvā dīnāra śatāni chatvārita dhānakā aṁḍikā raktikā yavā kalā pāda mūdṛikā

cha | etad mūlyaṁ vada me tatra m . sya kiṁ

1 to 12* | mū 400 | dhā 1 | phalaṁ dī 50 dīnāra nāsti dhāṇe
1 1 | 1 | aṁ 1
4* 12* dhānakā 10 dhāṇe nāsti āṁ 4*
ra 1 1* bhā
1 aṁ 1 ||
4
ya 1 3* bhā
1
5
ka 1 6* bhā
1
4
pā 1
4*
mū 1
4*

[49 *recto.*] i. This is the end of the answer to the problem on 49 *verso.* See Part I, §§ 101 (*iv*) and 111.

ii. *Example.*—The first part is too broken up to make out, but it appears to refer to a gift connected with an image of Devī and worship by Śūkhyas. (cf. *Sūkhara,* the name of a Śaiva sect). [It is possible to read *Mūkhyair* for *Śūkhyair,* in which case the chiefs of some clan or territory are intended. K. N. D.]

The statement (omitting change-ratios) means—

1 *to*° : cost 400 :: 1 *dhā*°+1 *aṁ*°-, 1 *ra*°+1 *ya*°+1 *ka*°+1 *pā*°+1 *mū*° : 50 *dī*° +10 *dhā*°+1 *aṁ*° <or 12 *dhā*°[: [400 *dī*° :: $\frac{2441}{7650}$ *dhā*° : x and x =50$\frac{41}{48}$ *dīnāras*>= 50 *dī*° +10 *dhā*° +1 *aṁ*°.

M 10.

i. 55 recto.

to° 1 / 1	va° 5 / 1 / 3	to° 1 / dhā° 1 / 12* / aṁ° 1 / 4* / ra° 1 1* bhā° 1 / 4 / ya° 1 3* bhā° 1 / 5 / si° 1 2* bhā° 1 / 2 / ka° 1 2* bhā° 1 / 2 / pā° 1 / 4* / mū° 1 / 4*	pha° va° 6 . . . 7227 / 1200	śe° 9 / 10	guṇitaṁ

atha śaḍdrammako . . . jjarad, vidhānakais dramam śā . . . . viṁśati-

pālā hatai dhānakā | asyaiva skandha-

ii.

to° 1 / 1	va° 5 / 1 / 3	to° 1 / 1	dhā° 1 / 1	1* / 12	aṁ° 1 / 1	1* / 48	ra° 1 / 1	1* / 60	ya° 1 / 1	1* / 192				

si° 1 / 1	1* / 480	ka° 1 / 1	1* / 120	pā° 1 / 1	1* / 4800	mū° 1 / 1	1* / 19200

M 10. [55 *recto*.] Folio 55 is here misplaced: it should come before folio 49, which has the same knot as 44.

i. The first statement means—

1 *to*° : $5\frac{1}{3}$ years :: 1 *to*°+1 *dhā*°+1 *aṁ*°+1 *ra*°+1 *ya*°+1 *si*°+1 *ka*°+1 *pa*°+1 *mū*° : x, and $x=5\frac{1}{3}\times\frac{21681}{19200}=\frac{7227}{1200}=6\frac{9}{400}$ years $=6$ years, $8\frac{1}{10}$ days. But the answer given appears to be 6 years...... $\frac{9}{10}$.

ii. This is the same proportion with the change-ratios given in cumulative form. See Part I, § 104.

. . . . paṁchatriṁ satam | 55 verso.

divardha tolakasya divardha māsakasya .

divardha chāṇḍikā divardha yavasya kiṁ mūlyaṁ ||

[55 *verso*.] If 1 *tola* cost thirty-five *drammas* what will be the price of one and a half *tolās*, one and a half *māshakas* and one and a half *āṇḍikas* and one and a half *yavas*.

M 10—*contd.*

nyāsa	to° 1 1	35 1	1 to° 1 2 1 1* mā° 1 6 2 1 1* aṁ° 1 2 2 1 1* ya° 1 2 2	pha° draṁ° 58 śe°	31 128

punānyaṁ

to° 1 1	35 1	1 1 1* 1 1* 1 1* 1 1 12 1 48 1 192 2 2 2 2	phalaṁ 58 śe°	31 128

Statement.—(i) 1 *to*° : 35 :: $1\frac{1}{2}$ *to*°+$1\frac{1}{2}$ *mā*°+$1\frac{1}{2}$ *aṁ*°+$1\frac{1}{2}$ *ya*° : $58\frac{31}{88}$ *dram*° or <1 : 35 :: $319\frac{1}{2}/192$: $58\frac{31}{128}$>.
(ii) This is exactly the same proportion with cumulative change-ratios indicated. See Part I, §§ 104, 105.

M 11.

. . . . 44 verso.

nivī sapta-śatānāṁ kaχ kālaṁ ārjana bhakshane ||

nyāsa sthāpanaṁ kṛiyat*e*

ā° 1 *di*° 1 *bhā*° 1 1 2 3	bha° 8 di° 5 bhā° 1 pa° 32 bhā śū° 2 36 bhā° 1 1 1 1 1 1 1 3 2 4	1 1 1 360

bhāṇḍā 700 1

vyaya rāśi | 223/144 | āya rāśi | 280/61 | etat kāleṇa ārjana bhaksh

M 11. [44 *verso.*] . . . . the capital is seven hundred. What is the time of the consumption of the earnings.
The statement means—
Daily earning $\frac{1\frac{1}{2}}{1\frac{1}{3}}$; given for Bha(vānī) 8 in $5\frac{1}{3}$ days; given for *pa*(*ra-loka*) 1 in 32; given for Śū(lin) $\frac{2\frac{1}{2}}{4\times 36}$; $\frac{1}{360}$ years ; reserve 700.
<The daily earning is $\frac{9}{8}=\frac{162}{144}$>. The expenditure quantity is <$\frac{8}{5\frac{1}{3}}+\frac{1}{32}+\frac{2\frac{1}{2}}{4\times 36}=$> $\frac{223}{144}$. <The daily loss is $\frac{223-162}{144}=\frac{61}{144}$ so 700 will last $\frac{700}{360}\div\frac{61}{144}$> $=\frac{280}{61}$ years and ' in this time the earning will be consumed.'
Then 1 day : $\frac{223}{144}$ :: $\frac{280}{61}\times 360$: $2559\frac{1}{61}$ and this is the (total) expenditure in $\frac{280}{61}$=4 years, 7 months, $2\frac{28}{61}$ days.
Then the income, $1\frac{1}{2}$ days : $1\frac{1}{3}$ :: $\frac{280}{1}\times 360$: $1859\frac{1}{61}$, and $2559\frac{1}{61}-1859\frac{1}{61}=700$.

M 11—*contd.*

di	1	223	280	ūrdha chchhedaṁ 360 phalaṁ . . 2559 śe $\frac{1}{161}$ | esha
	1	144	61	

vyaye ||

. . . va° 4 mā° 7 *di*° 2 śe°

28
61

atha āy*a*	di°	1	1	280	. . . .
		1	1	61	
		2	3		

i.	2559	di	1	——	223	esha vyaya pramāṇaṁ 44 recto.
	1		1		144	
	61					

ii. udā° || eka daśārdham utpati sa tṛibhāga dina dvayāt

pūjārthaṁ sa tṛibhāgaṁ cha trayodaśa . *ta*tāś chayet

sāshṭa bhāga dinā trīṇi vāsudevasya chārchayet

pādoṇa trayodaśāṇāṁ cha ashṭa sārdha dināni chet ||

brāhmaṇā bhojane dadyā paraloka hitārthinaḥ

sa tṛibhāgam . jjaraṁ sa paṁcha bhāga dinattrayet

pa°

ardhaṁ sārdhaṁ dine

[44 *recto.* 1. Again $\frac{280}{61}$: 2559$\frac{1}{61}$ :: $\frac{1}{360}$: $\frac{223}{144}$. This is the expenditure measure. See Part I, § 96.

Example.—One produces ten and a half in two and one-third days. For the sake of religion he gives thirteen and one-third in three and] one-eighth days ; he offers for VĀSUDEVA one quarter less than thirteen in eight and a half days. Desiring reward in a future world he gave to Brāhmans for food one and one-third in three and one-fifth days .... two and a quarter in five days ....

M 12.

. . . . ārayet **43 recto.**

. . sārdha dvādaśam evā tra bhojanē madyam uttamet

sa tṛi bhāga trayastriṁśai diւaid vāṇijyakasya tu. |

bhāṇḍāre dvādaśa śata vajārāṇāṁ sthitāsya vai |

eshā vyayasamutpattau kaχ kālaṁ brūhi paṇḍita ||

karaṇa-vidhānena dvādaśa śatasya bhāṇḍāre sti ta .

10	2	bhā°	13	3	bhā°	13	8	bhā°	1	3	bhā°	1	1	bhā°	1	5	bhā°	2		1] *bhā°*
1	1		1	1		1+	1		1	1		2	1		1	1		[1	1	1
2	3		3	8		4	2		3	5			2			3		[4	4	4

12	33 bhā	1	1	bhāṇḍā	1200	guṇitāni
1	1	1	360		1	
2	3					

M 12. [43 *recto.*] and also twelve and a half in thirty-three and one-third days for the best wine for the consumption of merchants. In the treasure house was stored twelve hundred. Say, O Pandit, how long can this expenditure continue.

The statement means :—

Daily income $= \frac{10\frac{1}{2}}{2\frac{1}{3}} = \frac{9}{2}$.

Daily expenditure $= \frac{13\frac{1}{3}}{3\frac{1}{8}} + \frac{13\frac{1}{4}}{8\frac{1}{2}} + \frac{1\frac{1}{3}}{3\frac{1}{5}} + \frac{\frac{1}{2}}{1\frac{1}{2}} + \frac{1}{5\frac{1}{3}} + \frac{2\frac{1}{4}}{5} + \frac{12\frac{1}{2}}{33\frac{1}{3}} = \frac{1807}{240}$.

<The daily loss is therefore $\frac{1807}{240} - \frac{9}{2} = \frac{727}{240}$ and $\frac{1200}{727 \div 240} \times \frac{1}{360} = \frac{800}{727}$ is the period>.

. . . .

2	10	800	adha-chchhedaṁ 360 diva	. . teṇa saha	||	ya-piṇḍaṁ	**43 verso.**
1	1	727					
3	2						

2982	. . . adhunā vyaya piṇḍaṁ	di°	1	1807	800
486			1	240	727
727					

ūrdha-chchhedaṁ 360 *phalaṁ* diva	2982	puna	800	2982	1
	486		727	486	1
	727			727	

*adha-chchhe*daṁ 360 di . . phalaṁ pratidina . . .	1807	evaṁ sarva
	240	

trai-rāśikena | udā°

[43 *verso.*] *Proofs.*—$2\frac{1}{3} : 10\frac{1}{2} :: \frac{800}{727} \times 360 : 1782\frac{86}{727}$ the total amount earned and $1782\frac{486}{727} + 1200 = 2982\frac{486}{727}$.
Again $1 : \frac{1807}{240} :: \frac{800}{727} \times 360 : 2982\frac{486}{727}$; and lastly $\frac{800}{727} : 2982\frac{486}{727} :: \frac{1}{360} : \frac{1807}{240}$ the daily expenditure. Thus each item (can be tested) by the rule of hree.

M 13.

ārdha yukte trayo-daśa sārdhaṁ bhavati 42 recto.

40 bhā°	160	13
1	1	1
		2

. . . . . eshāṁ chchhedāṁ kṛitā jātā ekeṇa . .

. . . sārdha trayo-daśabhi kim iti

1	4	27
1	1	2

pha° 54 eshām . .

. ekena labdha chatvārish shaḍbhi saṁpadyate kathaṁ

1 4
1

. . eko *labhati* chatvāri *śaṇsardhasya* tu kiṁ bhavet . . . .

. . . .

M 13. [42 *recto.*] This contains portions of a solution that is not, at present, fully understood. The preliminary work is missing and then comes the following proportion 40 : 160 :: $13\frac{1}{2}$: 54, or cancelling by 40 we get 1 : 4 :: $\frac{27}{2}$: 54. The next part is missing but apparently was—

1 : 4 :: 6 : 24
1 : 4 :: 3 : 12
1 : 4 :: $\frac{9}{2}$: 18

. . . .

i. jātā 54 | śaḍbhi 24 | . . 12 | ardhā 18 | ekatraṁ 54 || 42 verso.

e . . . trai-rāśika karaṇa pratyeka mūlya vidhi ||

ii. aparaṁ vakshyāmi | vim*śānām* diva . . . kiṁ prathame khandhakeśu yo bhilikhita | apāsya prashṇā vidhi

20	1	1
	1	1	3
	2	

. . . guṇaye | guṇitā jātā

20	3	1
1	2	3

chhedaṁ

20	1	1
1	2	

bhāge . . . jātaṁ phalaṁ rū 10 || esha viṁsānāṁ diva . . *bha*vati | atra uparimāś khandhakasya esha guṇākāraṁ bhavati | . . . .

[42 *verso.*] i. A fragment: 24 + 12 + (24 + 12) ÷ 2 = 54 . . . . . . This *sūtrā* gives the three term solution with respect to one price.
ii. I shall instance another . . . . . . what is that which is written in the first term? The solution is a matter of intelligence. $20 \times 1\frac{1}{2} \times \frac{1}{3} = 20 \times \frac{3}{2} \times \frac{1}{3} = 20 \times \frac{1}{2} \times 1 = 10$ . . . . . . Now this is the calculation of the foremost term.

M 14.

50 verso.

i. dramme tra*pu*sa śataṁ labdhaṁ ardheṇa labhyate χ kati |

eka rāśis tu kalanā gaṇita prakṛiyā kuruḥ

1 dramme 100 trapusā 1 2	phalaṁ 50

ii. aparaṁ udā° || sārdha dvaye . yasardha divardhe labhyateχ kati

2
1
2
1
2
1
2
1
2

iii. sūtraṁ || ardhen opari saṁguṇya . . vardha krameṇa cha |

ardheṇa ūrdhaṁ guṇaye ma . . paṁcha saṁguṇe |

bhājaye labdha paṇyaṁ

M 14. [50 *verso*.] i. The solution is 1 *dramma* : 100 *trapusā* :: $\frac{1}{2}$: 50.
ii—iii. The problem is too mutilated to understand. The *sūtra* seems to apply to the problem, but it is not clear.

. . . . vaśishta putra 50 recto.

sikasyarthe putra pautra upayogyaṁ bhavatuḥ

likhitaṁ Chchhajaka putra gaṇaka rāje brāhmaṇena |

sarveshāṁ-m-eva śāstrāṇāṁ gaṇitaṁ mūrdhni tishṭati |

ādyāvasāne saṁsāre utpaṁnna . . . mahat

paśchā śṛishṭi tadā kartuṁ śivena paramātmana

. . yādyaṁ cha-m-utpaṁnnaṁ gaṇitaṁ sakhya kāraṇaṁ |

yach . . . . .

[50 *recto*.] At the top of this page is the remnant of a problem, too broken up to make out. The rest of the page is devoted to what appears to be a colophon. This is not all clear but what remains seems to state that the work was written by a certain Brāhmaṇ, a prince of calculators, the son of Chhajaka. It also refers to the importance of the science of calculation, which, it is said, we owe to Śiva.

M

MGIPC—L—III-9-6—2-6-33—550.